Principles of Stable Isotope Geochemistry

Zachary Sharp
The University of New Mexico

Upper Saddle River, New Jersey 07458

Library of Congress Cataloging-in-Publication Data
Sharp, Zachary.
 Principles of stable isotope geochemistry / Sharp Zachary.— 1st ed.
    p. cm.
  Includes bibliographical references and index.
  ISBN-13: 978-0-13-009139-0
  ISBN-10: 0-13-009139-1
  1. Isotope geology. I. Title.
  QE501.4.N9S422 2007
  551.9—dc22
                                                        2005032635

Acquisitions Editor: *Chris Rapp*
Project Manager: *Dorothy Marrero*
Editorial Assistant: *Sean Hale*
Executive Managing Editor: *Kathleen Schiaparelli*
Assistant Managing Editor: *Beth Sweeten*
Production Editor: *Donna Hibbs, PPA*
Director of Marketing: *Patrick Lynch*
Director of Creative Services: *Paul Belfanti*
Art Director: *Jayne Conte*
Senior Managing Editor, AV Management and Production: *Patricia Burns*
Managing Editor, AV Management: *Abigail Bass*
AV Production Editor: *Greg Dulles*
Art Studio: *Laserwords*
Manufacturing Manager: *Alexis Heydt-Long*
Manufacturing Buyer: *Alan Fischer*

About the Cover: A Metropolitan Vickers MS 2 mass spectrometer, bought by the Ecole de Géologie, Nancy, France in 1958 for analyses of Pb, and later Rb/Sr at the CNRS, Nancy. Similar mass spectrometers were used for oxygen isotope analyses. The mass spectrometer consists of a copper vertical tube pumped by a diffusion pump (bottom center). The flight tube cuts through the central vacuum tube, allowing it to be pumped essentially at both ends. This is the predecessor of the early VG Micromass spectrometers. Photograph by Andreas Pack.

© 2007 Pearson Education, Inc.
Pearson Prentice Hall
Pearson Education, Inc.
Upper Saddle River, NJ 07458

All rights reserved. No part of this book may be reproduced in any form or by any means, without permission in writing from the publisher.

All uncredited photos are the property of the author.

Pearson Prentice Hall™ is a trademark of Pearson Education, Inc.

Printed in the United States of America

10 9 8 7 6 5 4 3 2 1

ISBN 0-13-009139-1

Pearson Education Ltd., *London*
Pearson Education Australia Pty. Ltd., *Sydney*
Pearson Education Singapore, Pte. Ltd.
Pearson Education North Asia Ltd., *Hong Kong*
Pearson Education Canada, Inc., *Toronto*
Pearson Educación de Mexico, S.A. de C.V.
Pearson Education—Japan, *Tokyo*
Pearson Education Malaysia, Pte. Ltd.

Dedicated to James R. O'Neil, my mentor and friend.

CONTENTS

PREFACE xi

ABOUT THE AUTHOR xiii

1 **INTRODUCTION** 1
    1.1 Historical Background 1
    1.2 Scope of the Discipline 4
        1.2.1 What Are Stable Isotopes? 5
        1.2.2 Which Elements and Why? 7
    1.3 Abundances of the Rare Isotopes of Light Elements 7
    1.4 Characteristics of Elements That Undergo Significant Isotopic Fractionation 7
    1.5 Applications in the Earth Sciences 9
    1.6 Isotope Effects 10
        1.6.1 Kinetic Isotope Effects 10
        1.6.2 Equilibrium Isotope Effects 11
    References 13

2 **TERMINOLOGY, STANDARDS, AND MASS SPECTROMETRY** 15
    2.1 Overview 15
    2.2 Isotopologues, Isotopomers, and Mass Isotopomers 15
    2.3 The Delta Value 17
    2.4 Isotope Exchange Reactions 20
    2.5 The Fractionation Factor 21
    2.6 $10^3 \ln \alpha$, Δ, and the ε Value 22
    2.7 Reference Standards 24
        2.7.1 Hydrogen 25
        2.7.2 Carbon 28
        2.7.3 Nitrogen 28
        2.7.4 Oxygen 29
        2.7.5 Sulfur 30
    2.8 Isotope Ratio Mass Spectrometry 30
        2.8.1 The First Isotope Ratio Mass Spectrometers 30
        2.8.2 Modern Conventional Mass Spectrometers 31
        2.8.3 Gas Chromatograph Isotope Ratio Mass Spectrometry (GC-IRMS) 33
        2.8.4 Gases Measured in Isotope Ratio Mass Spectrometry 33

2.8.5 Relations between Measured and Desired Isotopic Ratios 35
2.8.6 Ion Microprobe Analyses of Stable Isotope Ratios 36
References 39

3 EQUILIBRIUM ISOTOPIC FRACTIONATION 40

3.1 Introduction 40
3.2 Theoretical Determination of Stable Isotope Fractionation Factors 41
    3.2.1 Free Energy of Reaction 41
    3.2.2 The Internal Energy of a Molecule 42
    3.2.3 Vibrational Partition Function 43
    3.2.4 Translational and Rotational Partition Function 45
    3.2.5 The Complete Partition Function Ratio 46
    3.2.6 Extension to More Complex Molecules 46
    3.2.7 Relationship to Temperature 46
    3.2.8 "Empirical" Theoretical Methods 47
3.3 Experimental Determination of Fractionation Factors 47
    3.3.1 Introduction 47
    3.3.2 Mineral–Water Exchange Reactions 49
    3.3.3 Mineral–Calcite Exchange Reactions 51
    3.3.4 Mineral–CO_2 Exchange Reactions 51
    3.3.5 The Three-Phase Approach 52
3.4 Empirical Determination of Fractionation Factors 52
3.5 Other Potential Factors Controlling Isotope Partitioning 53
    3.5.1 Pressure Effect 53
    3.5.2 Composition and Structure 54
3.6 So Which Fractionation Factors Are Correct? 56
    3.6.1 An Example from Quartz–Calcite Fractionation 56
References 60

4 THE HYDROSPHERE 64

4.1 Overview 64
4.2 Natural Abundances of the Isotopologues of Water 65
4.3 Meteoric Water 67
4.4 The Meteoric Water Line 68
    4.4.1 General Features of the GMWL 69
    4.4.2 Variations in Slopes and Intercepts of Local MWLs 69
    4.4.3 Meteoric Waters in Arid and Semiarid Environments 70
4.5 The Deuterium Excess Parameter 71
4.6 Evaporation and Condensation 74
    4.6.1 Evaporation 74
    4.6.2 Condensation: Closed-System (Batch) Isotopic Fractionation 75
    4.6.3 Condensation: Open-System (Rayleigh) Isotopic Fractionation 78
4.7 Factors Controlling the Isotopic Composition of Precipitation 80
    4.7.1 Temperature 80
    4.7.2 Distance or Continentality Effect 82
    4.7.3 Latitude Effect 83
    4.7.4 Altitude Effect 83
    4.7.5 Amount Effect 84
    4.7.6 Seasonal Effects 86

4.8 Groundwater 86
4.9 Geothermal Systems 88
4.10 Basinal Brines and Formation Waters 89
4.11 Glacial Ice 91
    4.11.1 Underlying Bases for Glacial Paleoclimatology 92
    4.11.2 Determining the Age of Glacial Ice 93
    4.11.3 Thinning of Ice Layers 94
    4.11.4 The Example of Camp Century, North Greenland 94
    4.11.5 Example of the GRIP Summit Core: Flickering Climates 97
References 100

5 THE OCEANS 103

5.1 Overview 103
5.2 Oxygen Isotope Variations in Modern Oceans 104
    5.2.1 Salinity–$\delta^{18}O$ Relations in Shallow Marine Waters 104
    5.2.2 Salinity–$\delta^{18}O$ Relations in Deep Ocean Waters 105
5.3 Depth Profiles in Modern Oceans: $\delta^{18}O(O_2)_{aq}$ and $\delta^{13}C(\Sigma CO_2)$ 108
5.4 Isotopic Compositions of Ancient Oceans 109
    5.4.1 Primitive Oceans 109
    5.4.2 Secular Changes in $\delta^{18}O$ of Marine Sediments 111
5.5 Seawater–Basalt Interactions: Buffering the $\delta^{18}O$ Value of the Ocean 112
    5.5.1 Low-Temperature Alteration 112
    5.5.2 High-Temperature Alteration 113
    5.5.3 Evidence from Drill Core Material 114
    5.5.4 Evidence from Obducted Material 114
5.6 Buffering the $^{18}O/^{16}O$ Ratio of Ocean Water 116
    5.6.1 Summing the Processes Affecting the $^{18}O/^{16}O$ Ratio of Seawater 116
    5.6.2 Model Calculations 116
    5.6.3 Unresolved Controversy 117
References 118

6 BIOGENIC CARBONATES: OXYGEN 120

6.1 Introduction 120
6.2 The Phosphoric Acid Method 121
    6.2.1 A Major Breakthrough 121
    6.2.2 Acid Fractionation Factors 123
    6.2.3 Applicability 124
6.3 The Oxygen Isotope Paleotemperature Scale 125
6.4 Factors Affecting Oxygen Isotope Paleotemperatures 129
    6.4.1 Variations in $\delta^{18}O$ of Ocean Water in Space and Time 130
    6.4.2 Vital Effects 131
    6.4.3 Diagenesis 133
    6.4.4 Ecology of the Organism 138
6.5 Applications of Oxygen Isotope Paleothermometry 139
    6.5.1 The Quaternary 139
    6.5.2 The Paleogene and Neogene (Cenozoic) 140
    6.5.3 Older Samples 140

6.6 Application to Continental Carbonates 141
References 145

7 CARBON IN THE LOW-TEMPERATURE ENVIRONMENT 149
7.1 Introduction 149
7.2 The Carbon Cycle 150
    7.2.1 Carbon Isotope Budget of the Earth 153
7.3 Carbon Reservoirs 153
    7.3.1 Mantle 153
    7.3.2 Plants 153
    7.3.3 Organic Carbon in Sediments 157
    7.3.4 Methane 159
    7.3.5 Atmospheric CO_2 160
7.4 $\delta^{13}C$ Values of Carbonates 161
    7.4.1 Introduction 161
    7.4.2 General Characterization of Carbonates 162
    7.4.3 The Vital Effect 162
    7.4.4 Carbonate Speciation Effects 165
    7.4.5 Controls on the $\delta^{13}C$ Value of Marine Carbonates over Long Timescales 165
    7.4.6 Variations in the $\delta^{13}C$ Value of Marine Carbonates at Short Timescales 169
7.5 $\delta^{13}C$ Studies of Terrestrial Carbonates 171
References 174

8 LOW-TEMPERATURE MINERALS, EXCLUSIVE OF CARBONATES 179
8.1 Introduction 179
8.2 Phosphates 179
    8.2.1 Analytical Techniques 180
    8.2.2 Applications to Marine Paleothermometry 181
    8.2.3 Application to Mammals: Theory 183
    8.2.4 Sample Applications 186
8.3 Cherts 189
    8.3.1 Application to Precambrian Chert Deposits 189
    8.3.2 Application to Phanerozoic Cherts 190
    8.3.3 Diagenesis 191
    8.3.4 Application to Recent Sediments 193
    8.3.5 Other Silica Applications 194
8.4 Clay Minerals 195
    8.4.1 Early "Bulk" Sample Studies 195
    8.4.2 Grain-Size Considerations 196
8.5 Iron Oxides 199
References 201

9 NITROGEN 206
9.1 Introduction 206
9.2 The Nitrogen Cycle 207

9.3 Nitrogen Isotope Fractionation 208
    9.3.1 Nitrogen Fixation 209
    9.3.2 Mineralization 209
    9.3.3 Assimilation 210
    9.3.4 Nitrification 210
    9.3.5 Denitrification 210
9.4 The Characteristic $\delta^{15}N$ Values of Various Materials 211
    9.4.1 Plants and Soil 212
    9.4.2 Other Terrestrial Reservoirs 212
    9.4.3 Nitrogen in the Oceans 213
9.5 Nitrogen Isotope Ratios in Animals 216
References 219

10 SULFUR 222

10.1 Introduction 222
10.2 Analytical Techniques 223
10.3 Equilibrium Fractionations and Geothermometry 225
10.4 Sulfate and Sulfide Formation at Low Temperatures: The Sedimentary Sulfur Cycle 228
10.5 Secular Variations in Sulfur 231
    10.5.1 Long-Term Variations 231
    10.5.2 Alternative Approaches: Barite and Trace Carbonates 232
    10.5.3 Time Boundaries 233
    10.5.4 Archean Sulfates: Clues to the Early Atmosphere 234
    10.5.5 Sulfur Isotope Anomalies: Mass-Independent Fractionation 235
10.6 Sulfur Isotope Ratios in the Terrestrial Environment 238
10.7 Oxygen Isotope Variations in Sulfates 236
References 239

11 IGNEOUS PETROLOGY 242

11.1 Introduction 242
11.2 The Mantle 243
    11.2.1 Oxygen 243
    11.2.2 Carbon 247
    11.2.3 Nitrogen 251
    11.2.4 Hydrogen 252
    11.2.5 Sulfur 254
11.3 Emplacement of Plutonic Rocks: Interactions with the Crust and Hydrosphere 255
    11.3.1 Normal Igneous Rocks 256
    11.3.2 Shallow-Level Hydrothermal Alteration by Meteoric Water: Low $\delta^{18}O$ Plutonic Rocks 256
    11.3.3 High-$\delta^{18}O$ Igneous Rocks 258
11.4 Calculating Fluid/Rock Ratios 259
11.5 Other Processes: Degassing, Assimilation, and Fractional Crystallization 261
    11.5.1 Magmatic Volatiles 261
    11.5.2 Assimilation–Fractional Crystallization (AFC) Processes 263
References 266

12 METAMORPHIC GEOLOGY 272

- 12.1 Introduction 272
- 12.2 Stable Isotopes as Geochemical Tracers 273
    - 12.2.1 Closed System: Protolith Identification and Alteration 273
    - 12.2.2 Open Systems: Volatilization and Fluid Infiltration Processes 274
- 12.3 Fluid Sources and Fluid–Rock Interaction 280
    - 12.3.1 Oxygen and Hydrogen 280
    - 12.3.2 Carbon 282
    - 12.3.3 Sulfur 283
- 12.4 Scales of Equilibration During Metamorphism 284
    - 12.4.1 Regional-Scale Exchange 284
    - 12.4.2 Localized Exchange 285
- 12.5 Quantifying Fluid–Rock Ratios and Fluid Fluxes 286
    - 12.5.1 Simple Mixing Models: Zero-Dimensional Water–Rock Interaction Models 286
    - 12.5.2 One-Dimensional (Directional) Water–Rock Interaction Models 287
- 12.6 Thermometry 291
    - 12.6.1 Introduction 291
    - 12.6.2 Oxygen Isotope Thermometry in Metamorphic Rocks: Testing for Equilibrium 293
    - 12.6.3 Applications of Stable Isotope Thermometry 295
- 12.7 Retrograde Exchange: "Geospeedometry" 296
- 12.8 State of the Art 300
- References 302

13 EXTRATERRESTRIAL MATERIALS 309

- 13.1 Introduction 309
- 13.2 Classification of Meteorites 310
- 13.3 Oxygen Isotope Variations in Meteorites 310
    - 13.3.1 Introduction 310
    - 13.3.2 Discovery of an ^{17}O Anomaly 312
    - 13.3.3 Possible Explanations: Mixing of Two Distinct Reservoirs 313
    - 13.3.4 Mass-Independent Fractionation 316
- 13.4 Hydrogen 318
    - 13.4.1 Introduction 318
    - 13.4.2 Meteorites 319
- 13.5 Carbon 320
- 13.6 Nitrogen 322
- References 325

APPENDIX A Standard Reference Materials for Stable Isotopes 329

APPENDIX B Sample Calculation of the Correction Procedure for Adjusting Measured Isotope Data to Accepted IAEA Reference Scales 332

INDEX 334

PREFACE

Principles of Stable Isotope Geochemistry is written as a textbook to accompany a one semester course in stable isotope geochemistry. The book is tailored to an upper division or graduate level. There are 13 chapters, each dealing with a specific subtopic of the field. Other than Chapters 1 and 2—introduction and definitions—most of the remaining chapters can be read without reliance on any preceding ones. Therefore, a course can be customized to the interests of the instructor. It is also hoped that the book will serve as a general reference volume for researchers in the field.

Stable isotope geochemistry covers a bewildering array of applications. They are relevant to meteoritics, igneous petrology, metamorphic petrology, sedimentology, paleoclimate, paleontology, hydrology, tectonics, and atmospheric science, to name a few. If we add biology, anthropology, chemistry, and medicine, the scope grows exponentially. Therefore, by necessity, the range of topics covered in this book is broad, each covered in only a general manner. *Principles of Stable Isotope Geochemistry* has been organized in such a way that major concepts are explained and accompanied by numerous examples. In most cases, the first published examples are used for illustration, giving both a broad base of understanding and an appreciation for the historical development of the field. For any inadvertent omissions, my sincere apologies. I have tried to treat the very broad subject in a uniform manner.

When most geologists consider inorganic geochemistry, they think of variations in major, minor, and trace element abundances of solids and liquids (see Figure P1). In general, variations in *cation* concentrations or ratios are measured. The *anions or anion complexes*, such as oxygen, sulfur, nitrate, carbonate, sulfate, hydroxyl, etc., tend to be stoichiometric in natural inorganic solids, so that measurements of the abundance of anions is not particularly informative. For example, carbonates contain varying amounts of Ca^{2+}, Mg^{2+}, Mn^{2+}, Fe^{2+}, Sr^{2+}, etc., but have the same stoichiometric proportions of CO_3^{2-}. Fortunately, each of the anions or anion complexes (other than fluorine!) has multiple stable isotopes, and the variations in these ratios provide information that is often complimentary to that of the cation data. Just a few of the applications are given in Figure P1. New ones are being discovered all the time.

This book is dedicated to James R. O'Neil, my mentor and friend. Jim and I conceived and outlined this book over a decade ago, and spent many long hours in organization and development. Without Jim, the book would certainly never have been written. His imprint can be seen throughout the book. Jim's joy and enthusiasm are contagious, and he made for me the field of stable isotope geochemistry exciting and fun.

A good portion of the book was written during a sabbatical stay at the CRPG, CNRS in Nancy, France. My thanks to director Bernard Marty for inviting me to Nancy and to many others there for their ideas, including Pete Burnard, Marc Chaussidon, Etienne Deloule,

Christian France-Lanord, Guy Libourel, Andreas Pack, Laurie Reisberg, and Larry Shengold. I would also like to thank the following people for their help in reviewing the text, providing information and preparing illustrations: Paul Aharon, Ihsan Al-Aasm, Viorel Atudorei, Julie Bartley, Michael Cosca, Lee Cooper, John Dilles, Chuck Douthitt, Anthony Fallick, Carey Gazis, Stan Halas, Juske Horita, Rhian Jones, Peter Larson, Karen McLaughlin, Jean Morrison, Peter Nabelek, William Patterson, Adina Paytan, Arndt Schimmelmann, Torsten Vennemann, Moire Wadleigh, and David Wenner. Finally, thanks to my wife Sharon, and my two children Alana and Chloé for their support and understanding over the years.

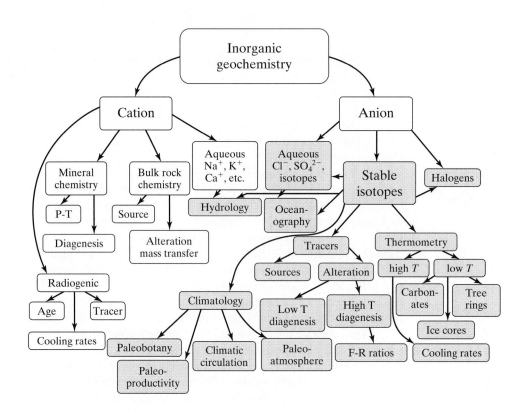

ABOUT THE AUTHOR

Photo by Chloé Sharp.

Zachary Sharp received a B.S. degree in Geology at the University of California at Berkeley and then moved eastward to continue his studies at the University of Michigan, where he received M.S. and Ph.D. degrees under the mentorship of Eric Essene. During this time, and with the encouragement of Eric, Zach worked in the stable isotope laboratories of John Bowman at the University of Utah and Jim O'Neil at the U.S. Geological Survey in Menlo Park, and made single crystal structure refinements at high pressure with Bob Hazen and Larry Finger at the Geophysical Laboratory, and high pressure phase equilibrium experiments with Steve Bohlen at SUNY Stony Brook. After graduation and the birth of his first daughter, he continued his eastward journey with a postdoctoral fellowship at the Geophysical Laboratory of the Carnegie Institution of Washington. There he developed the first laser extraction fluorination system for which he was later awarded the Mineralogical Society of America Award. After nearly two years in Washington, he continued eastward, this time taking a job as a research scientist and head of the stable isotope laboratory at the University of Lausanne in Switzerland. His second daughter was born in Switzerland, and after eight pleasant years there, he turned back to the West and returned to the United States as a professor in the Department of Earth and Planetary Sciences at the University of New Mexico. He has been there ever since, enjoying the sunshine and fruitful interactions with his wonderful colleagues.

CHAPTER 1

INTRODUCTION

1.1 HISTORICAL BACKGROUND

The first suggestion that physical chemical processes could cause isotopic fractionation of light elements in natural substances was made in 1925 by British scientists H. Briscoe and P. Robinson, who observed a variation in the atomic weight of boron in minerals from various localities (1925). They proposed that processes such as solution, crystallization, melting, and volatilization would likely cause such isotopic variations in nature. In the following year the eminent Russian scientist V. Vernadsky suggested that isotopic fractionation of the light elements should occur in living matter as well, but there were no experimental or natural data to support this hypothesis at that time. Variations in the hydrogen and oxygen isotope ratios of water in the hydrologic cycle of the Earth were recognized crudely as early as the mid-1930s on the basis of precise density measurements (Gilfillan, 1934). In that same decade, H. Urey[1] and his colleagues at Columbia University were conducting experiments and developing the theory of isotope exchange reactions, and A. Nier and his colleagues at the University of Minnesota were making mass spectrometric measurements of variations in the stable isotope ratios of several light elements in natural materials (Fig. 1.1).

Probably the first bona fide application of light stable isotope measurements to a major geochemical problem was published by F. Wickman (1941), who calculated the total amount of bitumen and coal in the Earth on the basis of carbon isotope analyses of these materials. The titles of several articles written in the 1930s and 1940s show clearly that the power of stable isotope measurements in resolving problems in earth science was recognized long ago by outstanding scientists throughout the world (Table 1.1).

Light stable isotope geochemistry as we know it today began in 1946. That year, Harold Urey traveled to several prominent universities in Europe to deliver a lecture sponsored

---

[1] H.C. Urey (1893–1981) won the Nobel Prize in Chemistry in 1934 for his discovery of deuterium. He is considered the father of modern stable isotope geochemistry.

FIGURE 1.1: Nier's 60° sector mass spectrometer, 1947. Reprinted from Rankama (1954) with permission from Elsevier.

TABLE 1.1: Selected early publications in stable isotope chemistry and geochemistry.

Year	Title	Reference
1932	A hydrogen isotope of mass 2 and its concentration.	Urey, H.C., Brickwedde, F.G., and Murphy, G.M., *Phys. Rev.* **40**, 1.
1934	The natural separation of the isotopes of hydrogen.	Dole, M., *J. Am. Chem. Soc.* **56**, 999.
1935	Isotopic exchange equilibria.	Urey, H.C., and Greiff, L.J., *J. Am. Chem. Soc.* **57**, 321.
1935	The relative atomic weight of oxygen in water and air.	Dole, M., *J. Am. Chem. Soc.* **57**, 2731.
1937	A micro-flotation method for the precise comparison of liquid densities and its application to a preliminary investigation of the distribution of heavy water in certain salt hydrates and to other matters.	Anderson, J.S., Purcell, R.H., Pearson, T.G., King, A., James, F.W., Emelius, H.J., and Briscoe, H.V.A., *J. Chem. Soc.*, p. 1492.
1939	Isotopic composition of rain water.	Teis, R.V., *Compt. Rend. Acad. Sci. U.R.S.S.* **23**, 674.
1941	Determination of the isotopic composition of [hydroxyl] waters in metamorphic rocks and minerals.	Vernadsky, W.I., Vinogradov, A.P., and Teis, R.V., *Compt. Rend. Acad. Sci. U.R.S.S.* **31**, 573.
1941	On a new possibility of calculating the total amount of coal and bitumen.	Wickman, F.E., *Geol. Fören. i Stockholm Förh.* **63**, 419.
1949	Natural variations in the isotopic content of sulfur and their significance.	Thode, H.G., MacNamara, J., and Collins, C.B., *Can. J. Res.* **27B**, 361.
1950	Isotopic composition of oxygen in silicate rocks.	Baertschi, P., *Nature* **166**, 112. Baertschi, P., and Silverman, S.R., *Geochim. Cosmochim. Acta* **1**., 4.
1951	Relative abundance of oxygen and carbon isotopes in carbonate rocks.	Baertschi, P., *Nature* **168**, 288.
1952	Variation in the relative abundance of carbon isotopes in plants.	Wickman, F.E., *Geochim. Cosmochim. Acta* **2**, 243.

annually by the Royal Society of London.² Urey presented results of semiempirical calculations of the isotopic fractionation of stable isotope ratios of the light elements among ideal gases and simple aqueous ions from spectroscopic data and the methods of statistical mechanics. At this lecture in Zurich in December of 1946, Paul Niggli asked if it might not be possible to determine the freshwater or marine origin of ancient deposits of limestone, coral, or shells from oxygen isotope analysis of the carbonate. At that time, it was already known that $^{18}O/^{16}O$ ratios of marine limestones were about 3 percent higher than those of the ocean and that ocean water was isotopically heavier than freshwater. Prompted by Niggli's remarks, Urey turned his attention to the temperature coefficient of the oxygen isotope fractionation between $CaCO_3$ and H_2O. On the basis of estimates made from his calculations, he concluded that this coefficient might be large enough to determine the temperatures of ancient oceans from oxygen isotope analyses of $CaCO_3$ in fossil shells. As Urey recounted the story later, "I suddenly found myself with a geological thermometer in my hands." And the games began.

The overall precision of isotopic measurements required to make temperature estimates that were meaningful in paleoclimatology was not attainable in the 1940s. At that time, the precision of mass spectrometric measurements of $^{18}O/^{16}O$ ratios was about a factor of 10 less than that required to determine temperatures to $\pm 0.5°C$. In addition, there was no reproducible technique for extracting CO_2 from $CaCO_3$, and there were no experimental data relating the oxygen isotope fractionation between calcium carbonate and water to temperature. Urey was not to be deterred from proceeding with this ambitious project, and he assembled an outstanding group of young scientists to work on it. The research team included postdoctoral fellow S. Epstein, doctoral students H. Craig and J. McCrea, doctoral student and later research associate C. Emiliani, paleontologist H. Lowenstam, and electronics engineer C. McKinney. By 1950, this group had successfully improved the precision of the Nier isotope ratio mass spectrometer (Nier, 1947) by the necessary factor of 10, developed reproducible analytical extraction methods for biogenic carbonates (McCrea, 1950), and established standards and protocols that are still being followed for the most part today. The development of the oxygen isotope paleotemperature scale (Urey et al., 1948; Epstein et al., 1951; Urey et al., 1951) has been heralded as one of the outstanding scientific achievements of the 20th century.

Concurrent with carrying out work on oxygen isotope analyses of carbonate shells, members of the Chicago group conducted survey studies of oxygen isotope variations in silicate rocks and minerals (Baertschi, 1950; Baertschi and Silverman, 1951), carbon isotope variations in nature (Craig, 1953), stable isotope ratios of natural waters (Epstein and Mayeda, 1953; Friedman, 1953) and oxygen isotope compositions of biogenic phosphates (Tudge, 1960). Important early stable isotope research was also conducted in Hamilton, Ontario, on sulfur isotope ratios of rocks and minerals (Thode, 1949), in Copenhagen on oxygen isotope variations in natural waters (Dansgaard, 1954), in Moscow on oxygen (Vinogradov and Dontsova, 1947) and on sulfur (Trofimov, 1949) isotope ratios of rocks and minerals. T. Hoering and his group in Arkansas were the first to investigate isotopic variations of nitrogen and chlorine in natural substances (Hoering, 1956; Hoering and Moore, 1958; Hoering and Parker, 1961). Within a few short years, it was recognized that oxygen isotope fractionations between cogenetic minerals were large enough to register temperatures of

---

²This lecture culminated in the now-classic paper entitled *The Thermodynamic Properties of Isotopic Substances*, published in *Transactions of the Royal Society of London* (1947).

formation of high-temperature rocks (Clayton and Epstein, 1958) and that hydrogen and oxygen isotope measurements of rocks and minerals were powerful petrologic tools (Taylor and Epstein, 1962).

From these beginnings, stable isotope research has blossomed to the point where thousands of isotope ratio mass spectrometers are in operation in laboratories all over the world. Stable isotope measurements are being made to resolve problems in many diverse fields, including geochemistry, climatology, hydrology, plant physiology, ecology, archaeology, forensic medicine, meteorology and atmospheric science, meteoritics, palaeobiology, and bacteriology.

Almost all the early achievements in the field of isotope geology were made by gifted chemists and physicists who developed the theory and techniques, improved the mass spectrometers, and thought and wrote about many fundamental scientific questions of geologic interest. It is well to keep in mind that, despite developments in isotope ratio mass spectrometry—most notably, very stable electronics and increased sensitivity of ion sources—and in techniques to analyze smaller samples, there has been no *significant* improvement in the overall precision of modern stable isotope analyses over those made in the early 1950s. The errors associated with chemical extraction procedures continue to be a limiting factor. It is instructive to read the early papers: One comes to the disturbing realization that many questions being examined with the aid of stable isotope measurements today were already addressed by these early workers. Similar or identical conclusions reached from analyses of several or even only a few carefully selected materials 30 to 40 years ago are reappearing in recent publications that may contain hundreds of analyses, many of them obviously superfluous. It is worthwhile, both from a historical standpoint and as proper scientific procedure, to be aware of pertinent observations and conclusions published in the older literature. In this spirit, efforts are made in this book to cite primary references whenever possible.

1.2 SCOPE OF THE DISCIPLINE

Stable isotope measurements have an extremely wide range of applications, and the principles employed are relatively easy to grasp. There are gross similarities between some of the approaches and scientific goals of stable isotope geochemistry and those of major and minor element geochemistry and radiogenic isotope geochemistry. For example, estimates of formation temperatures are often made on the basis of cation partitioning between coexisting mineral phases (e.g., garnet–pyroxene). Similarly, differences in $^{18}O/^{16}O$ ratios between coexisting minerals vary with temperature, just as Fe/Mg ratios vary between garnet and pyroxene. Therefore, by measuring ratios of $(^{18}O/^{16}O)_{garnet}$ to $(^{18}O/^{16}O)_{pyroxene}$, scientists can estimate temperatures of formation of rocks containing this mineral pair, and they can do so independently of pressure and activities of other phases.

As a second example, trace element concentrations and rare-earth element abundance patterns are frequently used to determine the origin and subsequent histories of rocks (e.g., ocean island basalts versus mid-ocean ridge basalts, and the degree of diagenesis in marine carbonates). Analogously, fluid histories can be ascertained from oxygen and hydrogen isotope compositions of waters, and fluid–rock ratios can be determined in modern and fossil hydrothermal systems. Paleotemperature estimates can be made from the oxygen isotope compositions of marine carbonates, and information about paleoproductivity can be estimated from their carbon isotope compositions. But applications extend beyond the boundaries of traditional geochemistry: Photosynthetic pathways, migration patterns of birds, and stratospheric chemical reactions can be all be evaluated with the use of stable isotope ratios. The applications are endless.

1.2.1 What Are Stable Isotopes?

In the most simple description, atoms consist of **protons, electrons,** and **neutrons**. Protons are positively charged particles, electrons are negatively charged, and neutrons have no charge. The mass of a neutron is about equal to that of a proton but, relative to protons and neutrons, the mass of electrons in an atom is negligible (Table 1.2). Both protons and neutrons are present in the nuclei of all atoms except protium, a form of hydrogen whose nucleus contains one proton but no neutrons.

An element is determined by the number of protons in its nucleus. In a neutral atom, the number of protons is balanced by an equal number of electrons, which are present as a negatively charged cloud around the nucleus. The configuration of the electron cloud determines the gross chemical properties of an atom. For a given element, the number of protons (the **atomic number Z**) is always the same, but the number of neutrons (the **neutron number N**) may vary. The **mass number** A is the sum $Z + N$. The number of neutrons in the nucleus of an element does not affect the gross chemical properties of the element and its compounds, but differences in mass attendant on changes in N can cause subtle chemical and physical differences between compounds containing elements of varying N. In fact, these small differences make up the subject of this discipline.

An **isotope**[3] of a given element differs from another isotope of the same element by the number of neutrons in its nucleus. Most elements in the periodic table have two or more naturally occurring isotopes (either stable or radioactive), but 21 elements, including fluorine, aluminum, sodium, and phosphorus, are **monoisotopic**. The nucleus of the single natural isotope of fluorine contains 9 protons ($Z = 9$) and 10 neutrons (so that $A = 19$). Oxygen has three naturally occurring isotopes: ^{16}O, with 8 protons and 8 neutrons; ^{17}O, with 8 protons and 9 neutrons; and ^{18}O, with 8 protons and 10 neutrons.

Consider the three isotopes of hydrogen. **Protium** has one proton, one electron and a mass of ~1 atomic mass unit, or amu. The nucleus of **deuterium**, a second isotope of hydrogen, contains one proton and one neutron. Deuterium has chemical properties almost identical to those of protium, but a mass of ~2 amu (Fig. 1.2). **Tritium**, the third naturally occurring isotope of hydrogen, has one proton and two neutrons in its nucleus and thus has a mass of ~3 amu. Whereas both protium and deuterium are stable isotopes of hydrogen, the additional neutron in tritium imparts instability to the nucleus, so that tritium is radioactive with a half-life of 12.3 years. Neither protium nor deuterium will undergo spontaneous radioactive decay.

TABLE 1.2: Charge and mass of the proton, neutron, and electron.

Particle	Charge	Mass (g)
Proton	+1	1.6726×10^{-24}
Neutron	0	$1.6749543 \times 10^{-24}$
Electron	−1	9.109534×10^{-28}

---

[3]The word *isotope* was coined in 1913 by Frederick Soddy, an English scientist who was awarded the 1921 Nobel Prize in Chemistry for his investigations into the origin and nature of isotopes.

FIGURE 1.2: Cartoon of the three isotopes of hydrogen. All have one proton (shown as the dark central sphere) and one electron, but differ in the number of neutrons in the nucleus (the light-colored central spheres). The three isotopes have very similar chemical properties, but their bonds to other elements have slightly different strengths. This phenomenon gives rise to hydrogen isotope fractionation in physical and chemical reactions.

Strictly speaking, any **nuclide**[4] could undergo spontaneous decay, but the *probability* of such decay is negligible for these so-called stable isotopes. The three isotopes of hydrogen have very similar chemical properties, but different masses, and these slight differences in mass result in slightly different strengths of bonds to other elements. In turn, these slight differences in bond strengths are responsible for fractionation of the different isotopes between coexisting phases undergoing a physical or chemical reaction and provide the foundation for all of stable isotope geochemistry.

An example of important effects that can arise as a result of small differences in bond strengths is provided by the chemical and physical properties of the various isotopic forms, or **isotopologues**,[5] of water (Table 1.3). Although the physical and chemical properties of the isotopologues of water are clearly distinct and large, pure isotopologues are not found in nature. Isotopic variations that occur in our solar system are much smaller than the isotopic differences between artificially produced pure isotopologues.

TABLE 1.3: Chemical and physical properties of three of the nine isotopologues of water.

Property	$H_2^{16}O$	$D_2^{16}O$	$H_2^{18}O$
Boiling Point (°C)	100.00	101.42	
Freezing Point (°C)	0.00	3.82	
Density at 0°C (gm/cm^2)	0.999841	1.10469	
Vapor Pressure at 20°C (bars × 10^2)	2.3379	2.0265	2.3161
Temperature of Maximum Density (°C)	4.0	11.6	
Critical Temperature (°C)	374.1	371.5	
Critical Pressure (bars)	220.6	221.5	
Ionization Product K_w at 25°C	1×10^{-14}	0.3×10^{-14}	
Dielectric Constant at 20°C	80.36	79.755	
Surface Tension at 19°C (dynes/cm)	73.66	72.83	
Viscosity at 20°C (centipoise)	1.009	1.260	
Refractive Index η_d at 20°C	1.33300	1.32844	
Representative Solubilities at 25°C (g/g of water)			
NaCl	0.359	0.305	
BaCl$_2$	0.357	0.289	

(From Weast, 1970.)

---

[4]Truman Kohman of Carnegie Mellon University coined the word **nuclide** as a general term for a *specific* isotope. Including those artificially produced, there are more than 2500 known nuclides, most of which are radioactive.
[5]An isotopically substituted molecule, such as $^{12}CO_2$ vs $^{13}CO_2$. See Chapter 2 for further definitions.

1.2.2 Which Elements and Why?

It often comes as a surprise to learn that classical stable isotope geochemistry concerns, for the most part, variations in the stable isotope ratios of only six elements: H, C, N, O, S, and Cl. Note that these elements constitute the bulk of tissues in living organisms. Of late there has been a great deal of interest in the isotopes of Li, Be, B, Si, and Cl and, with recent advances in inductive coupled mass spectrometry and ion probe techniques, in nontraditional isotopes such as Fe, Mg, Ca, Cr, and Mo as well (Johnson et al., 2004). Even without considering these new isotopic systems, the extent of applications of isotope geochemistry for the elements H, C, N, O, and S is enormous, and this text is focused on stable isotope variation of the "traditional" light elements.

Except for certain stable isotope relations in extraterrestrial materials and gases in the upper atmosphere of Earth, stable isotope geochemistry deals mainly with those isotopic variations or effects that arise either from isotopic exchange reactions or from mass-dependent fractionations that accompany biological and physical chemical processes occurring in nature or in the laboratory. While ultimately *quantum mechanical* in origin, such isotope effects are governed by kinetic theory and the laws of thermodynamics. Natural variations in the stable isotope ratios of heavy elements of geological interest, such as Sr, Nd, and Pb, involve nuclear reactions and are governed by other factors, including the ratio of radioactive parent and daughter nuclides, decay constants, and time.

1.3 ABUNDANCES OF THE RARE ISOTOPES OF LIGHT ELEMENTS

Isotopic ratios of the six elements of primary interest to light stable isotope geochemistry are written conventionally as the ratio of the heavy (and rare) isotope to the light (and more abundant) isotope, as in $^{18}O/^{16}O$, $^{34}S/^{32}S$, and the like. Early workers wrote these ratios in the opposite sense and reported values of absolute ratios. For example, $^{12}C/^{13}C$ or $^{32}S/^{34}S$ ratios were reported as relatively large numbers: 91.16 and 22.51, respectively. With our present knowledge of the absolute stable isotope ratios of certain international reference standards, it is now possible to compare these old analyses with modern analyses of similar materials. There is no accepted convention for writing isotopic ratios of other elements of geochemical interest. Sometimes the heavier isotope is the more abundant isotope and is still written in the numerator, as in $^{11}B/^{10}B$. In other cases, the lighter isotope is written in the numerator, regardless of its relative abundance, as in $^{3}He/^{4}He$ or $^{6}Li/^{7}Li$.

The elements under discussion in this text have one dominant isotope, such as ^{16}O or ^{32}S, and one or more rare isotopes, like ^{17}O or ^{18}O, or ^{33}S, ^{34}S, or ^{36}S, whose *average* abundances range from fractions of a percent to a few percent. The isotopic abundances and relative atomic weights of elements whose isotopic ratios vary as a result of mass-dependent processes are given in Table 1.4. Note that boron and chlorine are exceptions to the general rule concerning the disparity in the abundances of the heavy and light isotopes of an element. In these two cases, the abundances of the *rare* isotopes ^{10}B and ^{37}Cl are relatively high, at 19.78 and 24.47 percent, respectively.

1.4 CHARACTERISTICS OF ELEMENTS THAT UNDERGO SIGNIFICANT ISOTOPIC FRACTIONATION

The elements named in the previous section share several characteristics that are not possessed by other elements whose isotopic ratios are not fractionated to any significant extent in nature or in the laboratory. These characteristics, enumerated next, are only *observed* characteristics

TABLE 1.4: Isotopic abundances and relative atomic masses of the pertinent elements in stable isotope geochemistry.

Symbol	Atomic number	Mass number	Abundance (percent)	Atomic weight ($^{12}C = 12.$)
H	1	1	99.985	1.007825
D	1	2	0.015	2.0140
Li	3	6	7.42	6.01512
		7	92.58	7.01600
B	5	10	19.78	10.0129
		11	80.22	11.00931
C	6	12	98.89	≡12.
		13	1.11	13.00335
N	7	14	99.63	14.00307
		15	0.37	15.00011
O	8	16	99.759	15.99491
		17	0.037	16.99914
		18	0.204	17.99916
Si	14	28	92.21	27.97693
		29	4.70	28.97649
		30	3.09	29.97376
S	16	32	95.0	31.97207
		33	0.76	32.97146
		34	4.22	33.96786
		36	0.014	35.96709
Cl	17	35	75.53	34.96885
		37	24.47	36.96590

Symbols for the main elements in the discipline are in boldface.

and are not rigorously tied to theoretical principles. Recent advances in the field of isotope geochemistry of nontraditional stable isotopes (e.g., Fe, Mo, Cd, Mg, and Ca) have shown that not all of the following rules apply to all elements, but the rules are guiding principles for the elements classically treated as the light stable isotopes (H, C, N, O, S):

1. *These elements have a relatively low atomic mass.* Significant mass-dependent isotopic variations in terrestrial materials have been sought, but not clearly demonstrated, in heavier elements such as Sn and Ag.
2. *The relative mass difference between the rare (heavy) and abundant (light) isotope is large.* Compare, for example, the values of 8.3 percent and 12.5 percent for the pairs $^{13}C-^{12}C$ and $^{18}O-^{16}O$, respectively, with the value of only 1.2 percent for $^{87}Sr-^{86}Sr$. The relative mass difference between D (deuterium) and H (protium) is almost 100 percent, and hydrogen isotope fractionations are accordingly about 10 times larger than those of the other elements of interest. Note that a large relative mass difference is by no means sufficient to promote isotopic fractionation: The $^{48}Ca/^{40}Ca$ ratio varies little in terrestrial rocks despite the large relative mass difference between the isotopes (only that of D–H is larger).
3. *These elements form chemical bonds that have a high degree of covalent character.* Elements such as K, Ca, and Mg, which occupy cation sites in minerals, form ionic bonds to other elements and exhibit little or no site preference that could give rise to significant isotopic fractionations. Mg^{2+}, for example, is almost always surrounded by the same atomic environment in nature: an octahedron of oxygen. Nonetheless, small

Ca isotope variations observed in biogenic carbonates may have an origin in mass-dependent fractionation.

4. *These elements can exist in more than one oxidation state (C, N, S), form a wide variety of compounds (notably O), and are important constituents of naturally occurring solids and fluids.* Some of the largest fractionations in nature arise from differences in the nature of the chemical bonds to elements in different oxidation states as in the carbon isotope fractionation between CO_2 and CH_4. Silicon exists in a number of naturally occurring compounds, but is almost always bonded to the same element (oxygen); consequently, its isotopic ratios vary little in nature. Hydrogen is bonded almost exclusively to oxygen in inorganic minerals as —OH groups, but its isotopic composition is influenced greatly by the other ions bonded to the OH oxygen (Mg^{2+}, Al^{3+}, Fe^{2+}, etc.).

5. *The abundance of the element and rare isotope are sufficiently high (ranging from tenths to a few atom percent) to ensure precise determinations of the isotopic ratios by mass spectrometry.* Measurements of isotopic ratios in materials at trace levels are difficult. Large amounts of material are needed, and problems with blanks and contamination are quite serious. With recent advances in the sensitivity of conventional isotope ratio mass spectrometers and the introduction of continuous-flow mass spectrometers, low abundance is less of an issue. The abundance of the rare isotope still separates certain elements, such as He, which has a rare isotope abundance of 0.000137%, from the elements commonly considered the "light stable isotopes."

1.5 APPLICATIONS IN THE EARTH SCIENCES

Stable isotope measurements have been applied successfully to the resolution of fundamental problems in the earth sciences, human sciences, biological sciences, and several subdisciplines of chemistry. Applications in the earth sciences can be broadly classified into four main types:

1. *Thermometry*	Formation temperatures of rock, mineral, and gas systems are determined on the basis of temperature-dependent fractionations of the isotopic ratios between two or more cogenetic phases. Stable isotope thermometry has played a major role in studies of paleoclimatology.
2. *Tracers*	Large reservoirs like the ocean, the mantle, meteoric waters, and organic matter have distinct stable isotope signatures that can be used to trace the origin of rocks, fluids, plants, contaminants, and food sources. Isotopic ratios can also be used as biomarkers.
3. *Reaction mechanisms*	Distinctions can be made between diffusion and recrystallization, open and closed systems, bacterial and thermogenic processes, and various metabolic pathways.
4. *Paleoclimatology*	Isotope ratios of minerals and of fluid and gas inclusions preserve information about conditions in the past. There is a remarkably broad range of materials that have been studied for paleoclimate reconstruction, including bones, gas inclusions in ice (as well as the ice itself), carbonates, cherts, clays, grasses, amber, coal, and eggshells, to name a few.

1.6 ISOTOPE EFFECTS

1.6.1 Kinetic Isotope Effects

Kinetic isotope effects are common both in nature and in the laboratory, and their magnitudes are comparable to, and sometimes much larger than, those of equilibrium isotope effects. Kinetic isotope effects are normally associated with fast, incomplete, or unidirectional processes such as evaporation, diffusion, dissociation reactions, and almost all biological reactions. Isotope effects attendant on diffusion and evaporation are explained by the different translational velocities possessed by the different isotopic forms of molecules as they move through a phase or across a phase boundary. Classical kinetic theory tells us that the average kinetic energy (K.E.) per molecule is the same for all ideal gases at a given temperature. Consider, for example, the molecules $^{12}C^{16}O$ and $^{12}C^{18}O$, which have molecular weights of 28 and 30, respectively. Equating their kinetic energies at some temperature T yields

$$\text{K.E.}(^{12}C^{16}O) = \text{K.E.}(^{12}C^{18}O) \tag{1.1}$$

and

$$\text{K.E.} = \tfrac{1}{2}mv^2, \tag{1.2}$$

where m denotes mass and v represents velocity. Substituting the masses of these isotopologues of CO, we find that the preceding equations reduce to

$$\tfrac{1}{2}(28)(v_{28})^2 = \tfrac{1}{2}(30)(v_{30})^2, \tag{1.3}$$

or

$$v_{28} = \sqrt{30/28}\, v_{30} = 1.035 v_{30}. \tag{1.4}$$

That is, regardless of T, the average velocity of $^{12}C^{16}O$ molecules is 3.5 percent greater than the average velocity of $^{12}C^{18}O$ molecules in the same system.

Such velocity differences can lead to isotopic fractionations in a variety of ways. For example, isotopically light molecules can preferentially diffuse out of a system and leave the reservoir enriched in the heavy isotope. On average, more $^{12}CO_2$ molecules than $^{13}CO_2$ molecules strike the surfaces of leaves and enter the stomates, an effect partially responsible for the low $^{13}C/^{12}C$ ratios of plants relative to other carbon-containing substances in nature. In the case of evaporation, the greater average translational velocities of isotopically lighter water molecules allow them to break through the liquid surface preferentially, resulting in an isotopic fractionation between vapor and liquid that is superimposed on the equilibrium isotopic fractionation. Attesting to this phenomenon is the fact that water vapor over the oceans or over a large lake has $^{18}O/^{16}O$ and D/H ratios that are significantly lower than the ratios that would occur at equilibrium. These lower ratios arise from kinetic isotope effects associated with evaporation.

Molecules containing the heavy isotope are more stable and have higher dissociation energies than those containing the light isotope (Fig. 1.3). As a consequence, it is easier to break bonds like $^{12}C-H$ and $^{32}S-O$ than to break bonds like $^{13}C-H$ and $^{34}S-O$. Kinetic isotope effects arising from differences in dissociation energies can be extremely large in dissociation and in bacterial reactions that occur in nature. Organic matter has distinctly low $^{13}C/^{12}C$ ratios, principally because the ^{12}C isotope is preferentially incorporated into organic

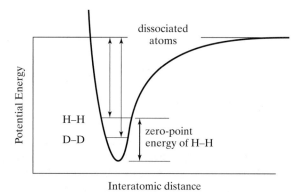

FIGURE 1.3: **Potential-energy diagram for interactions between hydrogen atoms.** At large interatomic distances, the atoms are dissociated. As the two atoms (H and H, H and D, or D and D) approach each other, they are acted upon by an attractive force, and the potential energy drops, reaching a minimum (the bottom of the potential-energy well) at the interatomic distance corresponding to a stable diatomic molecule. At very close interatomic distances, repulsive forces become larger than attractive forces, and the potential energy increases again. The horizontal lines represent the lowest vibrational energy levels for the molecules H_2 and D_2. The *zero-point energy* (shown for H-H as the thick double-arrowed vertical line) is the difference between the energy level at the bottom of the well and the first allowed vibrational energy level and is equal to the vibrational energy a molecule possesses at absolute zero. The energy needed to dissociate the H_2 molecule, depicted by thin double-arrowed vertical lines, is less than that required to dissociate the D_2 molecule, because D_2 has a lower zero-point energy. (See Chapter 3 for more details.)

tissues during photosynthetic reactions. The uniquely light carbon isotope composition of organic matter has been very useful in identifying material of biogenic origin. For example, low $^{13}C/^{12}C$ ratios of materials in ancient sediments have been used to identify the first appearance of life on Earth.

While it is important to be aware of kinetic isotope effects, they are relatively rare in high-temperature processes occurring on Earth. By contrast, transient processes can occur whereby differing rates of isotopic exchange between coexisting minerals themselves, or between the minerals and an external fluid, can result in assemblages that are grossly out of isotopic equilibrium. Such examples are explained, not by kinetic isotope effects, but rather by a series of equilibrium isotope exchange reactions that have not gone to completion.

1.6.2 Equilibrium Isotope Effects

Equilibrium isotope effects can be considered in terms of the effect of atomic mass on bond energy. When a light isotope in a molecule is substituted by a heavy isotope, the nuclear charges and electronic distributions remain unchanged. However, the internal energies of the different isotopes are slightly different due to mass differences, and there is consequently a subtle preference for the heavy isotope to be partitioned into one phase relative to another. For most materials, the fractionation varies regularly as a function of $1/T^2$, and the measured fractionation of two phases, such as two coexisting minerals (Fig. 1.4), or water and a precipitating phase (e.g., biogenic carbonate), can be used to estimate the temperature of formation. The temperature dependence of isotopic fractionation spawned the first major application of stable

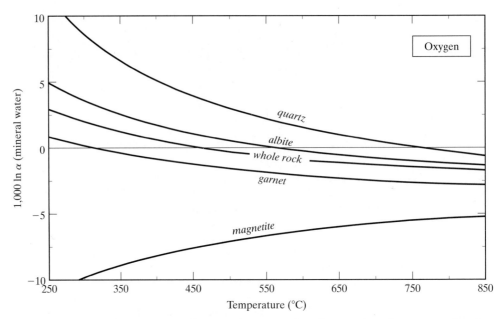

FIGURE 1.4: Oxygen isotope fractionation between selected minerals and water as a function of temperature. Under certain circumstances to be discussed later, knowledge of the $\delta^{18}O$ values of two cogenetic phases allows the temperatures of formation to be determined.
Note: $1{,}000\ln\alpha_{a-b} \cong \delta^{18}O_a - \delta^{18}O_b$ for phases a and b.

isotope chemistry to geological problems: the oxygen isotope paleotemperature scale (Urey et al., 1948; Epstein et al., 1951; Urey et al., 1951). Fractionations of not only oxygen isotopes, but those of hydrogen, carbon, nitrogen, and sulfur, have been used successfully over the years to place constraints on the formation temperatures of both high- and low-temperature systems in nature. These concepts are discussed in more detail in later chapters of this text.

PROBLEM SET

1. Comment on the following:
   a. Combined measurements of oxygen and phosphorus isotope ratios might be useful in tracing the movement of phosphate in the phosphorus cycle on Earth.
   b. Some of today's scientists cite only the most recent papers in a field because the latest information on the subject is probably more reliable and most meaningful.
   c. Applications of stable isotope measurements are almost unlimited in fields related to biomedicine.
2. Calculate the difference in average velocities at constant temperature between the molecules $^{13}C^{16}O^{16}O$ and $^{12}C^{16}O^{16}O$. What significance might this difference play in determining carbon isotope ratios of marine plants?
3. Both Si and Se are present in diatoms and seawater. At a given temperature, would you expect the isotopic fractionation between diatoms and water to be larger for $^{29}Si/^{28}Si$ or for $^{80}Se/^{78}Se$?
4. a. What do you conclude about the characteristics of bonds to H and D from the differences between solubilities of salts in D_2O and H_2O?
   b. Why should the boiling point of D_2O be higher than that of H_2O?

5. a. The atomic weight of fluorine is 18.9984032 ± 0.0000005. The atomic weight of sulfur is 32.065 ± 0.005. Why is the atomic weight of fluorine known to such higher precision than that of sulfur?
   b. Calculate the atomic weight of boron, as listed in the periodic table.
   c. What are the molecular weights of $^{12}C^{16}O^{16}O$, $^{13}C^{17}O^{16}O$, and $^{13}C^{18}O^{18}O$, to six significant figures?

ANSWERS

1. a. Phosphorus has only one isotope.
   b. Although it sometimes takes more work to track down the early papers, the information is often as valuable. It is also scientifically "correct" to credit the author who was the first to address a problem.
   c. Certainly, there are a great deal of applications of stable isotopes to biomedicine. Stable isotopes are a marvelous tracer for reactions, metabolic pathways, etc. The applications are endless.

2. Start with the equation $E = 1/2mv^2$, where E represents energy, m denotes mass, and v designates velocity. For two isotopomers, E is the same, so the velocity of one isotopomer relative to another is given by $m_1 v_1^2 = m_2 v_2^2$. The masses of $^{12}C^{16}O^{16}O$ and $^{13}C^{16}O^{16}O$ are 44 and 45, respectively, so that $v_{44} = v_{45}\sqrt{\frac{45}{44}} = 1.0113 v_{45}$. The velocity of $^{12}C^{16}O^{16}O$ is greater than that of $^{13}C^{16}O^{16}O$. Therefore, the probability of the lighter isotopomer being involved in photosynthesis in marine plants is higher, so that the $\delta^{13}C$ value of marine plants should favor the light isotope of carbon.

3. The relative mass difference between Si isotopes is vastly higher than between Se isotopes. Therefore, Si isotopes should fractionate much more strongly than Se isotopes.

4. a. Salts are less soluble in D_2O, because bonds to D are less **polar** than bonds to H.
   b. Because the molecule is heavier, its translational velocities are slower and the vapor pressure is lower. Also, the D–O bond is stronger than the H–O bond, and the polymeric bonding will also be slightly stronger.

5. a. Fluorine is monoisotopic. Its mass in amu is determined relative to that of ^{12}C and can be determined as accurately as techniques allow. In contrast, sulfur has multiple isotopes, and their relative abundance varies in nature, so different sulfur samples have different atomic weights, depending upon the abundance of the different isotopes.
   b. The atomic weight of an element is given by $\sum_i x_i MW_i$.
   c. From Table 1.4, the molecular weight of boron is $0.1978 \times 10.0129 + 0.8022 \times 11.00931 = 10.812$.

REFERENCES

Baertschi, P. (1950). Isotopic composition of the oxygen in silicate rocks. *Nature* **166,** 112–113.

Baertschi, P., and Silverman, S. R. (1951). The determination of relative abundances of the oxygen isotopes in silicate rocks. *Geochimica et Cosmochimica Acta* **1,** 4–6.

Briscoe, H. V. A., and Robinson, P. L. (1925). A redetermination of the atomic weight of boron. *Journal of the Chemical Society* **127,** 696.

Clayton, R. N., and Epstein, S. (1958). The relationship between O^{18}/O^{16} ratios in coexisting quartz, carbonate, and iron oxides from various geological deposits. *Journal of Geology* **66,** 352–373.

Craig, H. (1953). The geochemistry of the stable carbon isotopes. *Geochimica et Cosmochimica Acta* **3,** 53–92.

Dansgaard, W. (1954). The O^{18}-abundance in fresh water. *Geochimica et Cosmochimica Acta* **6,** 241–260.

Epstein, S., Buchsbaum, R., Lowenstam, H., and Urey, H. C. (1951). Carbonate–water isotopic temperature scale. *Journal of Geology* **62,** 417–426.

Epstein, S., and Mayeda, T. K. (1953). Variation of ^{18}O content of waters from natural sources. *Geochimica et Cosmochimica Acta* **4,** 213–224.

Friedman, I. (1953). Deuterium content of natural water and other substances. *Geochimica et Cosmochimica Acta* **4,** 89–103.

Gilfillan, E. S. J. (1934). The isotopic composition of sea water. *Journal of the American Chemical Society* **56,** 406–408.

Hoering, T., and Parker, P. L. (1961). The geochemistry of the stable isotopes of chlorine. *Geochimica et Cosmochimica Acta* **23,** 186–199.

Hoering, T. C. (1956). Variations in the nitrogen isotope abundance, *Nuclear Processes in Geologic Settings*, Natl. Research Council, Comm. Nuclear Sci., Nuclear Sci. Ser. Rept. no. **19,** Chapter 6, 39–44.

Hoering, T. C., and Moore, H. E. (1958). The isotopic composition of the nitrogen in natural gases and associated crude oils. *Geochimica et Cosmochimica Acta* **13,** 225–232.

Johnson, C. M., Beard, B. L., and Albarède, F. (2004). Overview and general concepts. In *Geochemistry of Non-Traditional Stable Isotopes*, Vol. 55 (ed. C. M. Johnson, B. L. Beard, and F. Albarède), pp. 1–24. Mineralogical Society of America, Washington, DC.

McCrea, J. M. (1950). On the isotopic chemistry of carbonates and a paleotemperature scale. *Journal of Chemical Physics* **18,** 849–857.

Nier, A. O. (1947). A mass spectrometer for isotope and gas analysis. *Review of Scientific Instruments* **18,** 398–411.

Rankama, K. (1954). *Isotope Geology*. New York: McGraw-Hill Book Co., Inc.

Taylor, J. H. P., and Epstein, S. (1962). Relationship between O^{18}/O^{16} ratios in coexisting minerals of igneous and metamorphic rocks, Part 2: Application to petrologic problems. *Geological Society of America Bulletin* **73,** 675–694.

Thode, H. G. (1949). Natural variations in the isotopic content of sulphur and their significance. *Canadian Journal of Research* **27B,** 361.

Trofimov, A. (1949). Isotopic constitution of sulfur in meteorites and in terrestrial objects. *Doklady Akademii Nauk SSSR* **66,** 181–184.

Tudge, A. P. (1960). A method of analysis of oxygen isotopes in orthophosphate; its use in the measurement of paleotemperatures. *Geochimica et Cosmochimica Acta* **18,** 81–93.

Urey, H. C., Epstein, S., McKinney, C., and McCrea, J. (1948). Method for measurement of paleotemperatures. *Bulletin of the Geological Society of America (abstract)* **59,** 1359–1360.

Urey, H. C., Epstein, S., and McKinney, C. R. (1951). Measurement of paleotemperatures and temperatures of the Upper Cretaceous of England, Denmark, and the southeastern United States. *Geological Society of America Bulletin* **62,** 399–416.

Vinogradov, A. P., and Dontsova, E. I. (1947). Izotopnyi sostav kisloroda alyumosilikatnykh gornykh porod. *Doklady Akademii Nauk SSSR* 83–84.

Weast, R. C. (1970). *CRC Handbook of Chemistry & Physics*. Cleveland: Ohio Chemical Rubber Company.

Wickman, F. E. (1941). On a new possibility of calculating the total amount of coal and bitumen. *Geol. Foren. i Stockholm Forh.* **63,** 419–422.

CHAPTER 2

TERMINOLOGY, STANDARDS, AND MASS SPECTROMETRY

2.1 OVERVIEW

Most of the accepted terms and symbols used in stable isotope geochemistry are precise and were developed by the earliest workers, who gave the matter considerable thought. Arguably, some of the terms could be improved, and some recent workers have unilaterally coined new symbols and expressions for reasons known only to themselves. Unfortunately, this practice has caused considerable confusion among new workers, and more and more improper usage is finding its way into the literature and into oral presentations. In this text, the terms established by the founders of our discipline will be used, both in homage to them and because these terms are, for the most part, logical and grammatically correct. In Table 2.1, a number of terms and phrases are presented that are considered to be mistakes, together with the reasons they are unacceptable and recommended alternatives. All examples were culled from the literature. Although some of these common mistakes can be seductive in their simplicity, they should be avoided, in part to preserve the historical purity of the discipline, but most importantly, because they are indeed mistakes and not simply a matter of style.

2.2 ISOTOPOLOGUES, ISOTOPOMERS, AND MASS ISOTOPOMERS

What words best describe different isotopic forms of molecules whose constituent elements have more than one stable isotope—as, for example, the six isotopic forms of carbon monoxide: $^{12}C^{16}O$, $^{13}C^{16}O$, $^{12}C^{17}O$, $^{13}C^{17}O$, $^{12}C^{18}O$, and $^{13}C^{18}O$? Urey called such molecules **isotopically substituted molecules**, a somewhat awkward and inexact expression. In the ensuing years, new, precise terms, including **isotopologue**, **isotopomer**, **mass isotopomer**, and **isotopic isomer**, were introduced to describe different types of isotopically substituted molecules. According to recommendations made in 1994 by the International Union of Pure and

TABLE 2.1: Common mistakes in terminology and phraseology.

Mistake	Recommended expressions	Explanation
Referring to the symbol δ as del $\delta^{18}O$, not $\partial^{18}O$	Since the time of the early Greeks, the name of this symbol has been, and remains, **delta**.	The word *del* describes either of two things in mathematics and science: an operator (∇) or the partial derivative (∂)
$\delta^{13}C$ composition	$\delta^{13}C$ value; carbon isotope composition	$\delta^{13}C$ values are numbers; a *composition of numbers* has no meaning.
Isotopically depleted water	^{18}O (or D) depleted water	A given sample of water is neither depleted nor enriched in isotopes.
Heavy (light) $\delta^{18}O$ values	High (low) $\delta^{18}O$ values	As numbers, δ values can be high or low, positive or negative, but not heavy or light.
Isotopically negative	Relatively low δ values	Isotopic ratios are neither negative nor positive; they are lower or higher than those of the standards.
Depleted $\delta^{13}C$ value Enriched (depleted) carbonates. Enriched (depleted) compositions Depleted carbon reservoir	Low $\delta^{13}C$ value (relative to another) isotopically heavy (light) carbonates (relatively) ^{18}O-rich or ^{13}C-poor carbonates reservoir of (isotopically) light carbon	$\delta^{13}C$ values are numbers; as such, they cannot be depleted or enriched. The words *enrich* and *deplete* are overused and much abused. These words should be reserved for describing a **process** that changes the content of the heavy isotope of the element in some substance.
Oxygen isotopes **in** chert; inferred from carbon isotopes; isotopes of soil water	Oxygen isotope **ratio (composition)** of chert; inferred from carbon isotope **measurements**; isotopic composition of soil water	Such written mistakes are a carryover from loose oral communication.
The isotopic composition of the water was $\delta^{18}O = -4.3‰$.	The $\delta^{18}O$ value of the water was $-4.3‰$.	A matter of redundancy.
The isotopic value changed.	The isotopic composition changed. The $\delta^{18}O$ value changed.	The phrase *isotopic value* is ambiguous. Does it denote R? δ? which element?
The isotopic signature of the rock was $\delta^{18}O = 5.7‰$.	The $\delta^{18}O$ value of the rock was 5.7‰. Thus, this rock has the oxygen isotope signature of the mantle.	The word **signature** should be used to describe the isotopic composition of a significant reservoir, such as the mantle, the ocean, or a major part of the system being studied, not to describe the isotopic composition of ordinary samples.
δ^{15}, δ^{18}, δ^{13}, etc.; $^{15}\delta$, $^{18}\delta$, $^{13}\delta$, etc.; $\delta-15$, $\delta-18$, $\delta-13$, etc.	$\delta^{15}N$, $\delta^{18}O$, $\delta^{13}C$, etc.	The introduction of new symbols that save one character of space is unnecessary at best and confusing at worst.
Sulfur was measured.	The sulfur **isotope composition** was measured.	Misleading because the reader may take the sentence to mean that the sulfur **content** of a rock or mineral was measured.
The ^{13}C content of ...	The $^{13}C/^{12}C$ ratio of ...	^{13}C content refers to how much ^{13}C there is in a rock. A sample of coal has a lot of ^{13}C (a high ^{13}C content), but a low $^{13}C/^{12}C$ ratio compared with most materials.

Applied Chemistry (IUPAC), **isotopologues** are molecules that differ from one another only in isotopic composition, as is the case for the carbon monoxide molecules mentioned above. Isotopologues *can have the same or different masses*. The word *isotopologue* is the appropriate term to describe such molecules that are encountered in stable isotope geochemistry and will be employed frequently in this text.

Isotopomers[1] (a contraction of *isotopic isomers*) are isotopologues that differ from one another only in the positions or locations of the isotopic elements. Thus, isotopomers always comprise the same number of each isotopic atom and thus *always have the same mass*. They differ from one another in the positions or locations of the isotopic elements and thus the connection to isomerism. Two different isotopic forms of acetaldehyde—$CH_2DCH=O$ and $CH_3CD=O$—provide an example of isotopomers. They have the same isotopic composition, but the D atom is bonded to the methyl group carbon in the first case and to the carboxyl carbon in the second. Isotopic forms of nitrous oxide ($^{15}N^{14}NO$ and $^{14}N^{15}NO$) and ozone ($^{16}O^{18}O^{16}O$ and $^{18}O^{16}O^{16}O$) are among the few isotopomers studied by stable isotope geochemists. In mass spectrometry, the expression *mass isotopomer*, normally denoting an organic compound, is used to describe a family of isotopologues that *have the same mass*. Because mass isotopomers are collected simultaneously on the same collectors of a mass spectrometer, they pose a problem in isotopic analysis. The molecules $^{13}C^{16}O$ and $^{12}C^{17}O$ are mass isotopomers that each have mass 29.

2.3 THE DELTA VALUE

Relative differences in isotopic ratios can be determined far more precisely than absolute isotopic ratios. McKinney et al. (1950) introduced the delta (δ) notation to report stable isotope data for all materials except interstellar dust, whose isotopic ratios and variations are frequently so large that absolute ratios are commonly (and reasonably) used in publications. (See Chapter 13.) The delta value is given by

$$\delta = \left(\frac{R_x - R_{std}}{R_{std}} \right) \times 1000, \tag{2.1}$$

where R is the ratio of the abundance of the heavy to the light isotope, x denotes the sample, and std is an abbreviation for standard. For the elements hydrogen, carbon, nitrogen, oxygen, sulfur, and chlorine, R is given by D/H, $^{13}C/^{12}C$, $^{15}N/^{14}N$, $^{18}O/^{16}O$, $^{34}S/^{32}S$, and $^{37}Cl/^{35}Cl$, respectively. The notation $^2H/^1H$ is strictly correct for hydrogen isotope ratios, but for historical reasons D/H is preferred by most workers.

Delta values are reported in per mil, or parts per thousand, and the symbol for per mil is ‰. A positive δ value means that the ratio of the heavy to the light isotope is higher in the sample than it is in the standard, and a negative δ value has the opposite meaning. A sample with a $\delta^{18}O$ value of +19.7‰ has an $^{18}O/^{16}O$ ratio that is 19.7 per mil, or 1.97 percent higher than that of the standard. Similarly, a negative δD value of −65.2‰ means that the D/H ratio of the sample is 65.2 per mil, or 6.52 percent lower than that of the standard. The δ value is what is reported on the basis of the intensities of ion signals measured in the isotope ratio mass spectrometer. Mass spectrometric analyses of pure gases other than H_2 are *reproducible* to ±0.01‰ or better. Excellent reproducibility like this does not represent the precision of an individual analysis of a natural sample, because errors are introduced from the collection and chemical preparation procedures employed. In assessing the precision of stable isotope analyses of waters and carbonates, early workers stated, without a clear explanation, that the overall precision

---

[1] Isotopomers can be either **constitutional isomers**, like the examples provided in the text, or **isotopic stereoisomers**, as in *cis-* and *trans-* forms of molecules that contain different isotopes of some element.

of δ values is about 0.1‰, or 10‰ of the δ value, whichever number is larger. That is, large δ values were considered inherently less precise than small δ values. Certainly, it can still be argued that the δ values of artificially enriched or unusual extraterrestrial samples are not known to the same precision as the δ values of samples close to that of the reference standard, but this is generally not an important factor in attempts to use stable isotope ratios to resolve geochemical problems. Such a consideration becomes important, however, when δ values are used to establish thermodynamic constants like equilibrium fractionation factors.

Oxygen isotope compositions are reported using the symbol $\delta^{18}O$, those of carbon by $\delta^{13}C$, and so forth. The symbol δ is the lowercase Greek letter **delta** and is commonly used in many disciplines to express a difference. The unrelated word *del* crept into the parlance of stable isotope geochemistry about 20 years ago, presumably as some kind of abbreviation for the correct word *delta*.[2] This incorrect usage should be abandoned because the word del has been used for centuries to denote either the symbol ∇, a mathematical operator, or the symbol ∂, the partial-derivative sign.

The delta notation provides a very convenient means of expressing the small **relative differences** in isotopic ratios between samples and standards that are measured by isotope ratio mass spectrometry. The effective precision of a stable isotope measurement is much higher than is immediately obvious from the apparent precision of a δ value, which in the best case is ±0.01‰. The absolute $^{18}O/^{16}O$ ratio of Standard Mean Ocean Water (SMOW), one of the international reference standards, is 2005.20×10^{-6}. (See later.) That is, this ratio is known to six significant figures, or to one part in 10^5. The $^{18}O/^{16}O$ ratio of a gas whose $\delta^{18}O$ value is +2.06‰ *relative* to SMOW ($\delta^{18}O$ of SMOW = 0.00‰ by definition) is 0.00200933, the value obtained by substituting 2.06 for δ in equation (2.1):

$$2.06 = \frac{^{18}O/^{16}O_{sample} - 2005.20 \times 10^{-6}}{2005.20 \times 10^{-6}} \times 1000. \tag{2.2}$$

A $\delta^{18}O$ value of +2.05‰ corresponds to an absolute ratio of 0.002009311, and so on. Given the absolute $^{18}O/^{16}O$ ratio of a reference standard, the absolute $^{18}O/^{16}O$ ratio of any other material can be determined or known to the same precision that we know the absolute ratio of the standard, in this case 1 part in 10^5. But note that a difference of 0.01‰ in $\delta^{18}O$ corresponds to a change of 0.00000002 in the absolute ratio. In other words, differences in absolute $^{18}O/^{16}O$ ratios of two substances can be detected at the remarkable level of two parts in the eighth decimal figure! Isotopic ratios of certain heavy elements commonly met in geochemistry can be determined very precisely simply because such ratios are close to unity (e.g., $^{87}Sr/^{86}Sr_{modern\ seawater}$ = 0.70918). The ability to determine relative differences in small isotopic ratios at the seventh or eighth decimal figure makes stable isotope measurements among the most precise measurements attainable in all of geochemistry.

Isotopic compositions of samples are measured relative to the isotopic composition of a reference gas, the **working standard**, in a mass spectrometer. To convert the δ value of sample X from one scale (reference standard A) to another scale (reference standard B), the following equation is used:

$$\delta_{X-B} = \delta_{X-A} + \delta_{A-B} + 0.001\delta_{X-A}\delta_{A-B}. \tag{2.3}$$

---

[2] I first encountered students using ∂ in place of δ because the symbol took only one stroke of a keyboard (on a Mac) instead of two for the δ. Needless to say, this was not a convincing reason to abandon the correct notation.

This simple calculation, analogous to converting temperatures on the Celsius scale to temperatures on the Fahrenheit scale, is made in every stable isotope laboratory in the world.[3] Laboratory working standards are calibrated relative to international reference standards, precious materials distributed to qualified workers by the International Atomic Energy Agency (IAEA) or the National Institute of Standards and Technology (NIST). In most stable isotope laboratories, there are supplies of gases such as CO_2, N_2, and H_2 contained in metal or glass tubes and tanks that are fitted with appropriate valves to allow aliquots of the gases to be taken for use as working standards or for calibration purposes. The δ values of the gases are well known from repeated measurements relative to the values of primary or secondary reference standards.

Suppose that the CO_2 working standard (WS) used in a given mass spectrometer has a $\delta^{13}C$ value of +4.75‰ relative to the international standard Pee Dee Belemnite (PDB; $\delta_{\text{WS-PDB}}$). If sample X has a $\delta^{13}C$ value of −22.32‰ relative to WS ($\delta_{\text{X-WS}}$), then the $\delta^{13}C$ value of X on the PDB scale is

$$4.75 - 22.32 + 0.001(4.75)(-22.32) = -17.68‰. \tag{2.4}$$

That is, δ values are converted from the working standard of the mass spectrometer to PDB, or to any international reference standard, simply by adding a scaling term $10^{-3}(\delta_{\text{X-WS}})(\delta_{\text{WS-PDB}})$ to the sum of the two delta values $\delta_{\text{X-WS}}$ and $\delta_{\text{WS-PDB}}$. Another equation (essentially the same equation, rearranged in a slightly different format) frequently used to calculate a change in scale like this has, for the case in question, the form

$$1.00475(-22.32) + 4.75 = -17.68‰. \tag{2.5}$$

The difference of 4.75‰ between the isotopic composition of the working standard and PDB must be added to the measured *raw* value (−22.32), but only after the raw value has been corrected for an expansion in scale, equivalent in this case to a multiplicative factor of 1.00475 (equal to $1 + 0.001 \times \delta_{\text{WS-PDB}}$). Note that there is a contraction or expansion in scales involved in these calculations, and this term is directly related to the magnitude of the difference in δ values between the two standards. The *size* of δ values changes from one scale to another. When converting between scales, one must apply both an additive and a multiplicative factor to the raw data. If the δ value of a working standard is 12.34‰, the multiplicative factor is 1.01234 (i.e., $1 + 0.001 \times \delta$) and the additive factor is 12.34. If the δ value of another working standard is −6.78‰, the multiplicative factor is 0.99322 ($1 + 0.001 \times -6.78$), which in this case contracts the scale, and the additive factor is −6.78. In order to minimize the size of the multiplicative factor, international reference standards were prepared with isotopic compositions that are as close as possible to the range of isotopic compositions of natural materials expected to be analyzed by most workers. In the same vein, a researcher can choose working standards whose isotopic compositions are close to those of the materials most commonly met in the research at hand. This practice results in only small improvements in precision, but is worthwhile in any case.

Over the years, stable isotope geochemists have tacitly developed a certain uniformity in the presentation of their data. Some notations used in the past have all but disappeared in the modern literature, but are noted here for the sake of completeness. In the early literature, you will see the expressions $\delta(O^{18}/O^{16})$, $\delta(D/H)$, and so forth, but these strictly more correct

---

[3] Certainly, many workers are not aware that such calculations are being made, as the conversion equations are hidden in the software packages provided with their mass spectrometers.

notations soon gave way to the simpler expressions $\delta^{18}O$ and δD. Prior to the mid-1970s, the mass number was always written as a right superscript of the symbol of the element, as in O^{18}, C^{14}, and U^{235}. It is for this reason that one usually *hears* the element name (or symbol) and number spoken in that order, as in "C-14 dating," "Sr-90 contamination," or "O-18 determinations." Subsequently, IUPAC officially changed the order in which mass number and symbol are written, to allow oxidation states and other identifying marks to be written to the right of the symbol for the element, something that is rarely, if ever, done. The notation δ^2H is frequently employed by hydrologists and is, in fact, more consistent than the notation δD, which is preferred by most workers in other fields and used in this text. In the early literature, δD and $\delta^{13}C$ values were often given in percent rather than per mil.

BOX 2.1 Why is ^{12}C the official reference mass for atomic mass units?

Prior to the 1970s, two conventions were used for determining relative atomic masses. Physicists related their mass spectrometric determinations to the mass of ^{16}O (i.e., ^{16}O has a mass of *exactly* 16 on the amu scale), the most abundant isotope of oxygen, and chemists used the weighted mass of all three isotopes of oxygen: ^{16}O, ^{17}O, and ^{18}O. At an international congress devoted to the standardization of scientific weights and measures, the redoubtable A.O. Nier proposed a solution to these disparate conventions whose negative consequences were becoming serious. He suggested that the carbon-12 isotope (^{12}C) be the reference for the atomic mass unit (amu). By definition, its mass would be *exactly* 12 amu, a convention that would be acceptable to the physicists. In accordance with this convention, the average mass for oxygen (the weighted sum of the three naturally occurring isotopes) becomes 15.9994 amu, a number close enough to 16 to satisfy the chemists.

2.4 ISOTOPE EXCHANGE REACTIONS

Isotope exchange reactions are equilibrium reactions in which isotopes of a single element are exchanged between two substances. The chemical makeup of reactants and products are identical. For the general case of an isotope exchange reaction between two substances A and B, where the subscripts 1 and 2 refer to molecules *totally substituted* by the light and heavy isotope, respectively, and a and b refer to the coefficients necessary to balance the reaction,

$$aA_1 + bB_2 = aA_2 + bB_1. \tag{2.6a}$$

An example is given by

$$^{12}CH_4 + {}^{13}CO_2 = {}^{13}CH_4 + {}^{12}CO_2. \tag{2.6b}$$

(In equation (2.6b), the coefficients of the reaction are all unity; see equation (2.9) for another example and section 3.2.1 for further details.) A methane isotopologue like $^{12}CH_3D$ (only one of the four hydrogen atoms is deuterium) is not in this class of molecules. The equilibrium constant for the preceding reaction is written in the usual way:

$$K = \frac{(A_2)^a(B_1)^b}{(A_1)^a(B_2)^b} = \frac{(A_2/A_1)^a}{(B_2/B_1)^b}. \tag{2.7}$$

The terms in parentheses are activities, but in practice, concentrations are normally used. The difference between concentrations and activities of isotopologues is normally negligible (i.e., the activity coefficient $\gamma \sim 1$), so that substituting concentrations for activities is valid. In rare cases, such as the case of concentrated brines, the small difference between activities and concentrations of hydrogen and oxygen isotopes becomes an issue.

The equilibrium constant K can be expressed in terms of partition functions Q, mathematical relations arising from statistical mechanics and containing all the energy information possessed by a molecule (Chapter 3). This formalism is used in the calculation of equilibrium constants for isotope exchange reactions from spectroscopic data and the methods of statistical mechanics. In terms of partition function ratios Q_2/Q_1,

$$K = \frac{(Q_2/Q_1)_A^a}{(Q_2/Q_1)_B^b}, \qquad (2.8)$$

where Q_2 and Q_1 are partition functions for molecules of A and B that are totally substituted by the heavy and light isotopes of the element, respectively.

For ease of mathematical manipulations, isotope exchange reactions are normally written such that *only one atom is exchanged*. Take, for example, the oxygen isotope exchange reaction between CO_2 and H_2O:

$$\tfrac{1}{2}C^{16}O_2 + H_2^{18}O = \tfrac{1}{2}C^{18}O_2 + H_2^{16}O. \qquad (2.9)$$

The equilibrium constant for this reaction is

$$K = \frac{(C^{18}O_2)^{1/2}(H_2^{16}O)}{(C^{16}O_2)^{1/2}(H_2^{18}O)}. \qquad (2.10)$$

Similarly, the isotope exchange reaction between SO_3 and O_2 is written

$$\tfrac{1}{3}S^{16}O_3 + \tfrac{1}{2}{}^{18}O_2 = \tfrac{1}{3}S^{18}O_3 + \tfrac{1}{2}{}^{16}O_2, \qquad (2.11)$$

and the various expressions for the equilibrium constant for this reaction are

$$K = \frac{(S^{18}O)^{1/3}\,({}^{16}O_2)^{1/2}}{(S^{16}O)^{1/3}\,({}^{18}O_2)^{1/2}} = \frac{(Q_2/Q_1)_{SO_3}^{1/3}}{(Q_2/Q_1)_{O_2}^{1/2}} = \frac{({}^{18}O/{}^{16}O)_{SO_3}}{({}^{18}O/{}^{16}O)_{O_2}}. \qquad (2.12)$$

Recall that $C^{18}O_2$ means that both oxygen atoms in the molecule are ^{18}O, but to all intents and purposes, such isotopologues do not exist in nature. Their occurrence is statistically improbable because of the low abundance of the rare isotopes in nature. This formalism is used simply because it is easier to make calculations for isotopologues that contain only one isotope of the element in question. The calculation of isotopic partition function ratios by the methods of statistical mechanics is discussed in Chapter 3.

2.5 THE FRACTIONATION FACTOR

The isotopic fractionation factor between two substances A and B is defined as

$$\alpha_{A-B} = \frac{R_A}{R_B}. \qquad (2.13)$$

In terms of δ values, this expression becomes

$$\alpha_{A-B} = \frac{1 + \dfrac{\delta_A}{1000}}{1 + \dfrac{\delta_B}{1000}} = \frac{1000 + \delta_A}{1000 + \delta_B}. \qquad [2.14]$$

If the isotopes are randomly distributed over all possible sites or positions in substances A and B, the fractionation factor α is related to the equilibrium constant K for isotope exchange reactions by the formula

$$\alpha = K^{1/n}, \qquad (2.15)$$

where n is the number of atoms exchanged, normally 1, as explained earlier. For the isotope exchange reaction between CO_2 and H_2O (equation (2.9)):

$$K = \alpha = \frac{(^{18}O/^{16}O)_{CO_2}}{(^{18}O/^{16}O)_{H_2O}}. \qquad (2.16)$$

Values of α are normally very close to unity, typically 1.00X. It is common to discuss isotopic fractionations in terms of the value of X, in per mil. As a true thermodynamic equilibrium constant, α is a function of temperature, so values of α are meaningful only when the temperature is specified. For example, the sulfur isotope fractionation between sphalerite (ZnS) and galena (PbS) is 1.00360 at 200°C. It is accepted parlance to state that, at 200°C, (1) the sphalerite–galena fractionation is 3.60 per mil, or (2) sphalerite concentrates ^{34}S by 3.60 per mil relative to galena.[4]

2.6 $10^3 \ln \alpha$, Δ, AND THE ε VALUE

It is a useful mathematical fact that $10^3 \ln(1.00X)$ is approximately equal to X. For the example of $^{34}S/^{32}S$ mentioned earlier, where $\alpha = 1.00360$, $10^3 \ln \alpha = 3.594$. That is, $10^3 \ln \alpha$ *is the fractionation*, but expressed more simply. It is sometimes called the **per mil fractionation**, but this terminology is not strictly correct, because α is unitless. This logarithmic function has added theoretical significance. For perfect gases, $\ln \alpha$ varies as $1/T^2$ and $1/T$ in the high and low temperature limits, respectively. In addition, smooth and often linear curves obtain when $10^3 \ln \alpha$ is plotted against $1/T^2$ for experimentally determined fractionation factors between mineral pairs or mineral–fluid pairs. As in any expressions or calculations in thermodynamics, T is absolute temperature in kelvins.

The fractionation expressed as $10^3 \ln \alpha$ is of prime importance in stable isotope geochemistry. This quantity is very well approximated by the Δ value:

$$\Delta_{A-B} = \delta_A - \delta_B \approx 10^3 \ln \alpha_{A-B}. \qquad (2.17)$$

That is, merely subtracting δ values is a good approximation to the per mil fractionation and identical to it within the limits of analytical error **when the individual values of δ_A and δ_B,**

---

[4]Note that the difference of 3.60‰ is strictly correct only if the $\delta^{34}S$ value of the galena is exactly zero.

TABLE 2.2: Comparison between values obtained from different expressions for isotopic fractionations.

δ_A	δ_B	α	Δ_{A-B}	ε_{A-B}	$10^3 \ln \alpha_{A-B}$
1.00	0.00	1.00000	1.00	1.00	1.00
5.00	0.00	1.00500	5.00	5.00	4.99
10.00	0.00	1.01000	10.00	10.00	9.95
12.00	0.00	1.01200	12.00	12.00	11.93
15.00	0.00	1.01500	15.00	15.00	14.89
20.00	0.00	1.02000	20.00	20.00	19.80
10.00	5.00	1.00498	5.00	4.98	4.96
20.00	5.00	1.00493	5.00	4.93	4.91
30.00	5.00	1.00488	5.00	4.88	4.87
30.00	20.00	1.00980	10.00	9.80	9.76
30.00	15.00	1.01478	15.00	14.78	14.67
30.00	10.00	1.01980	20.00	19.80	19.61

as well as Δ_{A-B}, are less than about 10. As the numbers in Table 2.2 indicate, however, it is important to calculate the exact function $10^3 \ln \alpha$ when the fractionations or the δ values are greater than 10. Δ is the uppercase symbol for the Greek letter delta and is frequently expressed orally as **"big delta."**

As has been mentioned, equilibrium constants for isotope exchange reactions (isotopic fractionation factors) can be calculated from isotopic partition function ratios Q_2/Q_1. The function $\ln(Q_2/Q_1)$ is called the **β-factor** and is frequently used in theoretical calculations. A few authors have used the symbol ε to designate an isotopic fractionation and define it as $\varepsilon = (\alpha - 1)10^3$. Again, for small values of ε, this function is almost identical to $10^3 \ln \alpha$ and is a better approximation than Δ (Table 2.2). It is recommended that ε not be used in stable isotope work, because that symbol is used for another purpose in the sister discipline of radiogenic isotope geochemistry.

The α function is used in a variety of analytical techniques. For example, the $\delta^{18}O$ value of a sample of liquid (l) water is determined by equilibrating the water with a small amount of CO_2 gas at some temperature, normally 25°C, and measuring the oxygen isotope composition of the equilibrated CO_2 gas in a mass spectrometer. At 25°C, the fractionation factor between CO_2 and H_2O_l is 1.04120. This statement is approximately equivalent to stating that CO_2 is 40.37‰ ($10^3 \ln \alpha = 40.37‰$) heavier than the water with which it was equilibrated. As an interesting aside, no accurate value for this fractionation factor was available to the early workers. Calculations by Urey suggested that it was about 39‰, but knowledge of the exact value was not necessary for purposes of analysis. It is important only that the fractionation factor be *constant* at a given temperature. It is always possible to determine *relative* differences in isotopic compositions of water, as long as samples and standards are treated in exactly the same way (with respect to temperature of equilibration, relative ratios of CO_2 and H_2O, time, etc.)

As another example, the isotopic compositions of carbon and oxygen in carbonates are determined on the CO_2 that is released during the reaction between the carbonate and phosphoric acid. All of the carbon, but only two-thirds of the oxygen, in the carbonate is transferred to the CO_2 gas. There is a temperature-dependent isotopic fractionation between the oxygen in the evolved CO_2 and the oxygen in the original carbonate: As long as the temperature of acid dissolution is held constant, the so-called **acid fractionation factor** will be constant as well. If we know the α value between evolved CO_2 gas and carbonate at the reaction temperature, we can calculate the $\delta^{18}O$ value of the carbonate itself. At 25°C, $\alpha(CO_2\text{-calcite})$ for the phosphoric acid reaction is 1.01025. That is, the liberated CO_2 is about 10‰ heavier than the calcite. From equation (2.15), we have

$$1.01025 = \frac{1000 + \delta_{CO_2}}{1000 + \delta_{calcite}}. \tag{2.18}$$

If the measured $\delta^{18}O$ value of the liberated CO_2 gas is 3.40‰, the $\delta^{18}O$ value of the calcite is −6.78‰.

The acid reaction can be done at any temperature, and the technique calibrated by analyzing standards at the same temperature. The value of α is then easily determined from equation (2.18) if one knows the $\delta^{18}O_{calcite}$ and measures the $\delta^{18}O_{CO_2}$ value on the mass spectrometer. Acid fractionation factors are different for different carbonates and become smaller at higher temperatures. For example, the acid fractionation factor for calcite at 50°C is 1.00931, about 1‰ smaller than it is at 25°C. These corrections are normally incorporated into the software programs used in the operation of modern mass spectrometers, but are easily done by hand as well.

2.7 REFERENCE STANDARDS

Very precise comparisons of isotopic compositions of materials can be determined in a given laboratory, but to compare data obtained in different laboratories, an internationally accepted set of reference standards is available to all workers in the field. The measured isotopic composition of any substance should be the same in all laboratories after calibrations have been made with these international reference standards. Beginning in the 1970s, committees of stable isotope geochemists convened periodically in Vienna to select standard materials and to establish protocols for calibrating mass spectrometer analyses and presenting stable isotope data (Coplen and Clayton, 1973; Coplen et al., 1983; Hut, 1987; Coplen, 1996). These reference materials (Table 2.3) are available from the National Institute of Standards and Technology (NIST) in Gaithersburg, Maryland (NIST, 100 Bureau Drive, Gaithersburg, MD, USA) and from the International Atomic Energy Agency (IAEA) in Vienna (P.O. Box 100, Wagramer Strasse 5, A-1400 Vienna, Austria; http://www.iaea.org/worldatom). International reference standards are in limited supply and are *not* intended for use as working standards. They are provided in small quantities to allow workers to establish larger supplies of secondary reference materials (solids, liquids, and gases) that in turn can be used on a daily basis as working standards, for calibrating extraction techniques, and so on.

The history of stable isotope reference materials is long and complex. The early Chicago group reported $\delta^{13}C$ and $\delta^{18}O$ values of carbonates relative to the carbon and oxygen isotope compositions of a powdered specimen of *Belemnitella americana* from the Upper Cretaceous Peedee formation of South Carolina. They called this calcite standard PDB (PeeDee Belemnite). When the original supply of this material became impoverished, another sample was prepared and named PDB II, a standard that was later replaced by PDB III. In each case, the new standard was carefully calibrated against the isotopic composition of the original sample of PDB. Despite the fact that the original supply of PDB is exhausted, PDB remains the standard used in reporting all carbon isotope analyses and most of the oxygen isotope analyses of low-temperature carbonates.

TABLE 2.3: δ values of some common stable isotope reference standards.

Reference standard	Substance	$\delta^{18}O$ (SMOW)	$\delta^{18}O$ (PDB)	$\delta^{13}C$ (PDB)	δD (SMOW)
VSMOW (Standard Mean Ocean Water)	water	$\equiv 0.00$			$\equiv 0.00$
SLAP (Standard Light Antarctic Precipitation)	water	$\equiv -55.50$			$\equiv -428.00$
GISP (Greenland Ice Sheet Precipitation)	water	-24.78			-189.73
NBS-19	calcite	28.64	$\equiv -2.20$	$\equiv +1.95$	
NBS-18	calcite	7.20	-23.00	-5.01	
IAEA-CO-1	calcite	28.39	-2.44	2.48	
IAEA-CO-8	calcite	7.54	-22.67	-5.75	
USGS-24	graphite			-15.99	
NBS-22	oil			-29.74	
IAEA-C-6	sucrose			-10.43	
NBS-28	quartz	9.58			
NBS-30	biotite	5.24			-65.7

		$\delta^{18}O$ (SMOW)	$\delta^{15}N$ (AIR)	$\delta^{34}S$ (CDT)
NSVEC	N_2 gas		-2.77	
NBS-14	N_2 gas		-1.18	
IAEA-N-1	$(NH_4)_2SO_4$		0.43	
IAEA-N-2	$(NH_4)_2SO_4$		20.32	
IAEA-NO-3	KNO_3	25.3	4.69	
IAEA-S-1	Ag_2S			$\equiv -0.30$
IAEA-S-2	Ag_2S			22.67
AEA-S-3	Ag_2S			-32.55
NBS-123	sphalerite			17.44
NBS-127	$BaSO_4$	8.7		21.1

A complete set is given in Appendix 1.

The Chicago group also reported oxygen isotope analyses relative to a standard that they called Mean Ocean Water, a sample of real ocean water that was prepared with the aim of representing the oxygen isotope compositions of all the major oceans in more-or-less correct proportions. Later, a fictitious reference standard, *defined* in terms of the isotopic composition of a sample of freshwater, was introduced and called SMOW (Standard Mean Ocean Water). To complicate matters, scientists in various laboratories throughout the world were developing their own working standards, occasionally obtaining materials of *known* isotopic composition from established laboratories. Predictably, chaos ensued, and there was widespread mistrust of the reliability of published isotopic analyses. Ultimately, relatively large supplies of standard materials were prepared for worldwide distribution, and standardization procedures were developed. In the hands of competent analysts, standardization procedures are now relatively routine, and stable isotope analyses made anywhere in the world are, for the most part, reliable.

2.7.1 Hydrogen

In much of the early literature on the abundance of deuterium in natural materials, a sample of Lake Michigan water was used as a reference standard. The δD value of the Lake Michigan standard is $-42.4‰$ on the SMOW scale. Today, all hydrogen isotope analyses are reported relative to SMOW, a logical geochemical reference material because ocean water is by far the largest terrestrial reservoir of water. By definition, the δD value of SMOW is equal to zero. SMOW has a D/H ratio that is higher than the ratios of most other materials on Earth, an interesting geochemical fact in itself. Thus, most δD values of natural materials on our planet are negative on this scale, in contrast to δD values of extraterrestrial substances, which can be extremely positive for reasons explained in Chapter 13.

In 1961, Harmon Craig *defined* the hydrogen and oxygen isotope ratios of SMOW relative to NBS-1 (Fig. 2.1), a sample of Potomac River water that was distributed by the National Bureau of Standards (now NIST) for laboratory calibration. The relation between D/H ratios of SMOW and Potomac River water NBS-1 was defined as follows (Craig, 1961):

$$(D/H)_{SMOW} = 1.050(D/H)_{NBS-1}. \qquad (2.19)$$

Note how δ values change when the scale changes. SMOW has a δD value of 50‰ relative to NBS-1 (on the NBS-1 scale), but NBS-1 has a δD value of −47.6‰ on the SMOW scale. Laboratory standards were calibrated relative to SMOW by the analysis of NBS-1. Another sample of water with a very low δ value, NBS-1A, was also distributed in those days and served as an additional check on the calibration. To resolve the problem that SMOW was only a defined standard, Harmon Craig and Ray Weiss distilled a large sample of ocean water and adjusted its hydrogen and oxygen isotope compositions to those of SMOW by carefully adding appropriate amounts of other waters of different isotopic compositions. Thus, they prepared a supply of their newly created SMOW that is now kept both in Vienna and in Gaithersburg for distribution by the I.A.E.A. and NIST, respectively. There is no difference in oxygen isotope composition between defined SMOW and the actual SMOW. There is a reputed minuscule difference between the hydrogen isotope composition of defined and actual SMOW, but it cannot be detected in most, if any, laboratories.

Many workers call the real sample of standard mean ocean water V-SMOW, where V stands for Vienna, the location of the IAEA. The purported intention of engaging in this practice is to assure the reader that the mass spectrometers and working standards used in the study were properly calibrated. In fact, the use of "V" in front of SMOW indicates that a mass spectrometer was calibrated in such a way that the δD and $\delta^{18}O$ values of SLAP are −428 and −55‰ relative to VSMOW (Coplen, 1995), a calibration that is certainly not made in every laboratory that uses VSMOW notation. The practice of adding V to accepted acronyms for international reference standards is unnecessary at best. The least ambiguous method of presenting stable isotope data is to report δ values relative to SMOW, PDB, CDT, etc., and to state what calibration corrections were actually made.[5]

The raw δD value of a sample whose D/H ratio is quite different—say, 20–30‰ or more—from that of the working standard will be different when measured on different mass spectrometers. The factor most responsible for this effect is the inevitable production of the molecule ion H_3^+ (the same mass 3 as DH^+) in the ion source of the machine. In order to resolve this problem, an isotopically light natural water from Antarctica was selected as an additional reference standard for use in determining the *stretching factor* for individual mass spectrometers. (See Appendix 2 for further discussion.) The stretching factor is especially important for hydrogen isotope measurements, because the variation in δD values of natural materials are about 10 times larger than variations in any other δ values. This standard was

---

[5]Harmon Craig coined the acronym SMOW and was strongly opposed to appending the letter V to it or to the acronym of any international reference standard. Read his arguments and the history of SMOW in the archives of ISOGEOCHEM (January 28, 1995).

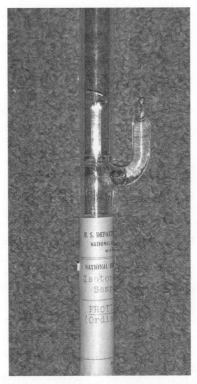

FIGURE 2.1: Picture of an ampoule (glass break-seal tube) containing NBS-1 standard, a reference standard that was formerly distributed by the National Bureau of Standards (now NIST). The label reads

Isotope Reference Sample #1.
PROTIUM OXIDE (Ordinary Water)

given the acronym SLAP (Standard Light Antarctic Precipitation) and has a very low δD value of $-428‰$, a value chosen as the most reliable value on the basis of a comparison study made in many of the major stable isotope laboratories in the world in the 1970s. In order to calibrate a machine for δD determinations, hydrogen isotope analyses of both SMOW and SLAP are analyzed relative to the working standard, and the difference obtained is multiplied by a factor, so that $\delta D_{SLAP\text{-}SMOW} = -428‰$ (Coplen, 1988).

Absolute ratios of D/H determined for both SMOW and SLAP and the absolute ratio of $^{18}O/^{16}O$ determined for SMOW are given in Table 2.4. The absolute values were determined

TABLE 2.4: Determinations of the absolute ratios of D/H and $^{18}O/^{16}O$ of the international reference standards SMOW and SLAP.

Ratio	SMOW	SLAP	Reference
D/H	$(155.76 \pm 0.05) \times 10^{-6}$	$(89.02 \pm 0.05) \times 10^{-6}$	Hagemann et al. (1970)
D/H	$(155.75 \pm 0.08) \times 10^{-6}$	$(89.12 \pm 0.07) \times 10^{-6}$	De Wit et al. (1980)
D/H	$(155.60 \pm 0.12) \times 10^{-6}$	$(88.88 \pm 0.18) \times 10^{-6}$	Tse et al. (1980)
$^{18}O/^{16}O$	$(2005.20 \pm 0.45) \times 10^{-6}$		Baertschi (1976)

Baertschi, P. (1976), *Earth Planet. Sci. Lett.* **31**, 341. De Wit, J.C., et al. (1980), *Geostand. Newslett*, **4**, 33–36. Hagemann, et al. (1970), *Tellus* **22**, 712–715. Tse, R.S., et al. (1980), *Anal. Chem.* **52**, 2445.

by mixing waters that were extremely pure samples of $^1H_2^{16}O$, $^1H_2^{18}O$, and $D_2^{16}O$. It is very difficult to prepare water that has no deuterium in it, but the best job possible was done at the time the determination was made. Through careful mixing, the D/H and $^{18}O/^{16}O$ ratios of these synthetic waters were *known*, but with significant uncertainty. SMOW and SLAP were then measured relative to the isotopic compositions of these waters to derive their absolute D/H and $^{18}O/^{16}O$ ratios. While it is desirable to have reliable determinations of the absolute ratios of these standards, keep in mind that knowledge of the absolute ratios is not necessary to conduct research in stable isotope geochemistry.

2.7.2 Carbon

Carbon isotope ratios are reported relative to the PDB standard described toward the beginning of this section, and by definition, the $\delta^{13}C$ value of PDB is zero. Several secondary carbonate standards (e.g., Carrara marble and Solenhofen limestone) were measured relative to PDB in the early years, and these standards are still in use in some older laboratories. Analysis of the international isotope reference standard NBS-19 is now the accepted means of calibrating to the PDB scale. NBS-19 was originally the TS (Toilet Seat) limestone working standard used in the laboratory of Irving Friedman at the U.S. Geological Survey and has a $\delta^{13}C$ value of $+1.95‰$ relative to PDB (Table 2.3). PDB is the only standard now universally accepted for reporting $\delta^{13}C$ values.

As in the case of SMOW, it has been suggested that the letter V be appended to the abbreviation PDB, but doing so would cause unnecessary confusion for newcomers to the field and create problems for editors of scientific journals. Notwithstanding the fact that V-PDB never existed, and that the Pee Dee formation is in South Carolina and not Vienna, the isotopic compositions of PDB and V-PDB are identical. The only thing new in the history of PDB is that there is now an accepted or defined $\delta^{13}C$ value for this time-honored standard relative to an available reference standard. As long as it is clearly stated that the appropriate calibration was made, $\delta^{13}C$ values reported on the PDB scale are reliable.

Carbon isotope geochemistry has remained a very active discipline since its inception, and thousands of carbon isotope analyses are reported every year. The carbon analyzed is present in a variety of substances, including the various carbonate minerals, organic matter in sediments, organic matter in meteorites, petroleum products, collagen extracted from plant material, graphite, carbonate in the apatite of bones and teeth, and carbon present in trace quantities in rocks (e.g., basalts) and minerals (e.g., goethite) and in archaeological and anthropological specimens. There are major differences in the extraction techniques used for these various carbonaceous materials, and the errors assigned to an analysis can depend on the complexities of the extraction method employed. In almost all cases, however, the carbon is put into the form of CO_2, and recently CO, for mass spectrometric analysis. In addition to various NBS carbonate standards, there are others, including sucrose, plastic, oil, and graphite standards (Table 2.3). Other than NBS-19, these are all secondary standards.

2.7.3 Nitrogen

Nitrogen isotope analyses are made from a variety of nitrogenous substances, including various nitrogen oxide gases in air and on particulates, nitrate in groundwaters, nitrate and ammonium in fertilizers, and nitrogen that is present in both terrestrial and extraterrestrial rocks and minerals. Reliable procedures, including bacterial mediation, are now available for extracting nitrogen from its various presentations in natural and synthetic materials.

The reference standard for nitrogen isotope analyses is atmospheric nitrogen and is called AIR. The $\delta^{15}N$ value of atmospheric nitrogen is almost constant everywhere on Earth and is 0‰ by definition (Mariotti, 1983). No AIR standard is distributed, so workers in each laboratory must prepare their own working standards. (See Chapter 9.) Solid secondary standards with recommended $\delta^{15}N$ values are available for calibration purposes, so that very pure tank N_2 can be used as a working standard. Nitrogen can be separated from air by combusting the O_2 with elemental copper, but contaminant Ar in the product can interfere with the nitrogen isotope analyses.[6]

2.7.4 Oxygen

Two international reference standards are used to report variations in oxygen isotope ratios: PDB and SMOW. SMOW was originally defined in terms of NBS-1 as

$$(^{18}O/^{16}O)_{SMOW} = 1.008(^{18}O/^{16}O)_{NBS-1}. \qquad (2.20)$$

SMOW has a $\delta^{18}O$ value of 8.00‰ relative to NBS-1, and NBS-1 has a $\delta^{18}O$ value of -7.94‰ versus SMOW. As with hydrogen, a stretching factor must be applied to oxygen isotope analyses, and this factor is calibrated by analyzing SMOW whose $\delta^{18}O \equiv 0$‰ and SLAP whose $\delta^{18}O = -55.5$‰. (See Appendix 2.) By calibrating in this manner, analyses of water reported on the SMOW scale are reliable. The stretching factor is much smaller for oxygen isotope analyses than for hydrogen isotope analyses and can generally be ignored.

Use of the PDB standard for reporting oxygen isotope compositions is restricted to analyses of carbonates of low-temperature origin (oceanic, lacustrine, or pedogenic) in studies of paleoclimate, paleoceanography, and carbonate diagenesis. As mentioned earlier, oxygen isotope compositions of carbonates are determined by analyses of CO_2 generated from them by reaction with 100 percent H_3PO_4 at some fixed temperature. Note that the *PDB standard is the solid carbonate*, not the acid-liberated CO_2 that is actually introduced to the mass spectrometer. Again, $\delta^{18}O$ of PDB is 0.00‰ by definition, and analysis of NBS-19 is the accepted means of relating oxygen isotope analyses to PDB. The $\delta^{18}O$ value of NBS-19 is $\equiv -2.20$‰ on the PDB scale. Two additional international secondary reference standards for carbonates are available (Table 2.3), and relating analyses to PDB no longer poses a problem for new workers.

Because the $\delta^{18}O$ value of PDB is 30.91‰ higher than that of SMOW (on the SMOW scale), the conversion between SMOW and PDB scales is given by the equation

$$\delta^{18}O_{SMOW} = 1.03091(\delta^{18}O_{PDB}) + 30.91. \qquad (2.21)$$

There is a difference of 0.28‰ (Fig. 6.1) between CO_2 in equilibrium with SMOW ($\alpha_{CO_2-H_2O} = 1.04120$) at 25°C and CO_2 liberated from PDB at 25°C ($\alpha_{CO_2-CaCO_3} = 1.01025$), and all these values were used in deriving equation (2.21). A substance with an $^{18}O/^{16}O$ ratio equal to that of PDB ($\delta^{18}O_{PDB} = 0$‰) has a $\delta^{18}O$ value of 30.91‰ on the SMOW scale. Marine carbonates have $\delta^{18}O$ values near zero on the PDB scale, while ocean waters have $\delta^{18}O$ values near zero on the SMOW scale. Due to this happenstance, it is not uncommon to see the two scales mixed in published reports, with data for carbonates reported on the PDB scale and data for waters reported on the SMOW scale. Their actual $\delta^{18}O$ values on the same scale, however, are approximately 30‰ apart! Care must be taken not to mix scales when presenting oxygen isotope data.

---

[6]Note that V-AIR has not yet been proposed.

2.7.5 Sulfur

Sulfur is extracted from aqueous, solid, or gaseous compounds and converted to either SO_2 or SF_6 for isotopic analysis. $\delta^{34}S$ values are reported relative to the CDT (Cañon Diablo Troilite) standard, a sample of meteoritic troilite (FeS) from Meteor Crater in Arizona. Once again, by definition, $\delta^{34}S(CDT) \equiv 0.0‰$. Sulfur isotope compositions of all samples of meteoritic troilite are about the same. With increased interest in analyzing smaller samples, it was discovered that the CDT standard is not as homogeneous as once thought, at least on a scale smaller than several milligrams (Beaudoin et al., 1994). The IAEA now distributes two synthetic Ag_2S standards with $\delta^{34}S$ defined values relative to CDT (Table 2.3). Sulfur isotope analyses are calibrated to CDT in each laboratory by analyses of these Ag_2S standards, whose $\delta^{34}S$ values differ by 22‰. CDT now joins the ranks of PDB in the sense that both are accepted international reference standards for reporting all sulfur and carbon isotope analyses, but neither is distributed and both are defined in terms of recommended δ values of secondary solid standards.

2.8 ISOTOPE RATIO MASS SPECTROMETRY

2.8.1 The First Isotope Ratio Mass Spectrometers

The basic components of a gas source isotope ratio mass spectrometer are (1) an inlet system, (2) ion source, (3) a flight tube, (4) an ion collector assembly, and (5) a recording system (Fig. 2.2). One of the primary goals of the early research group in Chicago was to increase the precision of the Nier mass spectrometer by a factor of 10 so that the small isotopic variations that occur in nature could be measured with the precision required for paleothermometry. They achieved this goal by making two fundamental changes to the existing machines: (1) substituting a vastly more sensitive (vibrating reed) electrometer to intensify the beam signals and (2) introducing an ingenious magnetic valve switching mechanism that allowed rapid switching from the standard gas to the unknown gas into the source of the mass spectrometer (Fig. 2.3). With rapid switching, the isotopic ratios of standard and unknown gases could be measured under almost identical conditions in the source, thus eliminating noise associated with variations in the stability of the electronics and other factors. Gases are introduced into the source of a mass spectrometer through metal capillary tubing. In early spectrometers, manually operated **mercury pistons** were used to adjust the volumes (pressures) of gas behind the capillary leaks, ensuring that the pressures of both standard and unknown gases were identical in the source. In most gas source machines, the gases pass through an approximately 1-meter-long capillary tubing that has a crimp at its end. The function of the crimped capillary is twofold: (1) to reduce the amount of gas that enters the source and (2) to assure **viscous flow**. With appropriate lengths of capillary tubing and associated back pressures that can be calculated for these lengths, viscous flow conditions are easily achieved, and no isotopic fractionation occurs when the gas exits from the capillaries into the source of the spectrometer. **Molecular flow** is avoided because, under these conditions, lighter molecules of gas preferentially enter the higher vacuum region of the mass spectrometer. A cluster of four changeover valves allows for rapid switching between sample and reference gases. In the earliest machines, gas flow was controlled simply by opening and closing ball-and-socket ground-glass joints (Fig. 2.3). These glass valves invariably leaked, but with careful grinding by hand, the leaks were amazingly small and constant. A small and easily measured leak correction was made to account for contamination of one gas by the other. The fate of the gas after it enters the source is about the same in both old and modern mass spectrometers and will be discussed in the next section.

Section 2.8: Isotope Ratio Mass Spectrometry   31

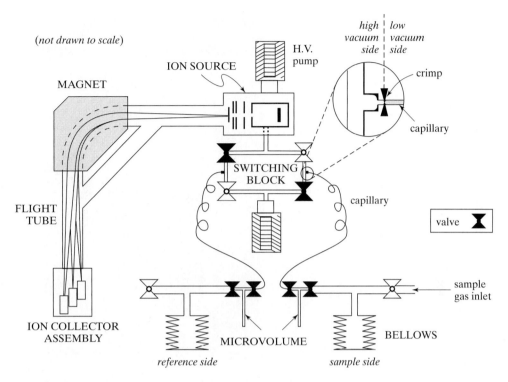

FIGURE 2.2: **Schematic of a typical modern mass spectrometer.** The isotopic composition of two gases (reference and sample) are measured relative to one another. The pressures of both gases in the source region are adjusted to the same value by compressing or expanding the bellows. The gas passes through a capillary about 1 meter long, with a crimp at the mass spectrometer end to help reduce the flow rate and to assure viscous flow. The capillary–crimp prevents a fractionation of the gas as it enters the high vacuum of the mass spectrometer. The bleed rate of the gas into the mass spectrometer is slow enough so that the reference gas will remain in the bellows system for a full day of measurement (except for hydrogen). The gas enters the source region, where it is ionized, focused into a coherent beam, and accelerated down the flight tube. The ion beams are deflected in a magnetic field in relation to the charge–mass ratio of the ion. The lighter ions are deflected more strongly than the heavy ones with the same charge. The ions enter the various collectors (Faraday cups), where the current that is developed is sent through resistors to produce voltages that are amplified and registered on a recording system (not shown). The intensities of these ion beams are proportional to the abundance of the isotopologue collected. The isotopic ratios and delta values are automatically calculated and recorded by the software that operates the machine.

2.8.2 Modern Conventional Mass Spectrometers

Modern mass spectrometers differ from the older machines principally in the gas inlet system, the sensitivity of the source, the use of multiple collectors, and more stable and rapid-response electronics. Gas pressure adjustments are made automatically employing a bellows system, and the changeover valves have Teflon® or gold seats that effectively eliminate the leakage problem. A schematic of a modern isotope ratio mass spectrometer is shown in Fig. 2.2.

After the volume of the bellows is adjusted to provide the desired pressure, the gas flows through the capillary into the ion source box, where it is bombarded by electrons emitted by a filament heated to high temperature. Outermost electrons are knocked off of a

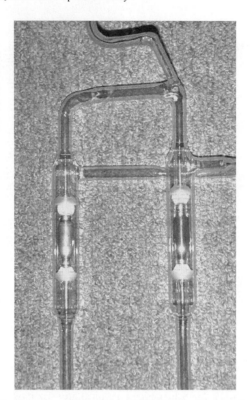

FIGURE 2.3: Glass changeover valves, circa 1967. Each valve consisted of two sets of ball-and-socket joints that were connected by a piece of glass containing a cylinder of metallic iron. The external glass tubes were surrounded by solenoid-activated electromagnets (not shown). The lower ball-and-socket joint was sealed by gravity when the magnet was off and the gas flowed into the source. When the magnets were turned on, the internal glass piece was pushed up, sealing the upper joint and opening the bottom joint, and the gas flowed into a waste pump. Adjustments were made to ensure that the rates of gas flow into the source and into the waste pump were identical. All modern mass spectrometers with dual-inlet systems use low-maintenance pneumatically controlled stainless-steel valves with teflon or gold seals.

fraction of the gas molecules, producing positively charged ions. These ions are accelerated by a high voltage potential and collimated into a coherent beam by a series of electric lenses. The beam passes through a strong magnetic field that deflects the ions in a circular trajectory according to their ratio of charge to mass. Some ions produced are multiply charged. Light ions are deflected more strongly than heavy ones of the same charge, thus causing a physical separation between ions of the different isotopologues of the molecule being analyzed. An analogous situation is a motor vehicle making a sharp turn on a mountain road. A small car can navigate a tight turn at high speeds much more easily than a truck can. The heavier isotopic species are represented by the truck and are deflected less strongly by the magnetic field. The ions pick up electrons at the collectors (Faraday cups), thereby creating electric currents that then pass through high-megohm resistors. The intensities of the voltages produced are proportional to the quantities of gas being collected and are magnified and recorded. Multiple collectors are used, so that ions of different masses are collected at the same time and the isotopic ratios can be monitored in time. Differences in the relative voltages of the sample and standard are then recorded as per mil deviations from the isotopic composition of the standard gas. That is, the δ value is the quantity directly measured by this type of mass spectrometer.

Only dual collectors were used on early isotope ratio mass spectrometers, so it was necessary to make corrections for collected ions with masses other than the desired ones. For example, there are four pertinent natural isotopologues of CO_2, with masses of 44 ($^{12}C^{16}O^{16}O$), 45 (the mass isotopomers $^{13}C^{16}O^{16}O$ and $^{12}C^{16}O^{17}O$), and 46 ($^{12}C^{16}O^{18}O$). Oxygen isotope ratios are calculated from 46/44 ratios, but in older mass spectrometers, mass 45 was collected and counted along with the main beam of mass 44. All modern isotope ratio mass spectrometers are equipped with multiple collectors so that the intensities of masses 44, 45, and 46 are measured separately and simultaneously. Other improvements in modern conventional isotope ratio mass spectrometers include more sensitive ion sources (smaller samples can now be measured with the same precision) and very stable electronics (less electronic noise and less downtime). The precision of repeated analyses *of the same gas* is higher on modern machines, but the chemical extraction procedures used in digesting and processing individual samples will always introduce errors. In general, there has been no significant improvement in precision over the years that can be exploited in the resolution of geochemical problems. The main advantages of the newer machines are ease of operation (speed and automation), ability to handle smaller samples, and significantly less downtime.

2.8.3 Gas Chromatograph Isotope Ratio Mass Spectrometry (GC-IRMS)

In recent years, the **continuous-flow method** of introducing gas has revolutionized isotope ratio mass spectrometry and opened up a panoply of research avenues (Matthews and Hayes, 1978). In place of analyzing gas by means of the dual-inlet method of measurement, samples of gas are entrained in a helium stream, purified in a gas chromatograph, and introduced directly into the source of the mass spectrometer. Initial applications of the GC were mainly to purify the gas. Early online applications included the now-routine high-temperature combustion of very small C- and/or N-bearing samples in an elemental analyzer (Fig. 2.4). Another important early application is compound specific analysis of organic molecules (Hayes et al., 1989). GC-IRMS (gas chromatography–isotope ratio mass spectrometry) technology has been modified so that it can be used in place of conventional extraction techniques for almost every material that has been analyzed conventionally. The advantages of GC-IRMS is that gases can be purified easily and efficiently, analysis is extremely rapid, and sample sizes are reduced by several orders of magnitude. Systems that combine gas chromatography and continuous-flow technology continue to find important applications in the earth and biological sciences. Recent improvements in analyses of water and carbonates effectively replace analytical methods that have been in use with only minor modification for over half a century. There is, however, a slight loss of precision relative to the traditional dual-inlet system, although the gap is reduced every year.

2.8.4 Gases Measured in Isotope Ratio Mass Spectrometry

Gases that are commonly introduced into isotope ratio mass spectrometers and the masses of the isotopologues measured are given in Table 2.5. Hydrogen isotope ratios are almost always determined from the 3/2 ratios of H_2, although other gases, including H_2O, have been used with limited success. Carbon isotope ratios are almost always determined from 45/44 ratios of CO_2 gas. Nitrogen isotope ratios are determined from 29/28 and 30/28 ratios of N_2. Because the concentration of $^{15}N^{15}N$ is negligible in nature, measurements are made of mass 30 only when materials artificially enriched in ^{15}N are being studied. Oxygen isotope ratios are most

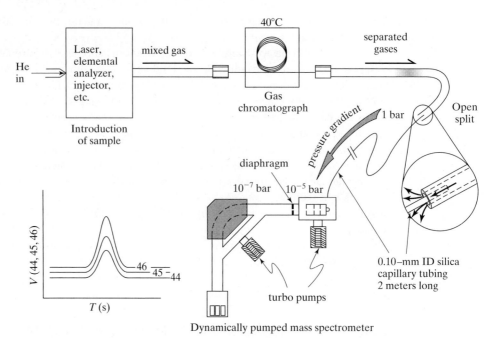

FIGURE 2.4: Schematic of a GC-IRMS system. Samples are introduced into a helium stream, and the different gases are separated on a gas chromatographic column. They pass through an open split into the mass spectrometer. The signal from the different isotopologues of a gas are recorded simultaneously. The integrated area of the peaks relative to a standard can be used to determine the isotope ratio in conventional delta format.

TABLE 2.5: Gases commonly measured in conventional gas source isotope ratio mass spectrometers.

Element	Gas	Masses of isotopologues measured
Hydrogen	H_2	2, 3 (interference from H_3^+)
Carbon	CO_2	44, 45, and 46
Nitrogen	N_2	28, 29 (and 30)
Oxygen	CO_2	44, 45, 46
	O_2 (fluorination)	32, (33), 34
	CO (pyrolysis)	28, 30
Sulfur	SO_2	64, 66
	SF_6	146, (147), 148, (150)

commonly determined from measurements of 46/44 ratios of CO_2, but more and more frequently, the gases CO and O_2 are being used. $^{18}O/^{16}O$ ratios are determined from 30/28 ratios of CO and 34/32 ratios of O_2, while $^{17}O/^{16}O$ ratios are determined from 33/32 ratios of O_2. CO is the gas of choice for emerging pyrolysis applications.

Sulfur isotope ratios can be measured on either SO_2 or SF_6. SO_2 is most commonly used as the sample gas because it is a simple matter to produce this gas from sulfur-bearing geological materials. There are two drawbacks to the use of SO_2: (1) Mass 66 is present as both $^{34}S^{16}O^{16}O$ and $^{32}S^{16}O^{18}O$, so that independent knowledge of the $^{18}O/^{16}O$ ratio is required to calculate the $^{34}S/^{32}S$ ratio, and (2) SO_2 is a relatively *dirty* and *sticky* gas that, in the relatively large quantities used in conventional mass spectrometry, can contaminate the source and collectors rather quickly. This problem is reduced significantly in continuous-flow systems, in which only minor amounts of SO_2 are introduced into the mass spectrometer. In contrast, SF_6 is a clean gas, and the monoisotopic nature of fluorine ensures that the different masses of SF_6 reflect variations in the isotopic ratios of sulfur only. The major drawbacks of using SF_6 are that this gas is relatively difficult to prepare in the laboratory and the relatively heavy masses of its isotopologues require a mass spectrometer with a very high voltage power supply and a large magnet.

2.8.5 Relations between Measured and Desired Isotopic Ratios

The relation between the intensities of measured masses in a mass spectrometer and the desired isotopic ratio of a given element is different for each gas. A few of the salient points pertinent to the analyses of three gases (H_2, N_2, and CO_2) will be made here to gain an appreciation of this important aspect of isotope ratio mass spectrometry.

After experimenting with various hydrogen-containing gases, Irving Friedman (1953) concluded that diatomic hydrogen was the most suitable for determining hydrogen isotope ratios, and H_2 remains the gas of choice to this day. The isotopic analysis of hydrogen poses a unique analytical problem. One of the ionic fragments produced in the source of the mass spectrometer is H_3^+, which has the same mass 3 as HD^+. Collection of this ion results in an apparent 3/2 ratio that is higher than the desired HD/HH ratio. The production of H_3^+ is proportional to the pressure of hydrogen gas in the source because that ion is produced by the reaction $H_2 + H_2^+ \rightarrow H_3^+ + H$. The H_3^+ contribution to mass 3 can be minimized, but not eliminated, by adding a repeller electrode to the source, using a high accelerating voltage and a low gas pressure. The contribution of H_3^+ is evaluated by measuring the 3/2 ratio as a function of pressure, a procedure that is done routinely and automatically on modern mass spectrometers. The correlation is linear on most modern machines (at low pressures, less H_3^+ is produced). Once the relationship between pressure and the contribution of H_3^+ is determined, a correction algorithm can be applied that removes the mass 3 contribution from the H_3^+ ion.

Hydrogen poses an additional problem because it diffuses back through the capillaries at a rate faster than the forward rate of viscous flow. As a result, the reference gas can fractionate with time. The magnitude of this effect is minimized by employing especially long capillaries.

Nitrogen isotope analysis is relatively straightforward. Measurements are made of two isotopologues of molecular nitrogen whose masses are 28 and 29. The isotopologue with mass 28 consists solely of $^{14}N^{14}N$, while that of mass 29 is a sum of the two possible orientations of ^{15}N and ^{14}N atoms in the molecule, $^{15}N^{14}N$ and $^{14}N^{15}N$. The desired mass ratio is then obtained from the relatively simple relation

$$\frac{29}{28} = \frac{^{15}N^{14}N}{^{14}N^{14}N} + \frac{^{14}N^{15}N}{^{14}N^{14}N} = 2 \times \frac{^{15}N}{^{14}N}. \qquad (2.22)$$

The molecular mass ratio 29/28 is almost identical to twice the atomic mass ratio of 15/14 and is the one used to determine nitrogen isotope ratios.

The nine isotopologues of CO_2 that are measured on a mass spectrometer are given in Table 2.6. As mentioned previously, isotopologues containing two rare isotopes like $^{17}O^{13}C^{16}O$ have negligible abundance in natural materials and are ignored. Deconvolution of the three masses 44, 45, and 46 to $^{13}C/^{12}C$ and $^{18}O/^{16}O$ ratios is seemingly very difficult, but a simple relation between the $^{17}O/^{16}O$ and $^{18}O/^{16}O$ ratios makes the task relatively simple. Given that $(^{17}O/^{16}O) \cong \frac{1}{2}(^{18}O/^{16}O)$ (see Box 13.1 for more details), the three masses 44, 45, and 46 of CO_2 can be used to determine both $^{13}C/^{12}C$ and $^{18}O/^{16}O$ ratios (Craig, 1957; Santrock et al., 1985).

TABLE 2.6: Masses and possible configurations of CO_2 isotopologues.

Mass	Isotopologue
44	$^{16}O^{12}C^{16}O$
45	$^{16}O^{13}C^{16}O$, $^{17}O^{12}C^{16}O$, $^{16}O^{12}C^{17}O$
46	$^{17}O^{13}C^{16}O$, $^{16}O^{13}C^{17}O$, $^{17}O^{12}C^{17}O$, $^{18}O^{12}C^{16}O$, $^{16}O^{12}C^{18}O$

2.8.6 Ion Microprobe Analyses of Stable Isotope Ratios

The ability to make precise in situ analyses of stable isotope ratios on micron-size samples has been a dream of geochemists for many years. With recent developments in laser fluorination techniques and ion microprobe technology, this dream is becoming a reality. Analyses of hydrogen, carbon, oxygen and sulfur isotope ratios can be made with a sensitive high-resolution ion microprobe (SHRIMP) and its various modified descendent instruments (Riciputi et al., 1998). Other than conventional isotope ratio mass spectrometry and GC/IRMS, this is the only spectroscopic means of analyzing these ratios precisely enough for applications in the earth and biological sciences. At the time of this writing, average (external) errors of ion microprobe analyses are generally no better than about 1‰, but are certainly high enough to resolve many important scientific questions. To make oxygen isotope analyses of minerals, a Cs^+, rather than the more common negative oxygen, ion beam is used to sputter off secondary $^{16}O^-$ and $^{18}O^-$ ions that can be extracted, collected, and counted to produce repeatable isotopic ratios. These beams can be focused down to about 25 μm or less and produce pits that are less than 3 μm deep, depending on the mineral being analyzed. With the passage of time, this technology will improve and be used to investigate geologic processes that take place on very small scales.

PROBLEM SET

1. a. A sample of sphalerite (ZnS) has a $\delta^{34}S$ value of 12.39‰ (the *raw* value) relative to the working standard of the laboratory. This working standard has a $\delta^{34}S$ value of -4.66‰ relative to CDT. What is the $\delta^{34}S$ value of the sphalerite relative to CDT?
   b. Graphitized diamond from Beni Bousera has a $\delta^{13}C$ value of -22.68‰ relative to the working standard. The $\delta^{13}C$ value of the working standard in the laboratory is -4.22‰ relative to PDB. What is the $\delta^{13}C$ value of the graphitized diamond relative to PDB?
2. The working standard used on a certain mass spectrometer has a $\delta^{18}O$ value of 38.72 per mil relative to SMOW. If a sample of CO_2 has a $\delta^{18}O$ value of -14.57 per mil relative to the this working standard,
   a. What is its $\delta^{18}O$ value relative to SMOW?
   b. What is its $\delta^{18}O$ value relative to PDB?
3. A $^{32}S/^{34}S$ ratio of 21.56 was reported for a sample of seawater sulfate in the early days of stable isotope geochemistry, when sulfur isotope compositions were reported as absolute ratios. Given that the absolute sulfur isotope ratio of Cañón Diablo Troilite (CDT) is 22.22, what is $\delta^{34}S$ (CDT) of this sulfate sample?
4. There was a computer crash after an important nitrogen sample (N_2) was analyzed on the mass spectrometer. No information was saved, other than the voltages of the masses 28 and 29 for the sample and the standard. (See the accompanying table.) Assuming that the reference gas has a $\delta^{15}N$ value of -0.57% relative to AIR, what is the $\delta^{15}N$ value of the sample relative to AIR?

	V[28]	v[29]
Standard	3.021	4.568
Sample	3.014	4.573

5. Vapor in oxygen isotope equilibrium with liquid water at 10°C has a $\delta^{18}O$ value of $-13.42‰$. What is the $\delta^{18}O$ value of the water? $\alpha^{18}O(l-v) = 1.0101$ at 10°C.

6. Quartz from the silica–carbonate rock in Trench 14 at the proposed disposal site for nuclear waste at Yucca Mountain, Nevada, has a $\delta^{18}O$ value of 24.4‰. Given that $10^3 \ln \alpha_{\text{quartz-water}} = 4.1(10^6 T^{-2}) - 3.70$,
   a. What would be the temperature of formation of this quartz if it formed in isotopic equilibrium with local groundwater whose $\delta^{18}O = -11.6‰$?
   b. What temperature would be obtained if you merely subtracted δ values instead of using the α value?
   c. What would the $\delta^{18}O$ value of water in equilibrium with this quartz need to be in order for the temperature of formation to be 300°C?

7. The term ε is commonly defined for phases a and b as $1000(\alpha_{a-b} - 1)$. For $\delta_b = \delta_a - 10$, show that the equality $\varepsilon_{a-b} = \delta_a - \delta_b$ is valid when δ_b is close to zero, but invalid if δ_a is very different from 0 (e.g., 100‰).

8. Correct the following sentences:
   a. Snow falling in the Sierra Nevada mountains is isotopically depleted. Del values of snow from this region have δD compositions of about -116.
   b. The carbon signature of the Hawaiian nodule was -2 per mil, a value which is heavier than most mantle rocks.
   c. Sulfur in these samples was analyzed and found to have no relation to the signature of carbon.

9. Write an expression for the sulfur isotope exchange reaction between pyrite and sphalerite.

10. Derive equation 2.3, $\delta_{X-B} = \delta_{X-A} + \delta_{A-B} + 0.001\delta_{X-A}\delta_{A-B}$, using the definitions of α and δ.

11. Give three isotopologues of water with the same mass.

12. In this chapter, it is stated that "A sample with a $\delta^{18}O$ value of $+19.7‰$ has an $^{18}O/^{16}O$ ratio that is 19.7 per mil, or 1.97 percent, higher than that of the standard." If the standard SMOW has an $^{18}O^{16}O$ ratio of 2005.20×10^6, calculate the $^{18}O/^{16}O$ ratio of the sample, using the stated expression. Show that you get the same answer as when you use the definition of delta, given in equation (2.1).

13. The Merriam–Webster dictionary gives the following definition of an isotope: "any of two or more species of atoms of a chemical element with the same atomic number and nearly identical chemical behavior but with differing atomic mass or mass number and different physical properties". What is wrong with this definition? How could you improve on it?

14. For a laboratory experiment, you wish to make a liter of water with a δD value of $+500‰$, starting with a mixture of normal water ($\delta D = -90‰$ and pure D_2O). How much D_2O must you add to 1 liter of water?

ANSWERS

1. a. Given the relationship $\delta_{X-B} = \delta_{X-A} + \delta_{A-B} + 0.001\delta_{X-A}\delta_{A-B}$ (equation (2.3)), $\delta^{34}S_{CDT}$ of the sample is $= 12.39 + (-4.66) + 0.001(12.39)(-4.66) = 7.67‰$.

2. For part (a), the problem is analogous to Problem 1(a). Given the relationship $\delta_{X-B} = \delta_{X-A} + \delta_{A-B} + 0.001\delta_{X-A}\delta_{A-B}$ (equation (2.3)), $\delta^{18}O$ of the sample is $= 38.72 + (-14.57) + 0.001(38.72)(-14.57) = 23.59‰$. For part (b), we need only find the working standard in terms of PDB instead of SMOW. To convert SMOW to PDB, we have $\delta^{18}O_{SMOW} = 0.970017\ \delta^{18}O_{PDB} - 29.98$. So the

working gas is $0.970017 \times 38.72 - 29.98 = 7.58‰$ vs. PDB. Then we use the equation $\delta_{X-B} = \delta_{X-A} + \delta_{A-B} + 0.001\delta_{X-A}\delta_{A-B}$ with the value of the working standard relative to PDB to finally solve the problem: $7.58 - 14.57 + 0.001(7.58)(-14.57) = -7.10‰$ vs PDB.

3. The $\frac{^{32}S}{^{34}S}$ value of the sample is 21.56. The $\frac{^{34}S}{^{32}S}$ value of the sample is 0.0464, and that of CDT is 0.0450. Given that the $\delta^{34}S$ value is given by equation (2.1), $\delta = \left(\frac{R_x - R_{std}}{R_{std}}\right)1000$, seawater sulfate has a $\delta^{34}S_{CDT}$ value $= \left(\frac{0.0464 - 0.0450}{0.0450}\right)1000 = 31.1‰$.

4. 2.83‰ vs AIR.

5. $\delta^{18}O_{water} = -3.46‰$

6. a. $\alpha_{qz\text{-water}} = \frac{1000 + \delta_{qz}}{1000 + \delta_w} = 1.0364$. Solving $10^3 \ln(1.0364) = 4.2(10^6 \times T^{-2}) - 3.70$ for T gives 326 K, or 53°C.

   b. In place of $10^3 \ln \alpha = 35.75$, we have $24.4 - (-11.6) = 36.00$. Solving for T gives 325 K, essentially the same as the temperature found in part (a).

7. For two phases a and b, $\alpha = \frac{1000 + \delta_a}{1000 + \delta_a}$ so that $\varepsilon = 1000\left(\frac{1000 + \delta_a}{1000 + \delta_a} - 1\right)$. If $\delta_b = 0$, $\delta_a = 10$. In this case, $\delta_b - \delta_a = 10$ and ε_{a-b} is exactly equal to 10. If $\delta_b = 100$, for example, then $\delta_a = 110$ and $\delta_b - \delta_a = 10$. But $\varepsilon = 9.09$.

8. a. The δD values of snow falling in the Sierra Nevada mountains are as low as $-116‰$ (or average $-116‰$, as the case may be).

   b. The carbon isotope composition of the Hawaiian nodule is $-2‰$, higher than value of normal mantle material. (Note that the $\delta^{13}C$ value of the sample still 'is' $-2‰$. There is no reason to use the past tense unless all of the sample was ground up and there is no more of this mantle nodule. Otherwise, we can all assume that the nodule has the same $\delta^{13}C$ value as when it was originally measured.

   c. There is no apparent correlation between the $\delta^{13}C$ and $\delta^{34}S$ values of these samples.

9. There are two sulfur atoms in the pyrite formula and only one in sphalerite..Therefore we write the reaction as follows:

$$Fe_{0.5}{}^{32}S + Zn^{34}S \Leftrightarrow Fe_{0.5}{}^{34}S + Zn^{32}S$$

In terms of thermodynamic properties such as volume or free energy, we would use 1/2 the value reported for FeS_2, because our formula is written as 1/2 of the total formula.

11. $HD^{17}O$, $DD^{16}O$, and $H_2^{18}O$.

12. For the first part, the sample should have an $^{18}O/^{16}O$ ratio $= 0.00200520 \times 1.0197 = 2044.70 \times 10^6$. Using the definition of delta, we solve the equation $\delta = \left(\frac{R_x - R_{std}}{R_{std}}\right) \times 1000$, where $\delta = 19.7$, and $R_{std} = 2005.20 \times 10^6$. Upon rearranging terms, it is found that the value is identical to that obtained by the first method.

13. The definition assumes that there is more than one isotope of a particular element. In the case of fluorine, for example, there is only one naturally occurring isotope. There is no requirement that two isotopes of an element must exist.

14. Starting with the definition of delta $\delta = \left(\frac{R_x - R_{std}}{R_{std}}\right) \times 1000$, and knowing that D/H of SMOW $= 155.76 \times 10^{-6}$, we determine the D/H ratio of our starting water to be 141.73×10^{-6} and the D/H ratio of water with $\delta D = 500‰$ to be 233.63×10^6. For one liter, the amount of H_2O

is $(1 - D_2O)$, or $\dfrac{D_2O}{H_2O} = \dfrac{D_2O}{1 - D_2O} = 233.63 \times 10^{-6}$, which translates to 0.234 ml of D_2O per liter of water. The same calculation for our starting water gives 0.1417 ml D_2O per liter of water. The difference is $0.233 - 0.142 = 0.091$ ml. That is, adding only 0.091 ml of D_2O to a liter of water with an initial value of -90% will increase the δD value to $+500‰$ vs. SMOW.

REFERENCES

Beaudoin, G., Taylor, B. E., Rumble, D., III, and Thiemens, M. (1994). Variations in the sulfur isotope composition of troilite from the Cañon Diablo iron meteorite. *Geochimica et Cosmochimica Acta* **58,** 4253–4255.

Clayton, R. N., Goldsmith, J. R., and Mayeda, T. K. (1989). Oxygen isotope fractionation in quartz, albite, anorthite and calcite. *Geochimica et Cosmochimica Acta* **53,** 725–733.

Coplen, T. B. (1988). Normalization of oxygen and hydrogen isotope data. *Chemical Geology* **72,** 293–297.

Coplen, T. B. (1995). Discontinuance of SMOW and PDB. *Nature* **375,** 285.

Coplen, T. B. (1996). New guidelines for reporting stable hydrogen, carbon, and oxygen isotope-ratio data. *Geochimica et Cosmochimica Acta* **60,** 3359–3360.

Coplen, T. B., and Clayton, R. N. (1973). Hydrogen isotopic composition of NBS and IAEA stable isotope water reference samples. *Geochimica et Cosmochimica Acta* **37,** 2347–2349.

Coplen, T. B., Kendall, C., and Hopple, J. (1983). Comparison of stable isotope reference samples. *Nature* **302,** 236–238.

Craig, H. (1957). Isotopic standards for carbon and oxygen and correction factors for mass-spectrometric analysis of carbon dioxide. *Geochimica et Cosmochimica Acta* **12,** 133–149.

Craig, H. (1961). Standard for reporting concentrations of deuterium and oxygen-18 in natural waters. *Science* **133,** 1833–1834.

Friedman, I. (1953). Deuterium content of natural water and other substances. *Geochimica et Cosmochimica Acta* **4,** 89–103.

Hayes, J. M., Freeman, K. H., Hoham, C. H., and Popp, B. N. (1989). Compound-specific isotopic analyses: a novel tool for reconstruction of ancient biogeochemical processes. *Organic Geochemistry* **16,** 1115–1128.

Hut, G. (1987). *Consultants' Group Meeting on Stable Isotope Reference Samples for Geochemical and Hydrological Investigations*. Vienna: International Atomic Energy Agency.

Mariotti, A. (1983). Atmospheric nitrogen is a reliable standard for natural ^{15}N abundance measurements. *Nature* **303,** 685–687.

Matthews, D. E., and Hayes, J. M. (1978). Isotope-ratio-monitoring gas chromotagraphy–mass spectrometry. *Analytical Chemistry* **50,** 1465–1473.

McKinney, C. R., McCrea, J. M., Epstein, S., Allen, H. A., and Urey, H. C. (1950). Improvements in mass spectrometers for the measurement of small differences in isotope abundance ratios. *Review of Scientific Instruments* **21,** 724–730.

Riciputi, L. R., Paterson, B. A., and Ripperdan, R. L. (1998). Measurement of light stable isotope ratios by SIMS: Matrix effects for oxygen, carbon, and sulfur isotopes in minerals. *International Journal of Mass Spectrometry* **178,** 81–112.

Santrock, J., Studley, S. A., and Hayes, J. M. (1985). Isotopic analyses based on the mass spectrum of carbon dioxide. *Analytical Chemistry* **57,** 1444–1448.

CHAPTER 3

EQUILIBRIUM ISOTOPIC FRACTIONATION

3.1 INTRODUCTION

We can classify isotope exchange reactions as either kinetic or equilibrium reactions. Kinetic reactions are irreversible and, by definition, cannot be treated with the methods of classical thermodynamics. Evaporation of water into unsaturated air cannot be reversed; the isotopic fractionation that occurs during evaporation is a combination of equilibrium fractionation and that related to the different translational velocities of the isotopologues of water. The extent of processes such as evaporation and diffusion can be calculated for certain conditions with the use of kinetic-based theories. Other kinetic isotope effects, such as those associated with bacterial metabolism, are extremely complex and mostly defy quantification (although qualitative models can be constructed). Products from bacterial reactions tend to be enriched in the light isotope, because the dissociation energies are lower and bonds are more easily broken. As the title of this chapter suggests, we will not consider kinetic-based reactions at this point, although discussions of kinetic effects are addressed at various places in the book. The remainder of this chapter is devoted to quantifying the fractionation associated with reversible equilibrium processes.

Many processes involving isotope exchange can be modeled by classical equilibrium thermodynamics, because they are near-equilibrium phenomena. High-temperature processes, such as crystallization, generally approach isotopic equilibrium, as do a number of low-temperature processes, including the precipitation of *some* carbonate, phosphate, and silica phases in water. Equilibrium fractionation between two phases is based on the differences in bond strength of the different isotopes of an element. The heavier isotope will form a stronger bond and will be concentrated in the phase with higher bond energy or "stiffness." Qualitative rules for equilibrium isotope fractionation are given by Schauble (2004):

1. Equilibrium fractionation between two phases generally decreases with increasing temperature, proportional to $1/T^2$.
2. The degree of fractionation is generally larger for elements whose mass ratio is large. Thus, the mass ratio is defined as $\dfrac{m_{\text{heavy}} - m_{\text{light}}}{m_{\text{heavy}} m_{\text{light}}}$, where m_{heavy} and m_{light} are the heavy and light isotopes, respectively.

3. The heavy isotope is preferentially partitioned into the site with the stiffest bonds (strong and short chemical bonds). Bond stiffness increases qualitatively for
    a. a high oxidation state, or high oxidation state in which the element is bonded,
    b. the lighter elements
    c. covalent bonds
    d. a low coordination number.

From rule 1, fractionation varies regularly with temperature, a relationship that forms the basis for stable isotope thermometry. In order to have a *useful* isotope thermometer, we need to be able to measure the isotopic composition of the phases with the necessary precision, determine that they are indeed in isotopic equilibrium (often a daunting task), have a quantification of the fractionation as a function of temperature, and have a mineral pair for which the fractionation changes significantly in response to temperature. In this chapter, we are concerned only with determining the fractionation factors. Historically, quantification of fractionation factors has been made using three methods: (1) theoretical calculations based on statistical mechanics, (2) experimental determinations based on measured fractionations of phases equilibrated under known laboratory conditions, and (3) empirical calculations based on measured fractionations of natural samples where independent temperature estimates can be obtained.

Each method of determining fractionation factors has benefits and limitations. At present, theoretical statistical mechanical calculations applied to minerals do not have the same precision as experimental determinations for complex phases. A number of approximations must be made regarding the energy state of a phase because the quantum states of the individual molecules in solids and liquids do not behave independently from one another; therefore, approximate solutions to Schrödinger's equations cannot be used. (See Denbigh, 1971, for a general introduction to the topic.) Also, the magnitudes of frequency shifts for the isotopically substituted molecule are not well known. However, the form of a theoretically derived curve often allows the extrapolation of experimentally determined fractionations beyond the temperature range of the experimental conditions. Statistical mechanical calculations may also be used for reactions that are difficult or impossible to duplicate in the laboratory or that simply have not yet been performed. Experimental methods allow us to control most variables, such as temperature, reaction time, and chemical and isotopic composition. Experiments are difficult to conduct, however, and many are impractical, impeded by kinetic limitations. Empirical estimates take advantage of the fact that nature provides us with very long term experiments. A metamorphic rock heated to 500°C for 100 million years is an experiment that cannot be duplicated in the laboratory! At the same time, however, it is difficult to constrain temperatures very precisely in natural systems, and problems with isotopic inheritance and retrograde resetting always need to be considered when empirical estimates are made.

3.2 THEORETICAL DETERMINATION OF STABLE ISOTOPE FRACTIONATION FACTORS

3.2.1 Free Energy of Reaction

Fractionation factors can be calculated with the methods of statistical mechanics. The basic principles are not complicated, but the mathematics is complex, so only the basic concepts are presented here. The reader is referred to the following articles for additional details: Denbigh (1971); Richet et al. (1977); O'Neil (1986); Criss (1999); Chacko et al. (2001); and Schauble (2004).

The fundamental concept of the equilibrium exchange reaction was introduced in Section 2.4. A typical reaction is of the form

$$aA_1 + bB_2 = aA_2 + bB_1, \qquad (3.1)$$

where A_1 and A_2 are the two isotopologues of the molecule A and similarly for molecule B. An example would be an exchange between CO and O_2, written as

$$C^{16}O + \tfrac{1}{2}{}^{18}O^{18}O = C^{18}O + \tfrac{1}{2}{}^{16}O^{16}O. \qquad (3.2)$$

Here, A_1 is $C^{16}O$, A_2 is $C^{18}O$, B_1 is ${}^{16}O^{16}O$, and B_2 is ${}^{18}O^{18}O$. Obviously, as written, this reaction does not occur in nature. We do not have individual phases $C^{16}O$ and $C^{18}O$; instead, there is only one inseparable, mixed $C^{18}O-C^{16}O$ phase. The reaction does make sense from a thermodynamic standpoint, however, because it is possible to assign activities to each of the components $C^{16}O$, $C^{18}O$, ${}^{16}O^{16}O$, and ${}^{18}O^{18}O$ on the basis of concentrations. In this case, the equilibrium constant K is defined as

$$K = \frac{\prod (a_i)^{n_i}\ \text{products}}{\prod (a_i)^{n_i}\ \text{reactants}} = \frac{[a({}^{16}O^{16}O)]^{1/2} a(C^{18}O)}{[a({}^{18}O^{18}O)]^{1/2} a(C^{16}O)}, \qquad (3.3)$$

and because the activity coefficients are close to unity, equation (3.3) can be simplified to

$$K = \frac{\left(\dfrac{{}^{18}O}{{}^{16}O}\right)_{CO}}{\left(\dfrac{{}^{18}O}{{}^{16}O}\right)_{O_2}} = \alpha_{CO-O_2}. \qquad (3.4)$$

The equilibrium for exchange reaction (3.2) is given by

$$\Delta G^o_{r,T} = -RT \ln(K). \qquad (3.5)$$

At 300 K, α_{CO-O_2} is 1.028, so the change in free energy for the reaction is ~ -69 J. Note that this energy change is minuscule compared with those associated with chemical reactions. For the reaction $\tfrac{1}{2}O_2 + CO = CO_2$, for example, the change in free energy (at 298 K) is $-257{,}200$ J.

3.2.2 The Internal Energy of a Molecule

We can calculate the energy of reaction (3.1) by considering the total energy of each molecule in the reaction. The total internal energy (E_{tot}) is the sum of all forms, including translational energy (E_{tr}), rotational energy (E_{rot}), vibrational energy (E_{vib}), electronic energy (E_{el}), and nuclear spin (E_{sp}). The last two terms are negligible, so we can say

$$E_{tot} = E_{tr} + E_{rot} + E_{vib}. \qquad (3.6)$$

At equilibrium, the ratio of molecules having energy E_i to those having zero-point energy (discussed in the next section) is given by

$$\frac{n_i}{n_0} = g_i e^{-E_i/kT}, \qquad (3.7)$$

where e is the exponential function, k is Boltzmann's constant (1.381×10^{23} J/K, or 0.6951 cm^{-1} K^{-1}), T is temperature in K, and g is a statistical term to account for possible degeneracy, or different states. The sum over all possible quantum states i accessible to the system is defined as the partition function

$$Q = \sum g_i e^{-E_i/kT}. \tag{3.8}$$

We can relate the partition function Q back to our equilibrium constant K (and ultimately our fractionation factor α) in equation (3.3) by the formula

$$K = \frac{\prod (Q_i)^{n_i}{}_{\text{products}}}{\prod (Q_i)^{n_i}{}_{\text{reactants}}} = \frac{Q(C^{18}O)[Q(^{16}O^{16}O)]^{1/2}}{Q(C^{16}O)[Q(^{18}O^{18}O)]^{1/2}}. \tag{3.9}$$

The total partition function Q_{tot} can be split up into the partition functions relating to the different forms of energy—translation, vibration, and rotation—as

$$Q_{\text{tot}} = Q_{\text{vib}} Q_{\text{rot}} Q_{\text{tr}}. \tag{3.10}$$

The end result is that each of these components can be solved with quantum mechanical techniques. With accurate spectroscopic data, the fractionation between phases can be computed as a function of temperature. All that needs to be done is to compute the different components of the partition functions to determine our fractionation factors.

3.2.3 Vibrational Partition Function

We start by considering the potential energy of a diatomic molecule, such as H$_2$. As a first approximation, the energy can be approximated as a simple harmonic oscillator, illustrated by the harmonic potential curve in Fig. 3.1. The two atoms will have an average distance between each other so as to minimize the energy. That is, they will tend towards the energy well in the figure. If we bring two H atoms from far apart towards one another, there is an attraction. If they are moved too close to one another, repulsive forces overwhelm the attractive forces, and the atoms are pushed apart. The average spacing is at the base of the energy well. The energy for a harmonic oscillator is then given by

$$E = (n + \tfrac{1}{2}) h\nu, \tag{3.11}$$

where n is the vibrational energy level ($n = 0, 1, 2$, etc.), h is Planck's constant (6.624×10^{-34} J·sec, and ν is the frequency (sec^{-1}). At low temperatures, $n = 0$, the *ground vibrational state*; at higher temperatures, higher energy levels are reached. But even at absolute zero, the vibrational energy is given by $E = \tfrac{1}{2} h\nu$, and the atoms move.[1] This is the **zero-point energy**, given by the difference between the bottom of the potential-energy well and the energy at the ground vibration state. The difference in the zero-point energy of H–H and D–D illustrates the difference in their bond strengths: The amount of energy needed to dissociate a D–D molecule is larger than that for an H–H molecule because the former resides lower in the

---

[1] The atoms must move, or they would violate the uncertainty principle, a fundamental law of quantum mechanics. If an atom had no motion, then we could tell exactly where it is and know its momentum, in violation of this law.

44 Chapter 3 Equilibrium Isotopic Fractionation

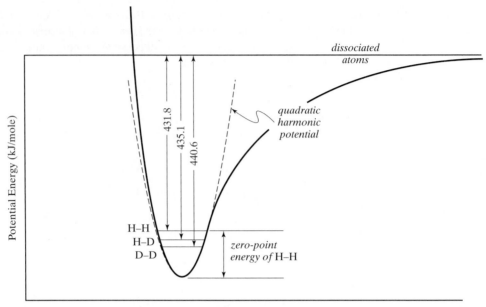

FIGURE 3.1: Potential-energy curve for diatomic hydrogen. Shown are the zero-point energies for the three isotopologues H–H, H–D, and D–D. Note that D–D sits lower in the potential energy well than H–D or H–H and has a higher dissociation energy. The result is that the D–D bond is stronger than the H–H bond.

potential well. More refined calculations of the vibrational energies take account of the deviation of the potential-energy curve from the simple harmonic oscillator, but they will not be considered here.

For a diatomic molecule a–b, the frequency is given by

$$\nu = \frac{1}{2\pi}\sqrt{\frac{k_s}{\mu}} = \frac{1}{2\pi}\sqrt{k_s\left(\frac{1}{m_a} + \frac{1}{m_b}\right)}, \quad (3.12)$$

where k_s is the effective spring constant (related to stiffness) and μ is the reduced mass of molecule a–b defined as $\frac{m_a m_b}{m_a + m_b}$. We can visualize this situation as two balls attached to either end of a rigid spring. They vibrate towards and away from each other with frequency ν. When the heavy isotope is substituted, the reduced mass changes, but the spring constant is unchanged. Imagine, then, our two-ball model with frequency reduced due to substitution by the heavier mass. It follows that the balls vibrate more slowly. The frequency ν is related to the wave number ω by $\nu = \omega c$, where c is the speed of light (2.998×10^{10} cm/sec). Wave numbers are measured from spectroscopic data or, for our purposes, taken from published tabulations. For $^{12}C^{16}O$, $\omega = 2{,}167.4$ cm^{-1}, and the spring constant (from equation (3.12)) is easily calculated. (See Problem 5.) The wave number for $^{12}C^{18}O$ is simply given by the relationship

$$\frac{\nu^*}{\nu} = \sqrt{\frac{\mu}{\mu^*}}, \quad (3.13)$$

where the asterisk refers to the isotopically substituted molecule. The frequency of $^{12}C^{18}O$ is calculated to be 2,115.2 cm^{-1}.

If we consider the simple harmonic oscillator, equation (3.8) becomes

$$Q_{vib} = \sum_i e^{-E_{vib}/kT} = \sum_{n=0}^{\infty} e^{-(n+1/2)h\nu/kT}. \tag{3.14}$$

Separating the terms gives

$$Q_{vib} = e^{-U/2} \sum_{n=0}^{\infty} e^{-(n)U}, \tag{3.15}$$

where $U = \dfrac{h\nu}{kT} = \dfrac{hc\omega}{kT}$. The following approximation can be applied (Criss, 1999)

$$\sum_{n=0}^{\infty} x^n = \frac{1}{1-x} \quad (\text{for } 0 < x < 1) \tag{3.16}$$

such that

$$Q_{vib} = e^{-U/2} \frac{1}{1 - e^{-U}}. \tag{3.17}$$

The E_{vib} values for $C^{16}O$ and $C^{18}O$ are 2.1527×10^{-20} J/mole and 2.10084×10^{-20} J/mole (from equation (3.11)), respectively. The ratio $Q_{vib}(C^{18}O)/Q_{vib}(C^{16}O)$, called **the vibrational partition function ratio,** is 1.1343, which we will return to in equation (3.20).

3.2.4 Translational and Rotational Partition Function

The translational partition function ratio is dependent on a number of terms, but ultimately, the translational partition function ratio of a diatomic molecule simplifies to

$$\frac{Q_{2tr}}{Q_{1tr}} = \left(\frac{M_2}{M_1}\right)^{3/2}, \tag{3.18}$$

where M_1 and M_2 are the molecular masses of the two molecules. Note that this formula is independent of temperature.

The rotational partition function is also a function of a number of terms that cancel out when we take the ratio of the two isotopically substituted molecules. For a diatomic molecule, the rotational partition function ratio is given by

$$\frac{Q_{2rot}}{Q_{1rot}} = \frac{\sigma_1 I_2}{\sigma_2 I_1}, \tag{3.19}$$

where σ is the symmetry number and I is the moment of inertia. For CO, in which only one molecule is exchanged, $\sigma = 1$. For O_2, in which ^{18}O can occupy one of two sites, $\sigma = 2$. $I = \mu r^2$, where μ is the reduced mass and r is the average interatomic distance.

3.2.5 The Complete Partition Function Ratio

Combining all partition function ratios into one complete term gives

$$\frac{Q_2}{Q_1} = \left(\frac{M_2}{M_1}\right)^{3/2} \frac{\sigma_1 I_2}{\sigma_2 I_1} \frac{e^{-U^*/2}}{e^{-U/2}} \frac{1-e^{-U}}{1-e^{-U^*}}. \tag{3.20}$$

(U^* is the isotopically substituted species.) This equation can be simplified with the Teller–Redlich spectroscopic theorem (see O'Neil, 1986, p. 9) to give the final solution

$$\frac{Q_2}{Q_1} = \left(\frac{m_2}{m_1}\right)^{3/2} \frac{\sigma_1 \omega_2}{\sigma_2 \omega_1} \frac{e^{-U^*/2}}{e^{-U/2}} \frac{1-e^{-U}}{1-e^{-U^*}}. \tag{3.21}$$

Here, m is the atomic mass and will cancel out, as will the term σ_1/σ_2. For $C^{16}O-C^{18}O$, we have $\omega_2/\omega_1 = 0.9759$, $\dfrac{e^{-U^*/2}}{e^{-U/2}} = 1.134$, and $\dfrac{1-e^{-U}}{1-e^{-U^*}} = 1$, giving a total partition function ratio of 1.1069. A similar calculation for O_2 gives a partition function ratio of 1.08304. Partition function ratios can be divided by one another to give respective α values. The α_{CO-O_2} is therefore $1.1069/1.08304 = 1.022$ at 25°C.

3.2.6 Extension to More Complex Molecules

Polyatomic gases, as well as solids, can be treated in a manner similar to the preceding, treating all possible modes of vibration in the summation terms. A number of assumptions and simplifications need to be made because the quantum states of molecules in liquids and solids (and high-pressure gases) are not independent from each other. Extensions of the foregoing models have been considered by a number of authors for relatively simple phases, such as calcite, quartz, and UO_2 (e.g., Bottinga, 1968; Hattori and Halas, 1982). Kieffer (1982) made some simplifying assumptions and was able to predict relative isotopic enrichments of more complex minerals. The largest uncertainty in calculating partition function ratios for complex solids is estimating the frequency shifts for the isotopically substituted molecule. Schauble (2004 and references therein) has predicted stable isotope fractionation for some elements other than those in the H–C–N–O–S system. Polyakov and Mineev (2000) used Mössbauer spectroscopy to estimate isotopic fractionation, and Driesner and Seward (2000) made simulations of salt effects on liquid–vapor partitioning. At the present time, however, statistical mechanical calculations do not have the precision necessary for accurate thermometry in most mineral systems. With ever more sophisticated computer programs and algorithms, however, this is almost certain to change in the future.

3.2.7 Relationship to Temperature

Bigeleisen and Mayer (1947) derived the following expression for partition function ratios:

$$\frac{Q_2}{Q_1} = 1 + \left(\frac{1}{2} + \frac{1}{U_2} + \frac{1}{e^{U_2}-1}\right)\Delta U. \tag{3.22}$$

Here, $U_i = \dfrac{hc\omega_i}{kT}$ and $\Delta U = U_1 - U_2$. If U is large, then $\dfrac{Q_2}{Q_1} \approx 1 + \frac{1}{2}\Delta U$. This would be the case when vibrational frequencies are high (as in reactions involving hydroxyl groups water) or temperatures are low. Under such conditions, $\dfrac{Q_2}{Q_1}$ will be proportional to $1/T$.

When $U < 5$, the terms in parentheses approaches a value of $U/12$. In this case, $\dfrac{Q_2}{Q_1}$ will be proportional to $1/T^2$. At room temperature and above, for anhydrous minerals (for which wave numbers are less than 1000 cm^{-1}), the $1/T^2$ relationship holds.

3.2.8 "Empirical" Theoretical Methods

Several methods of determining fractionation factors have been developed that take advantage of the empirical relationship between bond strength and relative isotope enrichment (Schütze, 1980; Richter and Hoernes, 1988; Smyth, 1989; Zheng, 1993; Hoffbauer et al., 1994). These techniques are based on ordering minerals according to their increasing anionic bond strength. Smyth's method involved calculating electrostatic site potentials for anionic sites, a function related to the bond strength of the oxygen in different crystallographic locations. Variations on this method include a consideration of the effects of the cation site and coordination. These latter methods are called "the increment method." They require that the relative bond strength data be calibrated to some independently (experimentally or theoretically) determined oxygen isotope fractionation relationship and then the correlations are extended to the entire data set. The general enrichment obtained with this method is mostly consistent with experimental data, but there are a number of notable exceptions, such as those for the aluminum silicate polymorphs (e.g., Sharp, 1995). The major advantage of the technique is that it can be applied to almost any mineral, is "internally consistent," and is easy to use. As a result, the increment method is widely used. It should be pointed out, however, that none of the methods mentioned here are based on any known physical or chemical laws relating isotope fractionation to anionic bond strengths, and they should be used with caution, as their results sometimes are in serious disagreement with other calibrations.

Savin and Lee (1988) devised an empirical bond-type approach for determining fractionation factors for phyllosilicates, particularly clay-forming minerals. They assume that oxygen in a given chemical bond has similar isotopic fractionation behavior, regardless of the mineral in which it is located. Once fractionations are assigned to each bond type (e.g., Si–O–Si, Al–O–Si, Al–OH), the fractionation for the entire mineral can be determined by summing the proportions of each bond. In general, the agreement between Savin and Lee's method and experimental and empirical calibrations is good. In the case of low-temperature clay minerals, such methods are necessary, because experimental data are limited by the sluggish reaction rates of minerals in the low-temperature range defining their stability field.

3.3 EXPERIMENTAL DETERMINATION OF FRACTIONATION FACTORS

3.3.1 Introduction

The first application of stable isotope thermometry was to the calcite–water system. McCrea (1950) synthesized calcite in water at room temperature and estimated fractionation factors using statistical mechanical methods. Epstein et al. (1951; 1953) were able to determine the equilibrium calcite–water fractionation at 29 and 31°C by drilling small holes into a living snail and bivalve (*Pinna* sp.). The organisms repaired their shells in an aquarium at constant temperature where the $\delta^{18}O$ value of ambient water was held to a constant and known value. The newly repaired material was removed and analyzed in order to determine the calcite–water fractionation at those specific temperatures.

More commonly, experiments are made by synthesizing or equilibrating two phases at high temperatures. O'Neil et al. (1969) expanded on McCrea's earlier work, both by synthesizing calcite at room-temperature conditions and by equilibrating calcite and water at high temperatures in hydrothermal bombs. Their results agreed with the earlier low-temperature calibration of Epstein et al.

The fractionation between phases m and n is defined as

$$1000 \ln \alpha_{m-n} = \frac{a \times 10^6}{T^2} + b \quad \text{(T in K)} \quad (3.23)$$

at high temperatures and

$$1000 \ln \alpha_{m-n} = \frac{a \times 10^6}{T} + b \quad \text{(T in K)} \quad (3.24)$$

at low temperatures, where a and b are constants. The temperature at which the crossover between equations 3.23 and 3.24 occurs is not known, but is probably below room temperature for most phases, as discussed in Section 3.2.7. Equations containing both $1/T^2$ and $1/T$ are probably not valid (Criss, 1991).

Bottinga and Javoy (1973) argued that equation (3.23) applies to all rocks. The constant b is 0 for fractionation between anhydrous phases, in which vibrational frequencies vary from 900 to 1,200 cm^{-1}. For water, the vibrational frequencies are far higher, ranging from $\omega_2 = 1647$ to $\omega_3 = 3939$ cm^{-1}. Therefore, $1,000 \ln \alpha$ does not vary linearly with T^{-2}, but *over a limited temperature interval* equation (3.23) should be valid. For anhydrous mineral–water exchange, the constant $b = 3.7$.[2] For fractionation between anhydrous minerals and hydrous minerals, the b term is proportional to the number of OH bonds in the phase. Others have questioned the validity of this term when applied to solid–solid equilibria (e.g., Chacko et al., 1996).

All experiments are made by isotopically equilibrating two phases. Most often, minerals are equilibrated with water, calcite, or CO_2 gas. Oftentimes, the reaction may not be of geological interest—for example albite–CO_2 gas fractionations may not have much geological importance—but the combination of two experiments can give a third fractionation that *is* of geological interest. For example, experimental fractionation factors have been measured separately for quartz–water and muscovite–water systems, but combining the results of these experiments gives the more interesting quartz–muscovite equation, a widely used isotope thermometer. Quite simply, the a and b terms can be subtracted from one another to eliminate the intermediate phase:

If $\quad 1000 \ln \alpha(\text{qz–water}) = 4.10 \times 10^6/T^2 - 3.70$
and $\quad 1000 \ln \alpha(\text{musc–water}) = 1.90 \times 10^6/T^2 - 3.10,$
then $\quad 1000 \ln \alpha(\text{qz–musc}) = 2.20 \times 10^6/T^2 - 0.6.$

---

[2]The value 3.7 was determined by averaging data from experimental studies. See original paper for details.

The choice of exchange medium (water, calcite, or CO_2) is determined on the basis of a number of factors, but all of the experiments have in common the requirement that the mineral and exchange medium be easily separable for later analysis.

3.3.2 Mineral–Water Exchange Reactions

Most early exchange experiments were made for mineral–water pairs. Ten to 20 mg of a finely ground solid and ~200 mg of water are sealed in a noble-metal tube and heated to reaction temperature at a confining pressure of 1–2 kbar.[3] The experiments are relatively easy to perform, the reaction rates are moderately rapid, and the initial isotopic composition of the water starting material can be varied.

To test for equilibrium, mineral–water exchange reactions are made by starting with waters that have δ values both higher and lower than the presumed equilibrium value. In this way, the equilibrium fractionation is approached from both directions (Fig. 3.2). Because the

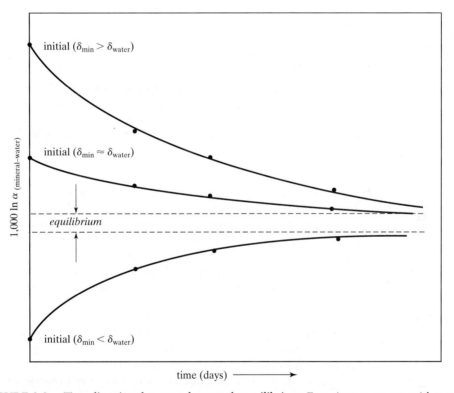

FIGURE 3.2: **Two-directional approach towards equilibrium.** Experiments are run with water that is lighter and with water that is heavier than the mineral. With increasing reaction time, samples tend towards equilibrium. Extrapolation to infinite time should give the equilibrium value.

---

[3]Some experiments have been made in piston–cylinder apparatuses at much higher pressures, with smaller sample charges. Although these experiments are more difficult to carry out, exchange rates are generally enhanced (Matthews et al., 1983), and certain mineral stability fields may be extended to higher temperatures at high pressures.

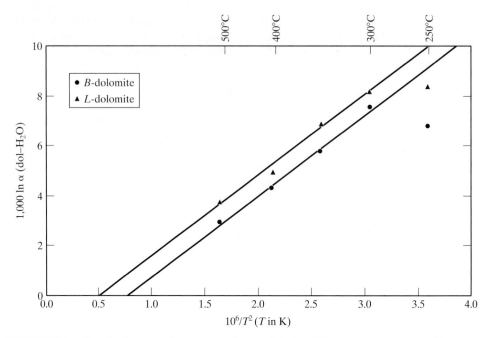

FIGURE 3.3: Results from exchange experiments for the dolomite–water system. Data over the range 300 to 500°C define a straight line when plotted against $1/T^2$. The slope is the a coefficient, the y intercept is the b coefficient. Experiments at 250°C did not appear to approach equilibrium. Two natural dolomite samples were used as starting materials and gave different results, for unknown reasons. After Northrop and Clayton (1966).

rates of reaction are independent of the direction of approach, the results can be extrapolated to infinite time to find the exact equilibrium value.[4]

In theory, experiments need be made only at a single temperature. From equation (3.23), assuming that $b = 3.7$, the constant a can be determined and extended over the entire temperature range. In practice, experiments are made at a series of temperatures, and the best-fit line through the data is used to calculate both a and b (Fig. 3.3).

Several important considerations arise regarding the mineral–water experimental method:

- At high temperatures, the mineral may partially dissolve in the aqueous fluid, recrystallizing *out of isotopic equilibrium* during quench. This concern places an upper limit on the temperatures at which these reactions can be run.
- The drive towards equilibrium is controlled mostly by recrystallization, a process with a higher free-energy drive than isotopic exchange alone. This means that the isotope fractionations follow recrystallization as a "passive partner," and there is no way to prove that equilibrium is indeed being approached.
- Reaction rates are sluggish at low temperatures (Fig. 3.3). For application in low-temperature environments, a large extrapolation of the data is usually necessary. One

---

[4]See Northrop and Clayton (1966) and O'Neil (1986) for the mathematical treatment of this approach.

approach to minimizing this problem is to use metastable starting materials, such as amorphous silica for quartz–water exchange. Recrystallization to the more stable form certainly enhances exchange rates, but there is a greater likelihood that the reaction products will not be in isotopic equilibrium.

3.3.3 Mineral–Calcite Exchange Reactions

Clayton et al. (1989) developed a new method for high-pressure exchange experiments using calcite as the exchange medium. Starting materials were finely admixed calcite and mineral. During the course of the experiment, recrystallization occurred and isotopic equilibrium was approached. The novel method, which has since been employed for a number of minerals, has two striking advantages over earlier experiments, but several potential problems as well.

The main advantage, of course, is that water is eliminated as a reactant. This removes the complication from the high O–H stretching frequencies characteristic of water, and the b term in equation (3.23) reduces to zero. Also, dissolution in aqueous solutions and recrystallization during quench are greatly reduced. Concerns about the quench material being analyzed as part of the run products are virtually eliminated.

However, there are several drawbacks to this method as well. It is clear that at all but the highest temperatures, isotopic equilibration occurs only because the solids undergo recrystallization. As in the mineral–water experiments, the driving force towards equilibrium is recrystallization, and there is no way to know whether the changes in stable isotope ratios are trending towards the equilibrium values or are controlled by kinetically based recrystallization. A second limitation is that exchange rates are far slower than those of the mineral–water system. This raises the lower limits on the temperatures at which significant exchange occurs and hence reduces the possible temperature range of calibration. Finally, diffusion in the calcite "exchange medium" is far slower than in water. As a result, there is concern that isotopic exchange is controlled in part by the diffusion rates of oxygen through calcite (Sharp and Kirschner, 1994). This kinetic control could drive the reaction toward isotopic fractionations that are not the true equilibrium values.

3.3.4 Mineral–CO_2 Exchange Reactions

The idea of measuring the fractionation between a solid and CO_2 gas was pioneered by O'Neil and Epstein (1966) and resurrected over two decades later (Chacko et al., 1991; Stolper and Epstein, 1991). Mineral–CO_2 exchange has the distinct advantage over other experimental methods in that no recrystallization of the solid phases occurs during the reaction. The exchange should be purely diffusional and therefore tend towards equilibrium. The experimental setup is simple as well. Finely powdered samples are loaded in silica or metal tubes, CO_2 is introduced, and the tube is sealed. The tube is heated during which time the CO_2 approaches oxygen isotope equilibrium with the solid.

The major limitation of this type of experiment is that exchange is very slow. Thus, only phases that have very high oxygen diffusion rates, such as carbonates, albite, and silica glass, are accessible to this approach.

The same technique has been used for hydrogen isotope fractionation. Hydrous minerals and molecular hydrogen undergo isotopic exchange surprisingly fast, so that direct exchange between a mineral and H_2 gas provide an exciting new method for determining hydrogen isotope fractionations (Vennemann and O'Neil, 1996).

3.3.5 The Three-Phase Approach

Several studies have used both calcite and water (with or without CO_2) as the exchange medium (Zheng et al., 1994b; Hu and Clayton, 2003). With judicious proportions of H_2O and CO_2, exchange experiments for mineral phases that would otherwise be unstable (e.g., hydrous phases, Zheng et al., 1994a) can be carried out. The presence of water also enhances reaction rates between the mineral and carbonate.

3.4 EMPIRICAL DETERMINATION OF FRACTIONATION FACTORS

Empirical determinations are made by measuring the fractionation between two natural phases, with temperatures either measured (for modern samples) or calculated (for example, with metamorphic rocks). Some of the most successful low-temperature calibrations—notably, the calcite–water (Epstein et al., 1953) and phosphate–water (Longinelli and Nuti, 1973) systems—have been made empirically. The isotopic composition of shells and coexisting water were measured and then compared with measured temperatures of growth. The original calcite–water and phosphate–water equations have withstood the test of time virtually unchanged.

Other low-temperature equations have been made for clay minerals, for which temperatures of formation are estimated from the depositional environment. Savin and Epstein (1970) estimated the oxygen and hydrogen isotope fractionation for kaolinite–water, montmorillonite–water, and glauconite–water at low temperature. Other low-temperature examples include gibbsite (Bird et al., 1994) and silica (Leclerc and Labeyrie, 1987). These empirical studies are particularly successful because exchange experiments are virtually impossible at room temperature; the only experimental avenue to low-temperature exchange is mineral synthesis.

Modern empirical estimates have also been made at higher temperatures, often taking advantage of unique and unusual conditions. Amorphous silica–water fractionation was determined from deposits of thermal waters from power plants (Kita et al., 1985), and quartz–water, calcite–water, and adularia–water fractionations were measured from the Broadlands geothermal field, New Zealand, in drill cores where water could be sampled and temperatures measured (Blattner, 1975).

Finally, fractionation factors have been made from metamorphic or igneous minerals, for which independent temperature estimates are available. In some cases, empirical estimates are the only option available to the isotope geochemist. For certain "refractory" phases,[5] the experimental approach is limited due to extremely sluggish reaction rates, so empirical estimates are our only option. A good example is the quartz–aluminum silicate system, where empirical estimates have been applied successfully to a number of metamorphic terranes (Sharp, 1995). The effect of complex chemical substitutions can also be estimated from natural assemblages (Taylor and O'Neil, 1977; Kohn and Valley, 1998), bypassing the huge amount of effort that would be required to make such measurements experimentally. In all cases, multiple samples from multiple localities should be analyzed to avoid any problems that might inadvertently exist within a single, potentially "anomalous" site. Consider the case

---

[5]Minerals fitting into this category include kyanite, garnet, zircon, corundum, staurolite (?), and certainly diamond and graphite for carbon.

of trying to determine the equilibrium fractionation between orthopyroxene and clinopyroxene from mantle xenoliths. The minerals equilibrated at high temperatures, over an inordinately long time, and cooled rapidly following eruption. One might think that if ever a mineral pair would be in equilibrium, this would be it. And yet we find that in some xenoliths, the clinopyroxene has a higher $\delta^{18}O$ value than coexisting orthopyroxene, and in other samples it is reversed (Perkins et al., 2005). Clearly, empirical estimates should not be made by analyzing a single rock and declaring that the fractionation factors for all of the measured phases are now known!

There are a number of advantages to making empirical estimates. First and foremost, the amount of time that a mineral has had to reach equilibrium with its surroundings far exceeds anything that could be accomplished in the laboratory. A metamorphic rock heated to 500°C for millions of years provides a nice contrast to the same system heated to 900°C in the laboratory for a period of days. Also, many of the potential pitfalls inherent in experimental studies, such as quench recrystallization and metastable equilibria, can be avoided by measuring natural materials.

As with all calibration methods, however, numerous concerns exist as well. The most serious of these are (1) knowing the precise temperature at which equilibrium was attained and (2) ensuring that no retrograde exchange occurred during cooling. The latter is particularly a concern for slowly cooled metamorphic rocks, in which some diffusional resetting is expected. The problem is illustrated in a consideration of the $\Delta^{18}O$ (quartz–feldspar) values commonly measured in igneous rocks (Chapter 11). Fractionations commonly range from 1.5 to 2.5‰, corresponding to temperatures of 430–640°C, clearly indicating that some postcrystallization has occurred.

3.5 OTHER POTENTIAL FACTORS CONTROLLING ISOTOPE PARTITIONING

3.5.1 Pressure Effect

The effect of pressure on the equilibrium constant is given by

$$\left(\frac{\partial \ln K}{\partial P}\right)_T = -\frac{\Delta V_R}{RT}, \qquad (3.25)$$

where ΔV_R is the volume change of the reaction. For an isotope exchange reaction such as equation (3.1), the ΔV_R term is close to zero, so that pressure will have a minimal effect on the fractionation between coexisting species. Hoering (1961) first demonstrated the insensitivity of fractionation to pressure when he measured the $^{16}O/^{18}O$ fractionation between and H_2O and HCO_3^- at 1 atmosphere and at 4 kilobars (both at 43.5°C). There was a change of 0.2 (± 0.2)‰ fractionation between 4 kb and 1 atm, which he concluded was negligible. Later, calcite–water and quartz–water exchange experiments were made over a pressure range of 1 to 20 kbar, with no detectable pressure effect (Clayton et al., 1975; Matthews et al., 1983). Polyakov and Kharlashina (1994) devised a statistical mechanical method of estimating pressure effects. For most rocks exposed at the Earth's surface, the pressure effect will be near the detection limits of analysis. At very high pressures, however, the effects can be significant. Using this method, Sharp et al. (1992) found that quartz is particularly sensitive to pressure. For most minerals, the electrostatic site potentials increase slightly with pressure, but the reverse is found for quartz. As a result, $\Delta^{18}O_{qz-min}$ values will change by ~0.5‰ at 1200°C and 40 kbar. Fortunately, the unusual behavior of quartz becomes redundant, because coesite is the stable SiO_2 polymorph above ~27 kbar.

The effect of pressure is greater for graphite–diamond (Polyakov and Kharlashina, 1994), but the really striking pressure effects are seen for D/H fractionation between hydrous minerals and water. Figure 3.4 shows the effect of pressure on the brucite–water fractionation as a function of pressure and temperature. In their combined experimental–theoretical study, Horita et al. (2002) conclude that water is much more strongly affected than hydrous minerals, so hydrogen isotope fractionation pressure effects should exist for all water–hydrous mineral pairs.

3.5.2 Composition and Structure

Taylor and Epstein (1962) identified a simple relationship between chemical composition and isotopic enrichment, recognizing that bond strength—and, therefore, oxygen isotope enrichment—decreases from Si–O bonds, through Al–O bonds, to M^{2+}–O = M^{1+}–O bonds.

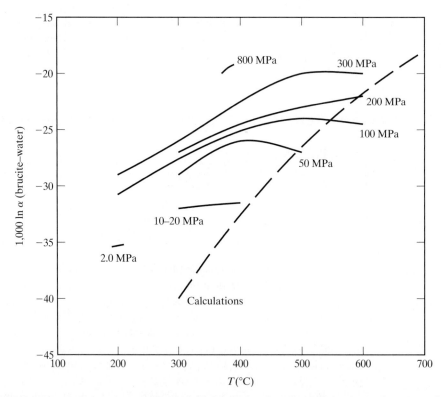

FIGURE 3.4: Hydrogen isotope fractionation between brucite and water as a function of temperature for various pressures. The dashed line labeled "Calculations" is the theoretical fractionation between brucite and water at 1 bar, based on statistical mechanical calculations. With increasing pressures there is an increase in brucite-water fractionations; the effect decreases with increasing temperature. Reprinted from Horita *et al.*, 2002 with permission from Elsevier.

Minerals follow this rule quite well. Quartz almost always has the highest $\delta^{18}O$ value, followed by feldspar and continuing down to the Si- and Al-free oxides, such as magnetite, rutile, and hematite. Rough estimates of relative isotopic enrichment are easily made by keeping this rule in mind. Consider olivine and clinopyroxene. Which one will concentrate ^{18}O relative to the other? Mg_2SiO_4 has a lower proportion of Si–O bonds than $MgSiO_3$ (or $CaMgSi_2O_6$) and, consequently, a lower $\delta^{18}O$ value. In general, the substitution of identically charged cations (e.g., Na $\Leftrightarrow$ K, Fe $\Leftrightarrow$ Mg, and Ca $\Leftrightarrow$ Mn) has a minimal effect on isotopic fractionation. There is no oxygen isotope fractionation between albite and potassium feldspar ($NaAlSi_3O_8$ vs. $KAlSi_3O_8$), or between almandine and pyrope ($Fe_3Al_2Si_3O_{12}$ vs. $Mg_3Al_2Si_3O_{12}$), and only a small effect of Ca $\Leftrightarrow$ (Mg, Fe) substitution. So, for minerals such as garnet, pyroxene, and biotite, composition can safely be ignored. This situation is completely opposite to that of cation exchange or mineral equilibria thermometers, wherein it is imperative to consider the effects of composition.

The largest effects are for coupled substitutions, such as NaSi $\Leftrightarrow$ CaAl in plagioclase ($NaAlSi_3O_8$ $\Leftrightarrow$ $CaAl_2Si_2O_8$) and NaAl $\Leftrightarrow$ Ca(Mg, Fe) in pyroxene ($NaAlSi_2O_6$ $\Leftrightarrow$ $CaMgSi_2O_6$). The temperature coefficient of fractionation (the term a in equation (3.23) is 0.94 for quartz–albite, increasing by $1.05x$ (x = fraction of anorthite in plagioclase), up to 1.99 for pure anorthite. Other substitutions that affect isotopic fractionation are F $\Leftrightarrow$ OH in phlogopite and Al^{3+} $\Leftrightarrow$ Fe^{3+} in garnet. (See Chacko et al., 2001, for more details.)

The effect of composition on hydrogen isotope fractionation has not been thoroughly studied, but in a seminal paper on the subject, Suzuoki and Epstein (1976) found that Al has the strongest affinity for deuterium, followed by Mg and Fe. They proposed a general equation to predict hydrogen isotope exchange between hydrous minerals and water given by[6]

$$1000 \ln \alpha_{\text{mineral}-H_2O} = \frac{-22.4 \times 10^6}{T^2} + 26.3 + (2X_{Al} - 4X_{Mg} - 68X_{Fe}) \quad (3.26)$$

where X refers to the portion of each element in the octahedral site. This equation generally predicts the correct degree of enrichment, but not necessarily the correct temperature dependence (Chacko et al., 2001).

There can be a large compositional effect for saline solutions relative to pure water. The effect is especially strong for $CaCl_2$ and $MgCl_2$ solutions and decreases towards $MgSO_4$ and (Na, K)Cl, being up to 6‰/molal and 1‰/molal for hydrogen and oxygen isotopes, respectively. For all but concentrated (Ca, Mg)Cl_2 and $MgSO_4$ solutions, however, the effects for both hydrogen and oxygen can be ignored.

The effect of polymorphism is also unimportant for the most part. There are several notable exceptions in which a polymorphic transition has a significant isotope effect, including graphite–diamond (Bottinga, 1969), calcite–aragonite (oxygen, Rubinson and Clayton, 1969), and perhaps quartz–coesite. For most polymorphic transitions, however, the effects are negligible. For example, no oxygen isotope fractionations have been seen between the different aluminum silicate polymorphs andalusite, kyanite, and sillimanite in which coexisting polymorphs are found to have nearly identical $\delta^{18}O$ values (Cavosie et al., 2002; Larson and Sharp, 2003).

---

[6]Equation (3.26) contained a printing error in the original publication. The constant 26.3 was originally given as 28.2 (Morikiyo, 1986).

3.6 SO WHICH FRACTIONATION FACTORS ARE CORRECT?

As with any thermometric estimate, be it based on cation exchange thermometry, solvus thermometry, or stable isotope thermometry, it is often difficult to know whether a calibration gives a meaningful answer or not. In some cases, most calibrations converge on a single result and the results are geologically reasonable. In other cases, the thermometers yield temperatures that are clearly unreasonable. So there are "good" and "bad" thermometers; our task is to determine which are which. Unfortunately, there are no firm rules defining which thermometers are the good ones. Some mineral pairs seem to be well behaved and give good results. There have been a number of independent calibrations of the quartz–magnetite thermometer. Most are in agreement with one another, and all are "field tested"; that is, they give reasonable answers when applied to natural rocks that have not been heated to temperatures in excess of ~600°C (Sharp et al., 1988). Most people would argue that quartz–magnetite is a good thermometer. For others, there are more serious discrepancies.

Bottinga and Javoy (1975) estimated fractionation factors for a number of minerals, combining empirical, experimental, and theoretical calibrations.[7] Their calibrations have been extremely successful and were used in innumerable studies. Their quartz–biotite fractionation, based mainly on empirical data, gives reasonable temperatures for igneous and many metamorphic rocks. Later, high-pressure quartz–calcite and biotite–calcite experiments were combined to give a new quartz–biotite calibration (Chacko et al., 1996) that is very different from the earlier Bottinga–Javoy curve. Applied to rocks, the experimental calibration invariably yields temperatures that are too low to be geologically reasonable. Chacko et al. explained the discrepancy in terms of the rapid diffusion rate of oxygen in biotite, concluding that exchange must continue down to low temperatures. In fact, the low temperatures are indeed expected: Diffusion rate data (Fortier and Giletti, 1991) support this conclusion. However, the Bottinga–Javoy calibration continues to give reliable temperature estimates. Is their estimate "fortuitously flawed"? At this point, it is difficult to say one way or the other.

3.6.1 An Example from Quartz–Calcite Fractionation

The fractionation factors for quartz–calcite should be among the best understood. After all, the calcite–water fractionation was the first fractionation to be determined, and multiple redeterminations have only reinforced the validity of the early estimates. There have been over a dozen quartz–water fractionation estimates and another half dozen theoretical, experimental, and empirical estimates involving quartz fractionation (Fig. 3.5). Unlike calcite fractionation, however, there is no convergence towards a single correct value. Early experiments were all made between quartz and water. A number of ingenious synthesis and exchange experiments were devised. Nevertheless, discrepancies exist. In some cases, they can be explained, but not always. Discrepancies also exist between the various theoretical (Shiro and Sakai, 1972; Kawabe, 1978; Kieffer, 1982), empirical (Blattner, 1975; Sharp and Kirschner, 1994), and experimental calibrations.

The new method of directly equilibrating quartz and calcite, devised by Clayton et al. (1989), appeared to eliminate all problems encountered in the earlier mineral–water experimental studies. Fractionations were measured directly, and problems with combining two

---

[7]The authors were aware of the effect of retrograde diffusion and were careful to take diffusion effects into account.

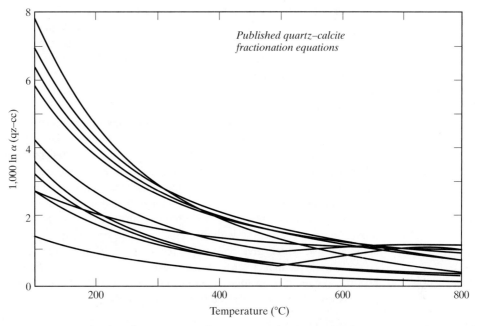

FIGURE 3.5: Published fractionations for quartz and calcite. The data represent theoretical, empirical and experimental curves. Quartz-water curves were combined with calcite-water of O'Neil *et al.* (1969). See Sharp and Kirschner (1994) for details.

different calibrations were eliminated. There were no problems with the high vibrational frequencies of water (Section 3.3.1) and no concerns of dissolution or evaporation. And yet, when applied to rocks, the new calibration method yielded temperature estimates that were generally far too low to be geologically reasonable. Sharp and Kirschner (1994) suggested that the direct-exchange experiments may suffer from kinetic effects associated with rapid recrystallization, whereby the measured fractionations are less than those which would be obtained under equilibrium conditions.

The problem of recrystallization was eliminated in a series of exchange experiments between solids and CO_2 gas (Chacko et al., 1991; Stolper and Epstein, 1991). Exchange between a mineral and CO_2 gas occurs by diffusion. There is no recrystallization, and no large free-energy processes driving isotopic exchange. So it would seem that exchange would converge towards the equilibrium value. Quartz–CO_2 exchange has not been measured. Oxygen diffusion rates in quartz are too slow. SiO_2 glass was used as a proxy for quartz.[8] The combined quartz (glass)–CO_2 and calcite–CO_2 fractionations give a quartz–calcite fractionation equation that is similar to the empirical calibration that "works" with the natural data. At present, the discrepancy between the direct-exchange experiments and the other calibrations is unresolved. A previously unrecognized "salt effect" on mineral–water exchange experiments has been proposed (Hu and Clayton, 2003), but obviously, this would not affect the mineral–CO_2 or empirical calibrations.

---

[8]Matthews et al. (1994) had demonstrated that there was no fractionation between albite and albite glass. By analogy, there should be no fractionation between quartz and quartz glass.

So which calibration is best? Which ones should be used? There are those who have the idea that anything theoretical—that is, anything that is "mathematically based"—must be the best. At the other extreme, there are those who would argue that experiments do not lie and must give the right answer. If nature and experiments don't agree, then there must be something wrong with the natural data. Yet, Nature has much to say, often in subtle ways. Just consider that Lord Kelvin used cooling theory to prove that the Earth was less than 20 million years old, even though Nature was giving lots of hints—thick sedimentary sequences and incredible evolutionary diversity—to the contrary. Something was being overlooked. That something, of course, turned out to be radioactivity. Its discovery brought immediate reconciliation between all the scientific groups.

So where do we stand? The stable isotope community has put at least 40 years of effort into gaining a better understanding of fractionation factors among minerals, fluids, and gases. Much has been learned in that time, with some ideas jettisoned and others retained. The fact that our methods have changed and continue to be modified does not mean that the utilization of stable isotope thermometers is in any way futile. Refinements are made and methods improve. One way of dealing with discrepancies is to try multiple calibrations, compare the results, and ask which give reasonable results. *Should* they give reasonable results? If they don't, why? What makes sense? What doesn't? With a judicious choice of calibrations and applications of multiple isotope thermometers, there is a very good chance that accurate temperature estimates can be recovered from geological materials.

PROBLEM SET

1. a. Why do anhydrous mineral–mineral fractionation data take the form $1000 \ln \alpha_{m-n} = \dfrac{a \times 10^6}{T^2}$, whereas mineral–water fractionations have the additional b term, given in $1000 \ln \alpha_{m-n} = \dfrac{a \times 10^6}{T^2} + b$?

   b. Why does the relationship $1000 \ln \alpha_{m-n} \propto \dfrac{1}{T^2}$ change to $1000 \ln \alpha_{m-n} \propto \dfrac{1}{T}$ at low temperatures?

2. Mineral–water, mineral–calcite, and mineral–gas exchange experiments have all been used in the experimental calibration of fractionation factors. What are the advantages and disadvantages of each?

3. Deuterium hydride is the name of the diatomic gas HD and is available commercially.

   a. What is the free-energy change for the formation of 1 mole of HD gas by the reaction $\tfrac{1}{2}D_2(g) + \tfrac{1}{2}H_2(g) \rightarrow HD(g)$, given that ΔG_f° for $H_2(g)$, $HD(g)$, and $D_2(g)$ are 0.0, -1.464, and 0.0 kJ/mole, respectively?

   b. What fraction of the gas will be HD at equilibrium?

   c. Why is HD (g) commercially available while HDO(l) is not?

4. What is the equilibrium δD value of water vapor (H_2O gas) in equilibrium with H_2 gas (at 25°C) if the δD value of the H_2 gas is -600%. Use the following information:

Phase	ΔG_f° (kJ/mol)
$H_2(g)$	0.0
$HD(g)$	-0.350
$H_2O(g)$	-54.634
$HDO(g)$	-55.719

5. Calculate the spring constant k_s for $^{12}C^{16}O$, given that, for $^{12}C^{16}O$, the wave number ω is 2167.4 cm^{-1}.
6. Ferrosilite ($Fe_2Si_2O_6$) is a pyroxene that is stable only at high pressures. If ferrosilite is heated at low pressure, it breaks down to fayalite (Fe_2SiO_4) and quartz (ignoring potential kinetic effects). Likewise, heating calcite to high temperatures produces lime (CaO) and CO_2. Comment on whether laboratory reactions in which these products are produced could be used for calibrating the fractionations between fayalite and quartz, on the one hand, and lime and CO_2, on the other.
7. Suppose the general enrichment of ^{18}O is more strongly concentrated in Si–O bonds, then Al–O bonds, and, finally, M^{2+}–O = M^{1+}–O bonds. For the following mineral triplets, which would you expect to have the higher $^{18}O/^{16}O$ ratio? (a) albite ($NaAlSi_3O_8$), microcline ($KAlSi_3O_8$), and anorthite ($CaAl_2Si_2O_8$); (b) mullite ($Al_6Si_2O_{13}$), sillimanite (Al_2SiO_5), and corundum (Al_2O_3).
8. The fundamental vibrational frequency of $^{16}O_2$ is 1580.4 cm^{-1}. Calculate the fundamental vibrational frequency of $^{18}O_2$.
9. Twelve mg of calcite ($\delta^{18}O$ = 24.32) and 50 mg of water ($\delta^{18}O$ = −3.65) are equilibrated in a sealed gold capsule at 350°C. At this temperature, α(cal–water) = 1.00490. What are the $\delta^{18}O$ values of the calcite and water after the equilibration?

ANSWERS

1. See discussion in Sections 3.2.7 and 3.3.1.
2. See discussion in Sections 3.2–3.4.
3. a. $\Delta G^\circ_{r,25°C} = \sum \Delta G^\circ_{f,25°C_{products}} - \sum \Delta G^\circ_{f,25°C_{reactants}} = -1.464$ kJ.
   b. We know that $\Delta G^\circ_r = -RT \ln(K)$, where $K = \dfrac{[HD]}{[H_2]^{1/2}[D_2]^{1/2}}$, and that, for initial fractions of H_2 and D_2 of 0.5, the final quantity of each is $0.5 - x$ and the amount of HD is $2x$. From part a, $\Delta G^\circ_r = -1464$ J and $K = 1.8055$. Finally, we solve the equation $1.8055 = \dfrac{[2x]}{[0.5-x]^{1/2}[0.5-x]^{1/2}} = \dfrac{[2x]}{[0.5-x]}$ to get $x = 0.237$. The percentage of HD in equilibrium with equal quantities of H_2 and D_2 is therefore 23.7%.
   c. HD, while not stable as a pure phase, is metastable. HDO will quickly react to equilibrium with H_2O and D_2O.
4. The exchange reaction is given by $H_2(g) + HDO(g) = HD(g) + H_2O(g)$. In this case, the free-energy change for the reaction is given by

$$\Delta G^\circ_{r,25°C} = \sum \Delta G^\circ_{f,25°C_{products}} - \sum \Delta G^\circ_{f,25°C_{reactants}} = -0.35 + -54.634 - (0.0 + -55.719)$$
$$= 0.735 \text{ kJ, or } 735 \text{ J}.$$

The equilibrium constant is $K = \dfrac{[H_2O][HD]}{[HDO][H_2]}$ and $\Delta G^\circ_r = -RT \ln(K)$, where $K = \alpha$ (equation (3.4)). Solving, we get $\alpha = 0.289$. From equation (2.14), $\delta^{18}O_{H_2O(g)}$ is +384‰. This is in fairly good agreement with a published α value of 0.2824 (Suess, 1949. *Zeitschr. Naturforschung.* 4, 328–332).

5. Equation (3.12) supplies the answer to this problem. Rearranging to solve for k_s gives $k_s = (2\pi\nu)^2 \left(\dfrac{m_a m_b}{m_a + m_b}\right)$ and $\nu = \omega c$. Given that c (speed of light) = 2.998×10^{10} cm/sec and $\omega = 2167.4$ cm^{-1}, we obtain $\nu = 6.4979 \times 10^{13}$ sec^{-1}. For $m_a = 12$ and $m_b = 16$, $k_s = 1.143 \times 10^{30}$. Question: What are the units of k_s?

6. In the first case, the two phases fayalite and quartz would be intergrown at a fine scale and probably physically impossible to separate. In both cases, although the reactions may go to completion, there is no guarantee that the phases will be in isotopic equilibrium, because the reactions are driven by more energetic chemical changes rather than isotopic ones.

7. Albite and microcline should be the same, because substitution between Na^+ and K^+ should have a negligible effect. Albite is in the proportions Na_2O: 1/2 Al_2O_3: 3/2 SiO_2, and anorthite is CaO: Al_2O_3: 2 SiO_2. Because there is a higher SiO_2/Al_2O_3 ratio in albite vs. anorthite, the former should concentrate more ^{18}O relative to the latter. (b). Sillimanite has the highest SiO_2/Al_2O_3 ratio, followed by mullite and corundum. They should concentrate ^{18}O in that order.

8. Equation 3.13 relates the vibrational frequency of the isotopically substituted molecule to the reduced mass, defined as $\frac{m_a m_b}{m_a + m_b}$. For $^{16}O_2$, the reduced mass is $\frac{16 \times 16}{16 + 16} = 8$. For $^{18}O_2$, the reduced mass is 9. The vibrational frequency of $^{18}O_2$ is calculated to be 1490.02.

9. The molecular weights of calcite ($CaCO_3$) and water are 100.087 and 18.015, respectively. The fraction of O_2 in calcite is 0.4796 and 0.888 for water. 12 mg of calcite has $12 \times 0.4796 = 5.75$ mg of O_2. For 50 mg of water, there are 44.41 mg of O_2. In general, the overall $\delta^{18}O$ value of a mixture is given by $\delta^{18}O_{mixture} = \frac{m_a \delta_a + m_b \delta_b}{m_a + m_b}$, where m_a is the mass (or number of moles) of O_2 in phase a and δ_a is the $\delta^{18}O$ value of phase a (similar notation for phase b). For our experimental conditions,

$$\delta^{18}O_{mixture} = \frac{5.75 \times 24.32 + 44.41 \times (-3.65)}{5.75 + 44.41} = -0.44\text{‰}.$$

The $\delta^{18}O$ value of our mixture does not change during the course of the reaction. At equilibrium, we can define $\delta^{18}O_{water} = x$ and $\delta^{18}O_{calcite}$ given by $1000\alpha - 1000 + \alpha \cdot \delta^{18}O_{water}$ (from definition of α). For $\alpha = 1.0049$, $\delta^{18}O_{calcite} = 4.9 + 1.0049 \times \delta^{18}O_{water}$. Plugging these numbers into the formula for the mixture (where $\delta^{18}O_{mixture} = -0.44\text{‰}$) gives

$$\delta^{18}O_{mixture} = -0.44 = \frac{5.75 \times (4.9 + 1.0049 \times \delta^{18}O_{water}) + 44.41 \times (\delta^{18}O_{water})}{5.75 + 44.41},$$

which gives a $\delta^{18}O_{water}$ value of -1.00‰ and a $\delta^{18}O_{calcite} = 3.90\text{‰}$.

REFERENCES

Bigeleisen, J., and Mayer, M. G. (1947). Calculation of equilibrium constants for isotopic exchange reactions. *Journal of Chemical Physics* **15**, 261–267.

Bird, M. I., Longstaffe, F. J., Fyfe, W. S., Tazaki, K., and Chivas, A. R. (1994). Oxygen-isotope fractionation in gibbsite: Synthesis experiments versus natural samples. *Geochimica et Cosmochimica Acta* **58**, 5267–5277.

Blattner, P. (1975). Oxygen isotopic composition of fissure-grown quartz, adularia, and calcite from Broadlands geothermal field, New Zealand, with an appendix on quartz–K–feldspar–calcite–muscovite oxygen isotope geothermometers. *American Journal of Science* **275**, 785–800.

Bottinga, Y. (1968). Calculation of fractionation factors for carbon and oxygen isotopic exchange in the system calcite–carbon dioxide–water. *Journal of Physical Chemistry* **72**, 800–808.

Bottinga, Y. (1969). Carbon isotope fractionation between graphite, diamond and carbon dioxide. *Earth and Planetary Science Letters* **5**, 301–307.

Bottinga, Y., and Javoy, M. (1973). Comments on oxygen isotope geothermometry. *Earth and Planetary Science Letters* **20**, 250–265.

Bottinga, Y., and Javoy, M. (1975). Oxygen isotope partitioning among the minerals in igneous and metamorphic rocks. *Reviews of Geophysics and Space Physics* **13**, 401–418.

Cavosie, A., Sharp, Z. D., and Selverstone, J. (2002). Co-existing aluminum silicates in quartz veins: A quantitative approach for determining andalusite–sillimanite equilibrium in natural samples using oxygen isotopes. *American Mineralogist* **87**, 417–423.

Chacko, T., Cole, D. R., and Horita, J. (2001). Equilibrium oxygen, hydrogen and carbon isotope fractionation factors applicable to geologic systems. In *Stable Isotope Geochemistry*, Vol. 43 (ed. J. W. Valley and D. R. Cole), pp. 1–81. Washington, DC: Mineralogical Society of America.

Chacko, T., Hu, X., Mayeda, T. M., Clayton, R. N., and Goldsmith, J. R. (1996). Oxygen isotope fractionations in muscovite, phlogopite, and rutile. *Geochimica et Cosmochimica Acta* **60**, 2595–2608.

Chacko, T., Mayeda, T. K., Clayton, R. N., and Goldsmith, J. R. (1991). Oxygen and carbon isotope fractionations between CO_2 and calcite. *Geochimica et Cosmochimica Acta* **55**, 2867–2882.

Clayton, R. N., Goldsmith, J. R., Karel, K. J., Mayeda, T. K., and Newton, R. C. (1975). Limits on the effect of pressure on isotopic fractionation. *Geochimica et Cosmochimica Acta* **39**, 1197–1201.

Clayton, R. N., Goldsmith, J. R., and Mayeda, T. K. (1989). Oxygen isotope fractionation in quartz, albite, anorthite, and calcite. *Geochimica et Cosmochimica Acta* **53**, 725–733.

Criss, R. E. (1991). Temperature dependence of isotopic fractionation factors. In *Stable Isotope Geochemistry: A Tribute to Samuel Epstein*, Vol. 3 (ed. H. P. J. Taylor, J. R. O'Neil, and I. R. Kaplan), pp. 11–16. San Antonio: The Geochemical Society.

Criss, R. E. (1999). *Prinicples of stable isotope distribution*. New York: Oxford University Press.

Denbigh, K. (1971). *The Principles of Chemical Equilibrium*. Cambridge, U.K.: Cambridge University Press.

Driesner T., and Seward, T. M. (2000). Experimental and simulation study of salt effects and pressure/density effects on oxygen and hydrogen stable isotope liquid–vapor fractionation for 4–5 molal aqueous NaCl and KCl solutions at 400 degrees C. *Geochimica et Cosmochimica Acta* **64**, 1773–1784.

Epstein, S., Buchsbaum, R., Lowenstam, H., and Urey, H. C. (1951). Carbonate–water isotopic temperature scale. *Journal of Geology* **62**, 417–426.

Epstein, S., Buchsbaum, R., Lowenstam, H. A., and Urey, H. C. (1953). Revised carbonate–water isotopic temperature scale. *Geological Society of America Bulletin* **64**, 1315–1326.

Fortier, S. M., and Giletti, B. J. (1991). Volume self-diffusion of oxygen in biotite, muscovite, and phlogopite micas. *Geochimica et Cosmochimica Acta* **55**, 1319–1330.

Hattori, K., and Halas, S. (1982). Calculation of oxygen isotope fractionation between uranium dioxide, uranium trioxide and water. *Geochimica et Cosmochimica Acta* **46**, 1863–1868.

Hoering, T. C. (1961). The effect of physical changes on isotopic fractionation. *Carnegie Institution of Washington Yearbook* **60**, 201–204.

Hoffbauer, R., Hoernes, S., and Fiorentini, E. (1994). Oxygen isotope thermometry based on a refined increment method and its application to granulite-grade rocks from Sri Lanka. *Precambrian Research* **66**, 199–220.

Horita, J., Cole, D. R., Polyakov, V. B., and Driesner, T. (2002). Experimental and theoretical study of pressure effects on hydrogen isotope fractionation in the system brucite–water at elevated temperatures. *Geochimica et Cosmochimica Acta* **66**, 3769–3788.

Hu, G., and Clayton, R. N. (2003). Oxygen isotope salt effects at high pressure and high temperature and the calibration of oxygen isotope geothermometers. *Geochimica et Cosmochimica Acta* **67**, 3227–3246.

Kawabe, I. (1978). Calculation of oxygen isotope fractionation in quartz–water system with special reference to the low temperature. *Geochimica et Cosmochimica Acta* **42**, 613–621.

Kieffer, S. W. (1982). Thermodynamics and lattice vibrations in minerals: 5. Applications to phase equilibria, isotopic fractionation, and high-pressure thermodynamic properties. *Reviews of Geophysics and Space Physics* **20**, 827–849.

Kita, I., Taguchi, S., and Matsubaya, O. (1985). Oxygen isotope fractionation between amorphous silica and water at 34–93°C. *Nature* **314**, 83–84.

Kohn, M. J., and Valley, J. W. (1998). Oxygen isotope geochemistry of the amphiboles; isotope effects of cation substitutions in minerals. *Geochimica et Cosmochimica Acta* **62**, 1947–1958.

Larson, T. E., and Sharp, Z. D. (2003). Stable isotope constraints on the Al_2SiO_5 'triple point' rocks from the Proterozoic Priest pluton contact aureole, New Mexico, USA. *Journal of Metamorphic Geology* **21**, 785–798.

Leclerc, A. J., and Labeyrie, L. (1987). Temperature dependence of oxygen isotopic fractionation between diatom silica and water. *Earth and Planetary Science Letters* **84**, 69–74.

Longinelli, A., and Nuti, S. (1973). Revised phosphate–water isotopic temperature scale. *Earth and Planetary Science Letters* **19**, 373–376.

Matthews, A., Goldsmith, J. R., and Clayton, R. N. (1983). On the mechanisms and kinetics of oxygen isotope exchange in quartz and feldspars at elevated temperatures and pressures. *Geological Society of America Bulletin* **94**, 396–412.

Matthews, A., Palin, J. M., Epstein, S., and Stolper, E. M. (1994). Experimental study of $^{18}O/^{16}O$ partitioning between crystalline albite, albite glass, and CO_2 gas. *Geochimica et Cosmochimica Acta* **58**, 5255–5266.

McCrea, J. M. (1950). On the isotopic chemistry of carbonates and a paleotemperature scale. *Journal of Chemical Physics* **18**, 849–857.

Morikiyo, T. (1986). Hydrogen and carbon isotope studies on the graphite-bearing metapelites in the northern Kiso District of central Japan. *Contributions to Mineralogy and Petrology* **94**, 165–177.

Northrop, D. A., and Clayton, R. N. (1966). Oxygen-isotope fractionations in systems containing dolomite. *Journal of Geology* **74**, 174–196.

O'Neil, J. R. (1986). Theoretical and experimental aspects of isotopic fractionation. In *Stable Isotopes in High Temperature Geological Processes*, Vol. 16 (ed. J. W. Valley, H. P. Taylor, Jr., and J. R. O'Neil), pp. 1–40. Mineralogical Society of America, Washington, D.C. Chelsea. MI

O'Neil, J. R., Clayton, R. N., and Mayeda, T. K. (1969). Oxygen isotope fractionation in divalent metal carbonates. *Journal of Chemical Physics* **51**, 5547–5558.

O'Neil, J. R., and Epstein, S. (1966). Oxygen isotope fractionation in the system dolomite–calcite–carbon dioxide. *Science* **152**, 198–200.

Perkins, G. B., Sharp, Z. D., and Selverstone, J. (2005). Oxygen isotope evidence for subduction and rift-related mantle metasomatism beneath the Colorado Plateau-Rio Grande Rift transition. *Contributions to Mineralogy and Petrology* (in press).

Polyakov, V. B., and Kharlashina, N. N. (1994). Effect of pressure on equilibrium isotopic fractionation. *Geochimica et Cosmochimica Acta* **58**, 4739–4750.

Polyakov, V. B., and Mineev, S. D. (2000). The use of Mössbauer spectroscopy in stable isotope geochemistry. *Geochimica et Cosmochimica Acta* **64**, 849–865.

Richet, P., Bottinga, Y., and Javoy, M. (1977). A review of hydrogen, carbon, nitrogen, oxygen, sulphur, and chlorine stable isotope fractionation among gaseous molecules. *Annual Review of Earth and Planetary Science* **5**, 65–110.

Richter, R., and Hoernes, S. (1988). The application of the increment method in comparison with experimentally derived and calculated O-isotope fractionations. *Chemie der Erde* **48**, 1–18.

Rubinson, M., and Clayton, R. N. (1969). Carbon-13 fractionation between aragonite and calcite. *Geochimica et Cosmochimica Acta* **33**, 997–1002.

Savin, S. M., and Epstein, S. (1970). The oxygen and hydrogen isotope geochemistry of clay minerals. *Geochimica et Cosmochimica Acta* **34**, 35–42.

Savin, S. M., and Lee, M. (1988) Isotopic studies of phyllosilicates. In *Hydrous Phyllosilicates*, Vol. 19 (ed. S. W. Bailey), pp. 189–223. Mineralogical Society of America, Chelsea, MI.

Schauble, E. A. (2004). Applying stable isotope fractionation theory to new systems. In *Geochemistry of Non-Traditional Stable Isotopes*, Vol. 55 (ed. C. M. Johnson, B. L. Beard, and F. Albarède), pp. 65–111. Washington, DC: Mineralogical Society of America.

Schütze, H. (1980). Der Isotopenindex—eine Inkrementenmethode zur näherungsweise Berechnung von Isotopenaustauschgleichgewichten zwischen kristallinin Substanzen. *Chemie der Erde* **39**, 321–334.

Sharp, Z. D. (1995). Oxygen isotope geochemistry of the Al_2SiO_5 polymorphs. *American Journal of Science* **295**, 1058–1076.

Sharp, Z. D., Essene, E. J., and Smyth, J. R. (1992). Ultra-high temperatures from oxygen isotope thermometry of a coesite–sanidine grospydite. *Contributions to Mineralogy and Petrology* **112**, 358–370.

Sharp, Z. D., and Kirschner, D. L. (1994). Quartz–calcite oxygen isotope thermometry; a calibration based on natural isotopic variations. *Geochimica et Cosmochimica Acta* **58**, 4491–4501.

Sharp, Z. D., O'Neil, J. R., and Essene, E. J. (1988). Oxygen isotope variations in granulite grade iron formations; constraints on oxygen diffusion and retrograde isotopic exchange. *Contributions to Mineralogy and Petrology* **98**, 490–501.

Shiro, Y., and Sakai, H. (1972). Calculation of the reduced partition function ratios of α-, β-quartz and calcite. *Bulletin of the Chemical Society of Japan* **45**, 2355–2359.

Smyth, J. R. (1989). Electrostatic characterization of oxygen sites in minerals. *Geochimica et Cosmochimica Acta* **53**, 1101–1110.

Stolper, E., and Epstein, S. (1991). An experimental study of oxygen isotope partitioning between silica glass and CO_2 vapor. In *Stable Isotope Geochemistry: A Tribute to Samuel Epstein*, Vol. 3 (ed. H. P. Taylor, Jr., J. R. O'Neil, and I. R. Kaplan), pp. 35–51. San Antonio: The Geochemical Society.

Suzuoki, T., and Epstein, S. (1976). Hydrogen isotope fractionation between OH-bearing minerals and water. *Geochimica et Cosmochimica Acta*. **40**, 1229–1240.

Taylor, B. E., and O'Neil, J. R. (1977). Stable isotope studies of metasomatic skarns and associated metamorphic and igneous rocks, Osgood Mountains, Nevada. *Contributions to Mineralogy and Petrology* **63**, 1–49.

Taylor, H. P., Jr., and Epstein, S. (1962). Relationship between O^{18}/O^{16} ratios in coexisting minerals of igneous and metamorphic rocks: Part 1. Principles and experimental results. *Geological Society of America Bulletin* **73**, 461–480.

Vennemann, T. W., and O'Neil, J. R. (1996). Hydrogen isotope exchange between hydrous minerals and molecular hydrogen: 1. A new approach for the determination of hydrogen isotope fractionation at moderate temperature. *Geochimica et Cosmochimica Acta* **60**, 2437–2451.

Zheng, Y. F. (1993). Calculation of oxygen isotope fractionation in anhydrous silicate minerals. *Geochimica et Cosmochimica Acta* **57**, 1079–1091.

Zheng, Y.-F., Metz, P., and Satir, M. (1994a) Oxygen isotope fractionation between calcite and tremolite: An experimental study. *Contributions to Mineralogy and Petrology* **118**, 249–255.

Zheng, Y. F., Metz, P., Satir, M., and Sharp, Z. D. (1994b). An experimental calibration of oxygen isotope fractionation between calcite and forsterite in the presence of a CO_2-H_2O fluid. *Chemical Geology* **116**, 17–27.

CHAPTER 4

THE HYDROSPHERE

4.1 OVERVIEW

The hydrosphere constitutes an astonishing 5.9 weight percent of the crust of Earth, and a significant amount of water is present in hydrous and nominally anhydrous minerals in the mantle as well. Water and aqueous solutions play a major role in the formation and development of rocks and minerals, the development of physiographic features, meteorological patterns, and all life processes. Thus, it is understandable that more stable isotope analyses have been made of water, collected from its myriad of reservoirs on Earth, than of any other naturally occurring substance. Despite the pronounced reactivity of water, the stable isotope compositions of large and relatively well mixed bodies of water like the ocean or large lakes vary little in time or place. In contrast, those of waters in minor abundance, and especially those that form as a result of evaporation (e.g., meteoric origin), vary widely. This behavior reflects a simple mass-balance principle that we shall meet over and over again in this text: During an interaction between two reservoirs of the same element, the isotopic composition of the element will change more in the smaller reservoir than it will in the larger reservoir and by an amount that is proportional to the relative sizes of the reservoirs. Finally, note that the wide range of $\delta^{18}O$ and δD values of meteoric water are fundamentally related to fractionation associated with phase change. As stated by Gat (1996), "In the water cycle, the most significant process in this respect is that of phase changes, from vapor to liquid or ice and vice versa."

Variations in the hydrogen and oxygen isotope compositions of modern natural waters are presented in Table 4.1. Ocean water is a logical reference material (SMOW = Standard Mean Ocean Water) for stable isotope analyses of natural waters, because it constitutes over 97% of the hydrosphere, is the source of essentially all atmospheric moisture, appears to be buffered to a relatively constant $^{18}O/^{16}O$ ratio by both high- and low-temperature interactions with mantle-derived rocks on the ocean floor, and controls the stable isotope compositions of authigenic minerals that form in it and the tissues of life-forms that inhabit it.

TABLE 4.1: Representative isotopic compositions and approximate volumes of natural waters.

Reservoir	Volume (%)	δD (‰ vs. SMOW)	$\delta^{18}O$ (‰ vs. SMOW)
Ocean	97.2	0 ± 5	0 ± 1
Deep Atlantic			$+0.05$
Deep Pacific			-0.15
Deep Antarctic			-0.40
Ice Caps and Glaciers	2.15	-230 ± 120	-30 ± 15
Groundwater	0.62	-50 ± 60	-8 ± 7
Fresh surface water	0.017	-50 ± 60	-8 ± 7
Atmospheric water	0.001	-150 ± 80	-20 ± 10

(After Criss, 1999.)

Although gross variations in stable isotope ratios of natural aqueous fluids were known from relatively precise density measurements made over 60 years ago, it was not until the publication of two landmark papers in 1953 (Epstein and Mayeda, 1953; Friedman, 1953) that we began to understand the relations between the isotopic compositions of natural waters and the physical chemical processes that modify them: *evaporation, condensation, mixing, and exchange reactions*. Epstein and Mayeda measured only oxygen isotope ratios and Friedman measured only hydrogen isotope ratios, but trends in both isotopic ratios of natural waters are, with subtle but important exceptions, almost the same. Relations governing variations in the isotopic compositions of meteoric waters were put on a quantitative basis by Dansgaard (1964) and will be examined in detail subsequently. An additional excellent summary of fractionation processes in the meteoric water cycle is given by Gat (1996).

The beauty of stable isotope measurements of natural waters is the fact that hydrogen and oxygen isotopes comprise the water molecules themselves and thereby constitute *built-in* tracers for water. They are most often conservative tracers, particularly for low-temperature surface waters, but the original isotopic compositions of natural waters can be changed by physical processes and by chemical reactions with rocks and other fluids.

4.2 NATURAL ABUNDANCES OF THE ISOTOPOLOGUES OF WATER

Because there are two stable isotopes of hydrogen—protium (^{1}H or H) and deuterium (^{2}H or D)—and three stable isotopes of oxygen (^{16}O, ^{17}O, and ^{18}O), there are nine possible isotopologues of water: $H_2^{16}O$, $H_2^{17}O$, $H_2^{18}O$, $HD^{16}O$, $HD^{17}O$, $HD^{18}O$, $D_2^{16}O$, $D_2^{17}O$, and $D_2^{18}O$. It is a simple matter to calculate the approximate abundance of each of these forms on Earth; using the average terrestrial abundance of each of the isotopes. The average terrestrial abundances of hydrogen and oxygen isotopes *in atom percent* are as follows (Coplen et al., 2002):

$$^{16}O = 99.7621 \quad H = 99.9844$$
$$^{17}O = 0.03790 \quad D = 0.01557$$
$$^{18}O = 0.20004$$

The abundance of a particular isotopologue is determined by multiplying together the average abundances, expressed as a ratio (e.g., 0.000374 for ^{17}O), *of each atom*

TABLE 4.2: Average natural abundances of the nine isotopologues of water vapor.

Isotopologues of Water	Average Abundance (%)
$H_2^{16}O$	99.73098
$H_2^{18}O$	0.199978
$H_2^{17}O$	0.037888
$HD^{16}O$	0.031460
$HD^{18}O$	0.0000006
$HD^{17}O$	0.0000001
$D_2^{16}O$	0.00000002
$D_2^{17}O$	0.00000000001
$D_2^{18}O$	0.00000000005

making up the compound, taking into account symmetry considerations when appropriate. For example,

$$\mathbf{H_2^{16}O} \qquad (0.999844)^2(0.997621) = 0.997310. \qquad (4.1)$$

That is, $H_2^{16}O$ accounts for 99.7% of all water on Earth. In contrast,

$$\mathbf{HD^{16}O} \qquad (2)(0.99928)(0.000156)(0.99759) = 0.0003146 \qquad (4.2)$$

constitutes only 0.03% of all water (note the *caveat* in the next paragraph), and it is the variation about this average value that is measured in hydrogen isotope geochemistry.

The squared term in equation (4.1) arises from the fact that there are two atoms of hydrogen in the water molecule and each atom in the molecule must be accounted for in the product of the abundances. The factor of 2 that appears in equation (4.2) is required because of the symmetry of the HDO molecule. In the strictest sense, this symmetry factor is valid only for gaseous or monomeric HDO.[1] Monomeric water vapor in nature exists almost exclusively in the atmosphere and arises from the evaporation of standing bodies of water and as emanations from volcanoes, fumaroles, hot springs, and the like. Thus, the symmetry consideration has no obvious practical implications. Nonetheless, it is instructive to introduce the concept of molecular symmetry at this point.

Oxygen is centrally symmetric in the gaseous water molecule; thus, there are two possible spatial configurations of $HD^{16}O$ (Fig. 4.1). In other words, it is equally probable for these two forms to occur, so their natural abundances must be multiplied by a factor of 2 to account for this statistical probability.[2]

The average natural abundances of the various isotopologues of water on Earth are given in Table 4.2. Only four isotopologues have sufficient abundance in nature to have any practical

FIGURE 4.1: Illustration of the two equally probable forms of $HD^{16}O$ due to symmetry.

---

[1]Liquid water is a complicated and constantly changing mixture of isotopically variable water molecules that are weakly joined by hydrogen bonds into clusters of different sizes. In reality, HDO cannot exist in liquid water and should be considered merely as a *component* of water.

[2]This can be envisioned in the following way: Suppose that 4 out of 20 atoms of hydrogen are deuterium. Then the abundance of deuterium is 4/20. But we have only 10 H_2O molecules for 20 H atoms, so the abundance of HDO is actually 4/10.

consequence for isotope geochemistry: $H_2{}^{16}O$, $H_2{}^{17}O$, $HD^{16}O$, and $H_2{}^{18}O$ (molecular weights = 18, 19, 19, and 20, respectively). Forms of any natural substance containing two rare stable isotopes, such as $HD^{18}O$ or $D_2{}^{16}O$, are to all intents and purposes absent in nature. (See, however, Eiler and Schauble, 2004.)

4.3 METEORIC WATER

Meteoric water is liquid or solid water that falls or has fallen from the sky and includes rain, fog, hail, sleet, and snow. Meteoric water resides on Earth principally in glaciers, groundwater systems, rivers, and lakes. On the basis of reasonable estimates of mean amounts and isotopic compositions of precipitation around the world, global meteoric water has $\delta^{18}O = -8‰$ and $\delta D = -50‰$. The development of precipitation can be discussed in general terms, with isotopic effects assigned to each part of the process. Most worldwide precipitation on Earth occurs over the oceans. About 90% of oceanic water vapor condenses after minimal horizontal movement of the confining air mass and than falls back into the ocean. The remaining 10% is carried by winds over the continents, where air masses pick up additional water vapor from sources of freshwater on land. Supersaturation occurs upon cooling of the air masses, and much of the vapor condenses and precipitates on land. Transport of atmospheric vapor over the continents helps regulate heat balance on Earth and provides plants and animals with life-sustaining freshwater. Because of the extreme importance of these processes and others, much attention has been given to the study of stable isotope variations of meteoric waters.

Using generally accepted estimates of vapor sources over land, one concludes that only about one-third of continental precipitation is derived *directly* from oceanic water vapor. The remaining two-thirds of continental water vapor enters the atmosphere through the evaporation of large lakes and rivers and, significantly, by evapotranspiration of plants (Fig. 4.2). Relatively minor amounts of water are introduced into the atmosphere by volcanic activity and from extraterrestrial sources, such as cometary material. These waters have distinct isotopic compositions, but under normal circumstances, they can never be identified, simply because they immediately become part of the vast amount of vapor in Earth's atmosphere and enter the hydrologic cycle. They are mentioned primarily for the sake of completeness. In the long term, however, they may contribute somewhat to the δD value of terrestrial waters.

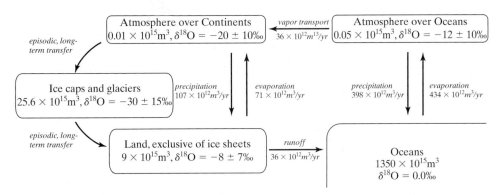

FIGURE 4.2: Fluxes for meteoric water cycle. The vast majority of freshwater is locked in the ice caps. Extensive transfer to and from the ice caps depends largely on climatic conditions. At present, warming is resulting in a diminution of the ice caps and an increase in flow to the oceans. If the ice caps were to melt completely, the $\delta^{18}O$ value of the oceans would decrease by ~0.6‰.

4.4 THE METEORIC WATER LINE

One of the most remarkable relations observed in the geochemistry of natural substances is the near-linear relation between δD and $\delta^{18}O$ values of the majority of waters of meteoric origin. Friedman (1953) first reported the covariance between δD and $\delta^{18}O$ values of natural waters, comparing his hydrogen isotope data with the oxygen isotope data reported by Epstein and Mayeda (1953). He proposed that the observed linear relation between δ values was controlled by the covariations of relative vapor pressures of $HD^{16}O$ versus $H_2^{16}O$ and $H_2^{18}O$ versus $H_2^{16}O$ that prevail during the various processes which characterize cloud dynamics. His relation and explanation were fundamentally correct, but the isotopic fractionations that occur upon initial evaporation and transport away from seawater and those attendant on subsequent processes of evaporation, precipitation, and exchange, as well as additions of water from sources on the continents, are to this day not understood in detail. (See Criss, 1999.) Because the correlation between $\delta^{18}O$ and δD values of natural waters was recognized early and, more importantly, because the equipment necessary to analyze both stable isotope ratios was rarely available in one institution, measurements were normally made of only one stable isotope ratio (D/H or $^{18}O/^{16}O$) of water in the early years.

In 1961 Craig (1961) published precise analyses made in a single laboratory of both isotopic ratios of many samples of meteoric water and defined the Meteoric Water Line (Fig. 4.3), often abbreviated simply as MWL. Related abbreviations commonly used are LMWL (Local Meteoric Water Line) and GMWL (Global Meteoric Water Line).

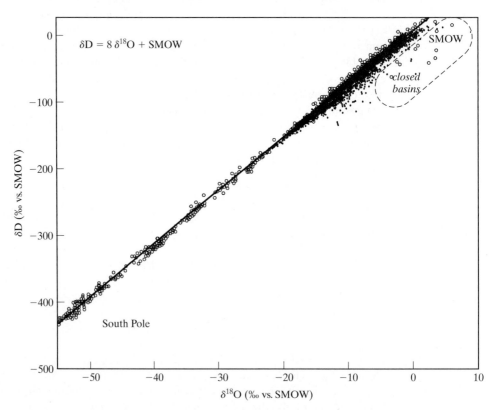

FIGURE 4.3: Global Meteoric Water Line. Shaded circles after Craig (1961); black spots after Kendall and Coplen (2001).

4.4.1 General Features of the GMWL

Despite the fact that ocean water is the ultimate source of all meteoric water, the datum for ocean water (δD and δ^{18}O = 0‰, by definition) does not lie on the GMWL. The explanation for this phenomenon lies in the fact that water vapor above the oceans is not in isotopic equilibrium with liquid water in the oceans. By its very nature, *evaporation is a nonequilibrium process*. In fact, δ^{18}O values of such vapor vary with latitude from about -15 to -10‰, values that are approximately 4‰ lower than the equilibrium values. The δD values of oceanic water vapor are lower than the equilibrium values as well, varying from about -110 to -70‰.

Regression of these early data for meteoric water, and many other data accumulated subsequently, results in the following expression, which represents the isotopic compositions of the majority of meteoric waters that fall in regions on Earth where climates are temperate:

$$\delta D = 8\delta^{18}O + 10. \tag{4.3}$$

Equation 4.3 represents the modern-day *global* meteoric water line very well.[3] If the stable isotope compositions of ocean water were different in the past, and global circulation patterns were roughly similar to those prevailing today, the GMWL at that time would be parallel to the modern GMWL. That is, the slope of the line would remain near a value of 8, but the intercept would be different.

Analyses of meteoric waters from closed basins or very dry areas yield results that lie to the right of the GMWL. This phenomenon arises from the interplay of kinetic and equilibrium isotopic fractionation factors that are appropriate to those meteorological conditions and different from those operating during evaporation and condensation processes in temperate regions. In this context, it is important to recognize that raindrops falling through very dry air can vaporize and possibly recondense at a later stage of the storm. (See Gat, 1996, for a thorough review of isotopic effects of meteoric water.)

Within the limits of analytical error, no data falling to the left of the GMWL were published for almost 20 years after the GMWL was defined. As a result, this region in δD–$\delta^{18}O$ space was considered a *forbidden region*. While we are now aware of a few important exceptions, such as the residual of evaporation water, this forbidden region continues to place constraints on possible processes undergone by meteoric waters.

4.4.2 Variations in Slopes and Intercepts of Local MWLs

Analyses of a given sample of water collected during a single storm may or may not lie on the MWL, but analyses of samples that are weighted averages of precipitation over a relatively long period (groundwater is a good example) at a specific site in temperate regions will almost certainly lie on or very near the MWL. Kendall and Coplen (2001) published stable isotope analyses of more than 4800 stream samples from 391 sites in the continental United States and demonstrated that such samples are normally very good proxies for modern local precipitation, corroborating earlier findings that LMWLs usually have slopes that are less than 8 (Fig. 4.4). In fact, LMWLs that characterize very large areas of the southern and western United States

---

[3]An *unweighted* GMWL generated from thousands of analyses of precipitation samples from the International Atomic Energy Agency network sites (Rozanski et al., 1992) is represented by an equation that can be considered a refinement of equation (4.3): $\delta D = 8.17\delta^{18}O + 10.35$. An analogous expression generated by Kendall and Coplen (2001) for samples from the 48 contiguous United States has a similar slope, but an intercept that is 1.4‰ lower: $\delta D = 8.11\delta^{18}O + 8.99$.

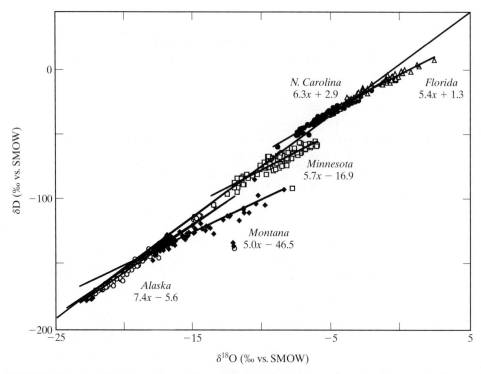

FIGURE 4.4: Local meteoric water lines for selected regions of the United States. In almost all cases, the slope of the LMWL is less than the global meteoric water line value of ~8. (After Kendall and Coplen, 2001.)

have slopes that are less than 6. The **GMWL represents a weighted average of LMWLs whose slopes are uniformly lower than 8 and whose intercepts vary widely, from values as negative as about −2 to values as positive as about +15.** It is not surprising that the many attempts to derive a general expression for the MWL from first principles have been met with frustration.

4.4.3 Meteoric Waters in Arid and Semiarid Environments

In regions where significant evaporation takes place, waters of meteoric origin can have quite unusual isotopic compositions. Data for waters in arid regions are commonly projected onto the figure for meteoric waters with a $\delta D/\delta^{18}O$ slope of ~5, in strong contrast to the slope of 8 for the GMWL. As has been mentioned, isotopic fractionations between liquid and vapor in these cases differ from the equilibrium fractionations due to kinetic isotope effects. In fact, $d\delta D/d\delta^{18}O$ slopes for evaporating waters are quite variable (typically 4–6) and depend on local conditions of humidity, isotopic composition of water vapor, and fraction of water evaporated.

For the simple case where the isotopic composition of evaporating water is in isotopic equilibrium with the water vapor, the following relation applies (Criss, 1999):

$$\frac{d\delta D}{d\delta^{18}O} \cong \frac{(1 - 1/\alpha^o_{\text{evap},D})(1000 + \delta D_w)}{(1 - 1/\alpha^o_{\text{evap,ox}})(1000 + \delta^{18}O_w)}. \tag{4.4}$$

In this equation, δD_w and $\delta^{18}O_w$ are the initial delta values of water prior to evaporation, and α^o_{evap} is the fractionation factor between water and vapor at zero humidity. At 20°C, values of α^o_{evap} are 1.0260 and 1.094 for the oxygen and hydrogen isotopologues, respectively. When the humidity changes, or the isotopic composition of the water changes in response to Rayleigh effects, the equations become substantially more complicated (Criss, 1999).

Under conditions of extreme evaporation in the Sahara Desert, extraordinary $\delta^{18}O$ and δD values of +31‰ and +129‰, respectively, for evaporated lake waters have been recorded (Fontes and Gonfiantini, 1967). Significant evaporation of natural waters also occurs in many regions of the world where the relative humidity is low, but nonetheless significant. In these cases, there is a limit to the amount of possible enrichment of heavy isotopes in the waters remaining after extensive evaporation. Once again, the explanation of the isotopic systematics of these systems lies in the concept of reservoir sizes. The $\delta^{18}O$ values of ocean water evaporating in the salt pans near San Francisco rise with increasing amount of evaporation, but level out at a value near 6‰ (Lloyd, 1966). In contrast to the situation in the Sahara Desert, where vapor leaves the system without the possibility of back exchange with the liquid, vapor in the San Francisco atmosphere ultimately becomes the dominant reservoir of water in that liquid–vapor system. That is, upon isotopic exchange between the remaining tiny amount of liquid in the evaporation pans and the larger reservoir of water in the atmosphere, the isotopic composition of the water is controlled by the vapor. At this point, the water cannot be further enriched in heavy isotopes, even though it is still evaporating. In fact, there are certain conditions of humidity and chemical composition of evaporating waters (brines) whereby the $\delta^{18}O$ values of the evaporating brines eventually start to become more negative.

Evaporation effects are commonly seen in lakes and rivers in semiarid environments. For example, meltwaters feeding Lake Tahoe have $\delta^{18}O$ values of about −16‰, whereas Lake Tahoe itself has a $\delta^{18}O$ value of about −5.5‰. Despite the relatively large volume of water in Lake Tahoe, it undergoes an amazing amount of evaporation during the year, and the $^{18}O/^{16}O$ and D/H ratios of the water increase dramatically in response to this process. In like fashion, $\delta^{18}O$ values of rivers downstream can be several per mil higher than those of the headwaters, due to evaporation effects as the water moves to lower and warmer elevations.

4.5 THE DEUTERIUM EXCESS PARAMETER

In spite of many complex processes operating in the hydrological cycle, the relative vapor pressures of the hydrogen and oxygen isotopologues of water normally vary sympathetically. Thus, there is an excellent linear relation between δD and $\delta^{18}O$ values of meteoric waters from all over the world, namely, the GMWL. The GMWL can be represented in a more general way than that given in equation (4.3) and for any time in Earth history. The slope of 8 is invariant, but the intercept is given by the variable d:

$$\delta D = 8\delta^{18}O + d. \tag{4.5}$$

The parameter d was given the name *deuterium excess parameter* by Dansgaard (1964) and is also referred to as the *deuterium excess value*, or simply, *deuterium excess*. A d value of 10‰ fits the data for *modern* worldwide samples rather well (Fig. 4.3).

The deuterium excess parameter is controlled predominantly by kinetic effects associated with the evaporation of water at the surface of the oceans or inland and increases with an increase in the *moisture deficit*, $1 - h$, of the oceanic air masses, where h is the relative

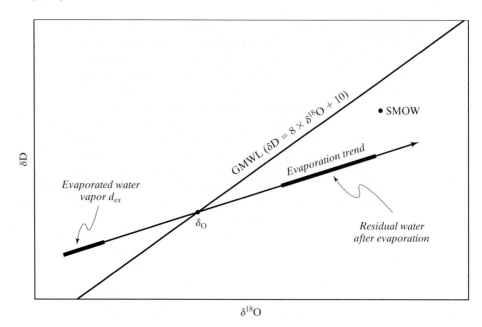

FIGURE 4.5: Deuterium excess parameter. Water starts on the global meteoric water line (GMWL) with a δ value of δ_O. During evaporation, the residual water moves to the right of the GMWL, because the $\delta D/\delta^{18}O$ slope during evaporation is less than 8. Vapor must lie to the left of the GMWL, due to simple mass-balance considerations. Upon condensation, vapor produces water at d_{ex}. Mixing of water d_{ex} with any other water on the GMWL will produce waters above the meteoric water line. (After Gat, 1996.)

humidity at the surface temperature of the water (Merlivat and Jouzel, 1979). The effect of evaporation is to drive the remaining water to higher $\delta^{18}O$ and and δD values, with a $\delta D/\delta^{18}O$ slope of less than 8 (Fig. 4.5). The removal of water with isotopic compositions that lie to the *right* of the meteoric water line results in a vapor that lies to the *left* of the meteoric water line, producing a deuterium excess.

The deuterium excess value of 10‰ that characterizes the modern GMWL corresponds to a mean relative humidity over the oceans of 81%. The mean relative humidity of air masses over the oceans is about 10% lower in winter (because the air masses are colder) than in summer, an effect that explains in part the seasonal shift from higher values of d in winter to lower values of d in summer precipitation over temperate continental regions. For waters undergoing intense evaporation in arid regions, kinetic effects are much stronger, and the value of d is correspondingly much higher. Precipitation falling in the eastern Mediterranean region can have a deuterium excess value as high as 18‰ (Gat and Carmie, 1970), and there is a difference of 10‰ in the deuterium excess parameter between meteoric waters that developed during glacial versus interglacial periods (Harmon and Schwarcz, 1981). This latter difference reflects extreme differences in humidity over the oceans under glacial and interglacial conditions. In the United States, the deuterium excess parameter varies strongly with geographic location, but is not explained by a single parameter. Instead, it is related to different source air masses, differences in temperature, aridity, and the contribution from lakes and evapotranspiration (Fig. 4.6).

Section 4.5: The Deuterium Excess Parameter

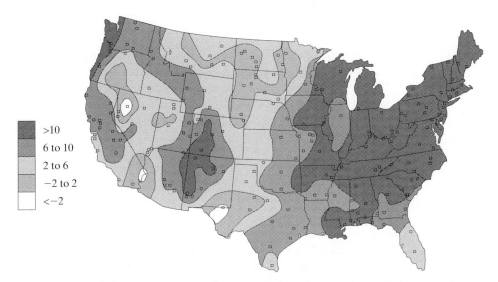

FIGURE 4.6: Spatial distribution of deuterium excess values in the United States. The most striking feature is the difference between samples east and west of the Mississippi River. There is no correlation with average mean annual temperature, amount of precipitation, or potential evapotranspiration (PET) loss, but a loose correlation appears with the difference of precipitation and PET loss (precipitation-PET). (From Kendall and Coplen, 2001.)

As mentioned above, LMWLs vary considerably around the earth in response to local and regional meteorological conditions. Dansgaard (1964) found that LMWLs in tropical and subtropical islands pass through SMOW (i.e., $d \sim 0$) according to the relation

$$\delta D = 4.6(\pm 0.4)\delta^{18}O + 0.1(\pm 1.6). \tag{4.6}$$

as shown in Fig. 4.7. This relation develops if the bulk of such precipitation arises from condensation of vapor that is initially generated from rapid evaporation of ocean water. Data for certain other island stations fall off the trend in equation (4.6) because the islands are located on the equatorial side of the subtropical high-pressure zone.

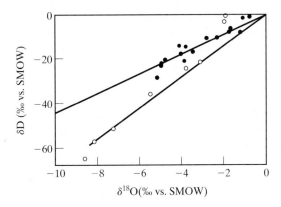

FIGURE 4.7: $\delta D/\delta^{18}O$ values for ocean island sites. Note that all tropical stations (filled circles) plot with a $\delta D/\delta^{18}O$ slope of 4.6. Subtropical stations (open circles) including high-latitude stations of Falkland, Adak, and Reykjavik, fall on the continental meteoric water line because they are strongly affected by Rayleigh processes. (After Dansgaard, 1964.)

4.6 EVAPORATION AND CONDENSATION

Natural waters undergo evaporation in a variety of settings, including clouds, large and small bodies of standing water, soil, and respiratory sites in animals and plants. **Evaporation is a kinetic process**. The stable isotope fractionations associated with evaporation depend on a number of factors and can be surprisingly large. **Condensation, by contrast, is an equilibrium process**. The fractionation factors attendant on condensation depend on temperature alone; consequently, the isotopic systematics of condensation processes are relatively easy to treat mathematically. Knowledge of the isotopic effects associated with evaporation and condensation in air masses and in other reservoirs of water, such as flowing streams, plants, and lakes, is fundamental to our understanding of the isotopic systematics of the hydrologic cycle.

4.6.1 Evaporation

We have already seen in Table 1.3 that vapor pressures of lighter molecules are higher than those of heavier molecules of the same chemical compound. Accordingly, the vapor pressures of the different isotopologues of water decrease in the order of decreasing molecular weight: $H_2^{16}O > HD^{16}O \cong H_2^{17}O > H_2^{18}O$. In simple terms, then, lighter isotopologues of water preferentially escape from the surface of the liquid into the vapor phase upon evaporation, and the opposite occurs upon condensation, with heavier isotopologues preferentially condensing out and leaving a vapor phase relatively enriched in isotopically lighter water.

The explanation is fairly straightforward in a general sense. Isotopically lighter forms of H_2O species that constitute liquid water (monomers and various hydrogen-bonded clusters) have higher translational velocities than heavier forms and are thus more likely to escape, or *rattle free*, from the surface of liquid water into the vapor phase. Evaporation is dominantly a unidirectional process when relative humidity is less than 100%, so that back isotopic exchange between the departing vapor and the remaining liquid is restricted. If the air above the ocean were to become saturated, and there were no winds, net evaporation would cease. At that point, isotopic exchange between vapor and liquid would become a natural consequence of the dynamic processes of evaporation and condensation that take place at the interface of any liquid and vapor in equilibrium with each other. Such exchange would then rapidly force an equilibrium distribution of isotopes between the two phases. This process does *not* occur over the oceans.

The failure to attain *complete*, or *equilibrium*, isotopic exchange between vapor and liquid causes vapor above the oceans to be isotopically lighter than that predicted from known equilibrium isotopic fractionations at temperatures of the ocean surface. Measured oxygen isotope fractionations between ocean water and nonequilibrium vapor vary with temperature and relative humidity, but a typical value is $\alpha_{\text{liquid-vapor}} = 1.013$. That is, vapor over the oceans has a $\delta^{18}O$ value of about $-13‰$ instead of an equilibrium $\delta^{18}O$ value of about $-9‰$ (Majoube, 1971).

At this juncture, we have no exact formulation for the difference of approximately 4 per mil between the observed kinetic fractionation between liquid water and water vapor ($\alpha_{\text{l-v}}$) and the equilibrium $\alpha_{\text{l-v}}$. During evaporation into unsaturated air, hydrogen and oxygen isotope fractionations depend not only on the vapor pressures of the different isotopologues of water, but also on the relative humidity of the air, turbulence in the liquid, and the relative rates of diffusion of hydrogen and oxygen isotopes from deeper layers in the liquid (Craig and Gordon, 1965). Evaporation into a *vacuum* depends strongly on the relative diffusion rates of the different isotopologues of water, but that condition is vastly different from the case of evaporation of ocean water into air.

One can calculate *effective* fractionation factors that take into account estimates of relative diffusion rates for the different isotopologues, but these fractionation factors are much too large and do not model well what is observed in nature. Actual diffusion rates are significantly lower than those predicted on the basis of simple kinetic theory, because the molecules in liquid water move as clusters, not as monomers. For example, the relative mass difference between $H_2^{18}O$ and $H_2^{16}O$ is $20/18 = 1.11$. But if water is considered as a polymerized chain of H_2O molecules, then the relative mass difference between two chains, each consisting of (for example) five molecules, with one of the two chains containing an ^{18}O isotope, is only $92/90 = 1.02$. Calculations using a five-molecule chain as our parameter for kinetic diffusion give a reasonable value for the fractionation between water and water vapor.

Measured $\delta^{18}O$ values of most vapor over the oceans are between -13 and $-11‰$, whereas equilibrium $\delta^{18}O$ values would be 3–4‰ more positive. The disequilibrium effect is less for hydrogen isotope fractionation because the difference in mass of the two hydrogen isotopologues is smaller than the corresponding difference in the mass of the pertinent oxygen isotopologues (i.e., $HDO/H_2O = 19/18$ vs. $H_2^{18}O/H_2^{16}O = 20/18$). The predicted lower sensitivity of the hydrogen isotopologues to diffusion effects is indeed observed in nature.

The departure from equilibrium in nature is maximum at latitudes of 18–26° (Fig. 4.8), where evaporation is highest and relative humidity is lowest (Craig and Gordon, 1965). Craig and Gordon (1965) proposed the following relation between the degree of disequilibrium evaporation and humidity:

$$\delta_l - \delta_v \approx \left[\left(1 - \frac{1}{\alpha'}\right)(2 - h) + \left(\frac{1}{\alpha'} - \frac{1}{\alpha}\right)\right]1000. \tag{4.7}$$

Here, δ_l and δ_v are the delta values of liquid water at the surface of the ocean and atmospheric vapor, respectively; h is the relative humidity; α is the equilibrium fractionation between liquid and vapor at a given temperature;[4] and α' is the effective fractionation factor, corrected for diffusion in the surface water.

4.6.2 Condensation: Closed-System (Batch) Isotopic Fractionation

Air masses cool as they rise and move away from their source. If the temperature of an air mass drops to the point of supersaturation, condensation and, eventually, precipitation occur. The $\delta^{18}O$ and δD values of the condensate will be substantially more positive than those of the parent vapor. As has been mentioned, condensation is dominantly an equilibrium process; thus, equilibrium fractionation factors are used to model the process. Under equilibrium conditions, hydrogen and oxygen isotope fractionation factors between liquid (l) and vapor (v) are nearly equal to the ratios of the vapor pressures of the pertinent isotopologues of water at a given value of T. For example, for the oxygen isotopologues,

$$\alpha_{l-v} = \frac{(^{18}O/^{16}O)_l}{(^{18}O/^{16}O)_v} = \frac{p'}{p} = pH_2^{18}O/pH_2^{16}O \tag{4.8}$$

---

[4]Always note the *direction* that a fractionation factor is stated. In this case, it is a liquid–vapor fractionation, not a vapor–liquid fractionation (i.e., α_{l-v}, not α_{v-l}).

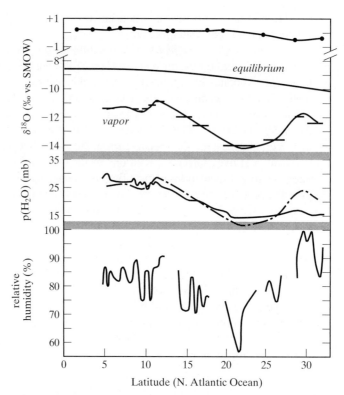

FIGURE 4.8: Oxygen isotope variations in water vapor collected on a north–south transect in the North Atlantic Ocean. The top curve depicts $\delta^{18}O$ values of surface water. The curve labeled *equilibrium* is the calculated $\delta^{18}O$ value of vapor in equilibrium with the surface water at the appropriate temperature. Calculated equilibrium $\delta^{18}O_v$ values decrease with latitude due to decreasing temperature; however, the relative humidity has a much greater effect on the $\delta^{18}O$ values of water vapor. The curve labeled $\delta^{18}O$ (*vapor*) is the measured $\delta^{18}O$ value of water vapor. Measured $\delta^{18}O$ values (repeated in this panel as a dotted line to show the form of the curve only) correlate well with measured $p(H_2O)$ and humidity, illustrating the nonequilibrium evaporation effect. Modified from Craig and Gordon (1965).

where p' is the vapor pressure of the heavy isotopologue and p is the vapor pressure of the light isotopologue. Determinations of α_{l-v} have been made both by vapor pressure measurements and by exchange experiments between liquid water and water vapor. At 20°C, measured values of $\alpha(D)_{l-v}$ and $\alpha(^{18}O)_{l-v}$ are 1.08351 and 1.00976, respectively (Kakiuchi and Matsuo, 1979; Horita and Wesolowski, 1994). The vapor pressures of $H_2^{16}O$ and $H_2^{18}O$ are 2.3379×10^{-2} and $2.0265 \times^{-2}$ bar (Table 1.3). Dividing the vapor pressure of $H_2^{18}O$ by $H_2^{16}O$ gives a value of 1.00941, in good agreement with the measured value.

Available fractionation factors indicate that water precipitating from a vapor at 25°C will have δD and $\delta^{18}O$ values that are, respectively, 73.6 and 9.26 per mil more positive than the δD and $\delta^{18}O$ values of the vapor.[5] Note that hydrogen isotope fractionations are almost

---

[5]To make such calculations, assign a value of 0 to δ_l and compute δ_v from the definition of α. In this case $\delta D_v = -79.4/1.0794 = -73.6‰$ and $\delta^{18}O_v = -9.35/1.00935 = -9.26‰$.

10 times larger than oxygen isotope fractionations, a phenomenon that is predicted on the basis of theoretical considerations and one that is observed frequently in nature.

If condensation were allowed to take place in a closed system at some fixed temperature, then the total liquid condensate at any point in the process would be in isotopic equilibrium with the vapor remaining in the vessel. For example, if $\delta^{18}O_{v,i}$ (the isotopic composition of the initial (*i*) vapor) were $-14.0‰$, then the first drop of liquid condensing would have $\delta^{18}O_l = -14.0 + 9.3 = -4.7‰$. Under equilibrium conditions, $\delta^{18}O$ values of the liquid and vapor at any time in the condensation process are easily calculated from a material-balance equation, assuming a constant value of α_{l-v} at 25°C.

As an example, we will calculate values of $\delta^{18}O_l$ and $\delta^{18}O_v$ when the fraction of vapor *remaining* (*F*) is 0.4. The material-balance equation for this particular closed system is

$$(F)\delta^{18}O_v + (1 - F)\delta^{18}O_l = \delta^{18}O_{\text{total water}} \tag{4.9}$$

The $\delta^{18}O$ value of total water in this closed system is always $-14.0‰$, the value of the initial vapor. Accordingly, the material-balance equation for $F = 0.4$ is

$$(0.4)\delta^{18}O_v + (0.6)\delta^{18}O_l = -14.0. \tag{4.10}$$

The expression for α_{l-v} at 25°C is

$$\alpha_{l-v} = \frac{1000 + \delta^{18}O_l}{1000 + \delta^{18}O_v} = 1.00935, \tag{4.11}$$

and rearranging terms yields

$$\delta^{18}O_l = (1.00935)(1000 + \delta^{18}O_v) - 1000. \tag{4.12}$$

Combining equations 4.9 and 4.11, we solve for the delta values:

$$\delta^{18}O_v = -19.61‰ \qquad \delta^{18}O_l = -10.09‰.$$

As expected, $\delta^{18}O_v$ becomes more negative than $\delta^{18}O_{v,i}$ as the isotopically heavier liquid condenses and preferentially removes the heavy isotopologues of water. Values of both δ_v and δ_l (either δD or $\delta^{18}O$) become more negative with decreasing *F* (Fig. 4.9). When *F* proceeds all the way to a value of zero (all vapor is converted to liquid), the $\delta^{18}O$ value of the liquid equals that of the initial vapor, namely, $-14‰$. The constant difference between δ_v and δ_l that is dictated by the equilibrium fractionation factor is maintained throughout the condensation process. Generalized equations relating δ_v and δ_l values to *F* at any point in the process *at constant T* are

$$\delta_l = \frac{\alpha\delta_{\text{tot}} + 1000F(\alpha - 1)}{\alpha(1 - F) + F} \tag{4.13}$$

and

$$\delta_v = \delta_l - 1000(\alpha - 1). \tag{4.14}$$

These equations are linear on a plot of δ versus *F* (δ_{v*} and δ_{c*} in Fig. 4.9).

In fact, the process of condensation is never isothermal. Condensation *requires* cooling, which means that α must increase as the process proceeds. The isotopic trajectory during condensation with falling temperature is shown by the dashed line in Fig. 4.9. The nonisothermal

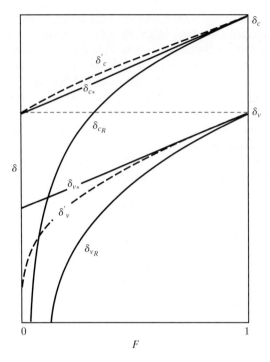

FIGURE 4.9: Correlation between isotopic composition and fraction of vapor remaining during equilibrium isothermal condensation of a liquid from vapor. At $F = 1$, there is only vapor, while $F = 0$ represents all liquid and no vapor. The curves $\delta_v{}^*$ and $\delta_c{}^*$ are the straight line trajectories of vapor and condensate, respectively, for equilibrium isothermal processes. The difference between the two is the equilibrium fractionation. The curves δ'_v and δ'_c are, respectively, vapor and condensate trajectories for condensation during cooling. Curves δ_{v_R} and δ_{c_R} are vapor and condensate trajectories for condensation under Rayleigh conditions. (After Dansgaard, 1964.)

model is a better approximation to nature, but is still not a valid model for condensation and precipitation. **The batch model for fractionation between condensate and vapor is not an appropriate model for stable isotope variations in precipitation on Earth.** It is presented here as an end-member case for the sake of completeness and because it illustrates the mass-balance principle. In addition, it provides examples of simple mathematical manipulations that are used frequently in stable isotope geochemistry.

4.6.3 Condensation: Open-System (Rayleigh) Isotopic Fractionation

Open-system Rayleigh (pronounced RAY-lee and not to be confused with Raleigh) fractionation is the other end-member fractionation model that complements the closed-system batch fractionation model just described. In the batch fractionation model, the condensed liquid and remaining vapor are always in contact with each other and in exchange equilibrium. Under Rayleigh conditions, condensate is continuously removed from the system, thus prohibiting back exchange between the two phases after their separation. Each increment of liquid condenses in isotopic equilibrium with the parent vapor, but is then forever removed from the system (i.e., the cloud) as precipitation. Because "heavy" water is continuously removed from the system, the δ value of the remaining vapor becomes progressively more negative with increasing degree of condensation, and the δ value of newly formed condensate becomes correspondingly lower and lower by an amount dictated by the fractionation factor at the temperature of condensation.

In effect, **Rayleigh fractionation in an air mass results in very large depletions of the heavy isotopes in precipitation**, particularly after the air mass has lost most of its vapor. Mathematically, as the fraction of vapor remaining approaches zero, the δ value of the vapor approaches the limiting value of $-1000‰$. Very low δD values of water vapor at very high altitudes are an example of an effective Rayleigh fractionation process. From observations made over many years, it appears that gross isotopic variations in precipitation on Earth are explained for the most part by Rayleigh fractionation in air masses. Nonetheless, we are far from understanding these processes in detail. (For further information on the Rayleigh equation and models developed to explain MWLs, see Jouzel and Merlivat (1984), Gat (1996), and Criss (1999)).

The Rayleigh equation is used to model many naturally occurring equilibrium processes other than precipitation in which a newly formed phase is removed or isolated from its "parent." Examples include crystals precipitating from a magma, decarbonation in a metamorphic system, and volcanic degassing, to name a few. In modeling precipitation, a single fractionation factor is normally used despite the obvious fact that temperatures of condensation in real systems must be decreasing. (See Dansgaard, 1964, for a more detailed discussion.) Also, true Rayleigh conditions are almost never met in real clouds, because vapor and condensate (both liquid and solid) coexist in clouds, and have an opportunity to undergo exchange reactions. For the stable isotope relations attendant on equilibrium precipitation of liquid water from water vapor in clouds, the relevant Rayleigh expression *with constant α* is

$$\left(\frac{R}{R_i}\right) = F^{\alpha-1}, \tag{4.15}$$

where R is the isotopic ratio (D/H or $^{18}O/^{16}O$), α is the fractionation factor *between liquid and vapor*, F is the fraction of vapor remaining, and i stands for the *initial* ratio.[6] In δ-notation, equation (4.15) reduces to

$$\delta_v = [\delta_{v,i} + 1000]F^{(\alpha-1)} - 1000, \tag{4.16}$$

where δ_v is the isotopic composition of the vapor for a given value of F and $\delta_{v,i}$ is the isotopic composition of the initial vapor.

The corresponding equation for liquid[7] water condensed at this value of F is

$$\delta^{18}O_l = \alpha(\delta^{18}O_v + 1000) - 1000. \tag{4.17}$$

The trajectories of vapor and liquid undergoing Rayleigh fractionation are shown in Fig. 4.9 as δ_{v_R} and δ_{c_R}.

Temperatures of condensation in real air masses are not constant, and the α value is continuously changing. The Rayleigh equation cannot be integrated with varying α values, unless the relationship between α (a function of temperature) and F is known. To address this problem, Dansgaard (1964) proposed the following expressions to model Rayleigh fractionation under conditions of varying temperature:

$$\delta_l = \frac{\alpha}{\alpha_i}(\delta_i + 1000)F^{(\alpha_m-1)} - 1000 \tag{4.18}$$

and

$$\delta_v = \frac{1}{\alpha_i}(\delta_i + 1000)F^{(\alpha_m-1)} - 1000. \tag{4.19}$$

---

[6] See Criss (1999, p. 106) for a derivation of the Rayleigh equation.
[7] Strictly speaking, the Rayleigh condensation model crudely explains the slope of 8 for the MWL, but only if the condensate is liquid. Slopes greater than 8 are obtained when ice is the condensate.

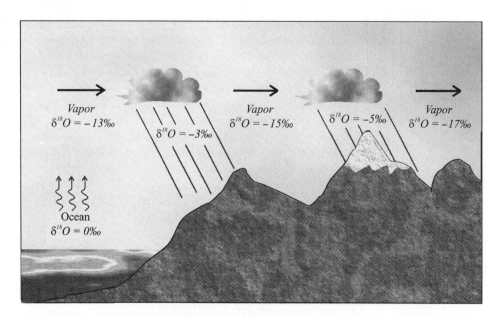

FIGURE 4.10: **Schematic of distillation effect for meteoric water.** A large kinetic fractionation occurs between ocean and vapor. Liquid condensing in clouds is in equilibrium with vapor and is heavier than the coexisting vapor. The remaining vapor becomes progressively lighter, leading to lower $\delta^{18}O$ and δD values farther from the ocean source.

In these equations, α, α_i, and α_m refer to the momentary condensation temperature T, at the initial temperature t_i, and at $(T + T_i)/2$, respectively.

Real air masses undergo a series of precipitation cycles during which both the vapor remaining and the precipitation become progressively lighter. A typical pattern of development of meteoric water precipitating from an air mass generated over ocean water is illustrated in Fig. 4.10. The water vapor over the ocean has a $\delta^{18}O$ value of about $-13‰$. Some vapor condenses and rains out a liquid phase with a $\delta^{18}O$ value of $-5‰$, 8 per mil heavier than the vapor. The removal of this relatively heavy water leaves behind vapor with even lower $\delta^{18}O$ values, so that the next cycle of rainout has a $\delta^{18}O$ of $-7‰$ (for example) instead of $-5‰$. As the air mass moves farther inland, and more and more precipitation cycles occur, the $\delta^{18}O$ values of vapor and condensate become progressively more negative, effectively following the Rayleigh curves of Fig. 4.9.

4.7 FACTORS CONTROLLING THE ISOTOPIC COMPOSITION OF PRECIPITATION

4.7.1 Temperature

Surface temperature, especially at high latitudes where precipitation forms near to the ground surface, is strongly correlated with the isotopic composition of meteoric water. This phenomenon is easily explained by the temperature dependence on α_{l-v}. Values of $1000 \ln \alpha_{l-v}$ for H and O vary strongly with temperature, but the ratio of the two has a nearly constant value of 8.1 ± 0.2 (Table 4.3, Fig. 4.11). **The dominant control on the isotopic composition of precipitation from a given air mass is the fraction of vapor remaining in that air mass.**

TABLE 4.3: Fractionation factors for H_2O (liquid–vapor) as a function of temperature. Although the fractionations are strongly dependent on temperature, the ratios of fractionation for H and O are relatively constant.

T (°C)	$1000(\alpha - 1)$ (D/H)	$1000(\alpha - 1)$ ($^{18}O/^{16}O$)	Ratio (of epsilons)
0	100.58	11.73	8.27
10	90.00	10.85	8.29
20	80.21	9.81	8.26
30	71.12	8.89	8.00

(Fractionation data from Kakivchi and Matsuo, 1979.)

A given air mass entering a region located at high latitudes, high altitudes, or significantly inland is cold and cannot hold as much water as it held or could have held in warmer regions that are closer to the ocean or at lower elevations and latitudes. Also, the fractionation between water vapor and condensate increase with decreasing temperature, so that the removal of water or ice has a more dramatic effect on lowering the isotopic composition of the remaining vapor. That is, precipitation in cold regions will more strongly lower the $\delta^{18}O$ and δD values of the remaining vapor because the α value is larger. Thus, temperature is indeed a control on the isotopic compositions of local meteoric water, but mainly in the sense that it controls the amount of water that a given air mass can hold. In fact, the apparent controlling factors or effects on the isotopic composition of meteoric water that will be discussed below are merely different expressions of this same thing.

Dansgaard (1964) first recognized the good correlation between the weighted mean isotopic composition of precipitation *in temperate climates* and mean annual surface temperature at the collection site. This correlation, shown in Fig. 4.12, provides the basis for many paleoclimate studies in which analyses are made of proxies for local precipitation in a given area in the past. Since that early publication, the correlation has been refined through thousands of analyses of meteoric water from all over the world. The *global* correlations between $\delta^{18}O$, δD, and mean annual surface temperature are given by

$$\delta^{18}O = 0.69 T_{average} - 13.6 \tag{4.20}$$

and

$$\delta D = 5.6 T_{average} - 100. \tag{4.21}$$

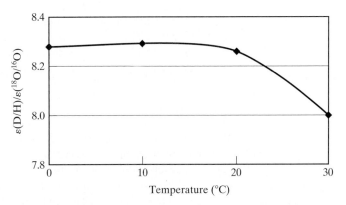

FIGURE 4.11: Plot of $\dfrac{\varepsilon(D/H)}{\varepsilon(^{18}O/^{16}O)}$ vs. T, where $\varepsilon = 1000\,(\alpha - 1)$. The ratio is nearly constant.

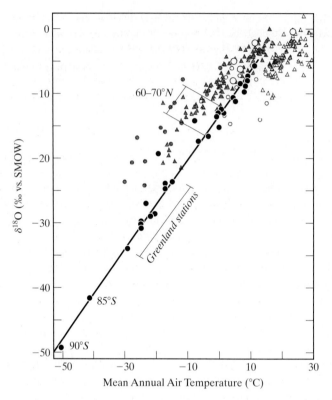

FIGURE 4.12: Effect of mean annual air temperature on the oxygen isotope composition of meteoric precipitation. The effect is largest and most linear at high latitudes, where condensation occurs close to the land surface. Circles are annual data from Dansgaard (1964); open diamonds are summer data, closed diamonds winter data, from Fricke and O'Neil (1999).

The temperature coefficient of $\left(\dfrac{d\delta^{18}O}{dT}\right) = 0.69$ is in good agreement with predictions made from experimental and theoretical considerations. The relationship is strongest at high latitudes, with far more scatter at mid- and equatorial regions (Fig. 4.12). Clearly, the temperature coefficient varies from place to place in response to a number of meteorological factors. For meaningful paleoclimate reconstructions in a given region, attempts must be made to evaluate this temperature coefficient in that region at the time of interest (Fricke and O'Neil, 1999). The combination of Rayleigh fractionation and temperature effects results in a strong latitudinal gradient in $\delta^{18}O$ and δD values of meteoric water. As an extreme example of the effect of latitude and temperature (not independent parameters) on the isotopic composition of precipitation, δD and $\delta^{18}O$ values of snow samples from the South Pole are as low as −495‰ and −62.8‰, respectively (Aldaz and Deutsch, 1967; Jouzel et al., 1987).

4.7.2 Distance or Continentality Effect

Precipitation becomes isotopically lighter as the parent air masses move farther from their sources and over the continents, simply because they have undergone more cycles of precipitation and F approaches ever-lower values. This continentality effect is associated with a number of other physical controls on F—most notably, the temperature decrease (ΔT) between the source of atmospheric vapor and the point of precipitation. The ΔT parameter is logically tied to the seasons, such that the continentality effect is greater during the colder months. In Eurasia, $\delta^{18}O$ values of meteoric water drop very regularly with distance eastward, the effect being more pronounced in wintertime (about $-3‰/1000$ km) than in summertime (about $-1.5‰/1000$ km). In summertime, the continentality effect is lower because temperatures (and hence α values) are less and because recycling of precipitation by evapotranspiration is more intense in the summer.

4.7.3 Latitude Effect

$\delta^{18}O$ and δD values of precipitation decrease with increasing latitude again because the degree of rainout of air masses increases and temperatures decrease with latitude. While decreases in delta values with latitude are regular, they are by no means linear, as many topographic and local meteorological conditions are operative as well in the process of rainout. Global-scale circulation patterns also differ over oceans and continents, as well as over eastern vs. western seaboards. The gross magnitude of the latitude effect in midlatitude regions is about $-0.5‰$ per degree of latitude. (Fig. 4.13).

4.7.4 Altitude Effect

The isotopic composition of water becomes lighter with increasing altitude, again because it is colder at higher elevations and air masses hold less water when they are cooled. The α values increase, so that the effect of Rayleigh fractionation is intensified. As an air mass is deflected upward by a mountain, it decompresses and cools adiabatically, whereupon more rainout occurs. The percentage of vapor remaining in the air mass will decrease rapidly if the relief is high. In fact, topographic relief has an intensely strong effect on the isotopic composition of the precipitation.

Both continentality and altitude effects are illustrated nicely by isotopic compositions of precipitation falling in and around the Sierra Nevada Mountains in western South America.

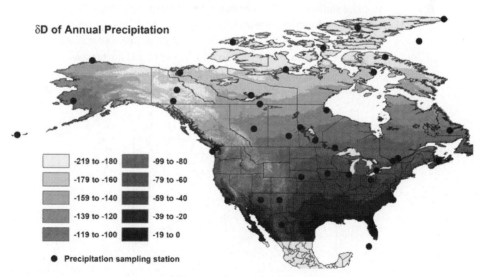

FIGURE 4.13: **Contours of δD values for meteoric water.** From Meehan, Giermakowski, and Cryan (2004), with permission from Taylor and Francis.

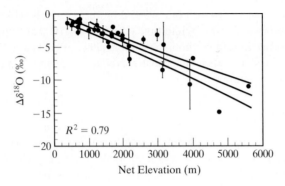

FIGURE 4.14: Changes in the $\delta^{18}O$ values of meteoric water relative to the local sea level $\delta^{18}O$ values (hence the $\Delta\delta^{18}O$ terminology). Data are compiled from published values the world over. (After Chamberlain et al. 1998.)

In this region, where prevailing winds are easterly, $\delta^{18}O$ values of precipitation drop slowly (about −1‰/1000 km) as air masses travel westward across the Amazon basin (the continentality effect), but drop rapidly to about −0.2‰/100 m as the remaining atmospheric moisture is carried up the Andes (the altitude effect). Poage and Chamberlain (2001) compiled data from many investigations of the altitude effect and concluded that a gradient of −0.26‰/100 m characterizes the effect at most places in the world for elevations up to about 5000 m (Fig. 4.14). Paleoaltitudes have been estimated from fluid inclusions, where low δD values of included water indicate extreme changes in altitude (e.g., for the Alps in the Neogene) in the past (Sharp et al., 2005).

4.7.5 Amount Effect

The amount effect is a negative correlation between mean δ values and the amount of monthly precipitation **in tropical regions**. Note that in higher latitude regions, this negative correlation disappears or can even show a weak positive correlation (Kendall and Coplen, 2001). For the tropical stations, δ values of rain falling at a given station are high in months with little rain and low during the rainy season (Fig. 4.15). In accordance with principles already established, the more rainout that occurs from a given air mass, the lower is the delta value of subsequent precipitation. But the behavior of water in convecting air masses is complicated, and explanations for the amount effect are given only in qualitative terms. With regard to convecting air mass systems, four processes should be considered in attempts to understand the amount effect:

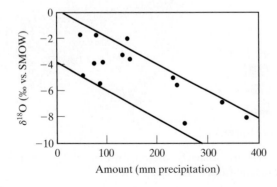

FIGURE 4.15: Amount effect at Binza, Congo, mm precipitation vs. $\delta^{18}O$ value. (After Dansgaard, 1964.)

(1) As air rises and cools to saturation, condensate falls through other droplets that formed below, and the condensate can exchange with them; (2) the droplets can grow larger by taking on more vapor as they fall; (3) upon exiting the cloud, the droplets can evaporate into dry air; and (4) droplets can exchange with vapor that is present in unsaturated air below the cloud.

When droplets evaporate into very dry air, isotopically light molecules vaporize preferentially and with a kinetic isotope effect. The liquid that reaches the surface in gentle rains will thus be relatively heavy. The lower the humidity, the greater is this effect. At times when the air is more humid, the probability of encounters between falling droplets and vapor molecules in the air increases, and exchange can occur between them. As a result of such exchange reactions, the liquid becomes richer in the heavier isotopes simply because of the positive direction of the isotopic fractionation between liquid and vapor. Thus, the two processes of evaporation and exchange explain the enrichment of heavy isotopes in gentle tropical rain. These same processes must *not* be operating during periods of intense tropical rainfall. At those times, rapidly ascending air masses result in deep cooling of the air mass, massive rainout, low values of F, and isotopically light precipitation. The $\delta^{18}O$ values of vapor over the tropical oceans are about -13 to -11‰, and mean $\delta^{18}O$ values of precipitation in the rainy months in tropical regions approach such values.

Isotopic effects associated with the processes undergone by water droplets falling through saturated or dry air deserve further consideration. When the amount of precipitation is low, the degree of cooling below the cloud mass is minimal. Under these conditions, the below-cloud air temperature is relatively high and significant evaporation can take place. Evaporation is most pronounced when rainwater falls through dry, hot air. The frequently beautiful natural phenomenon called *virga* (rain that never reaches the ground) develops under these conditions.

A water droplet falling through saturated air will undergo no evaporation, but will exchange with vapor in the surrounding air. At saturation, the rate of evaporation of a falling water drop equals the rate of condensation on the surface of the drop. Many studies have been made of the equilibration time between a raindrop and the vapor in its environment. Using HDO as a tracer, Friedman et al. (1962) determined that the rates of exchange between droplets and vapor are rapid for all but the largest raindrops (Table 4.4). The interplay between evaporation and exchange is obviously complex. It is natural to assume that rates of evaporation increase with a decrease in humidity, but there is a feedback mechanism that limits the degrees of evaporation and exchange processes. Large droplets fall rapidly and thus have a greater tendency to evaporate than do small droplets, but evaporation causes a diminution in the size of the droplet that, in turn, reduces its rate of descent and therefore the degree of evaporation. At the same time, because the water droplet is falling more slowly, it has more time to exchange with the water vapor in the air.

TABLE 4.4: Adjustment time τ for a water droplet of radius r.

r (cm)	τ (sec)	f(m)
0.01	7.1	5.1
0.05	9.2	370
0.075	164	890
0.10	242	1600
0.15	360	2900

f is the distance the droplet falls in time τ.
(After Friedman et al., 1962.)

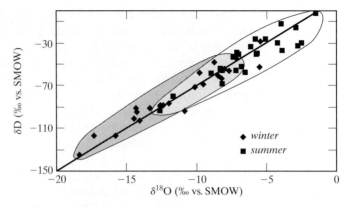

FIGURE 4.16: $\delta^{18}O$–δD values of individual storms in Albuquerque for 1998–1999. Although there is significant variation between individual storms, the averages for the summer and the winter months are clearly different.

4.7.6 Seasonal Effects

Many environmental parameters change with the seasons. Seasonal changes in temperature clearly affect the isotopic composition of precipitation, except in the tropics, where temperature is effectively constant all year long. In fact, all the factors discussed in this section are influenced by seasonal change. Both the continentality effect and the deuterium excess factor are higher in winter than in summer for reasons already explained. The amount of rain in one season can be far more than in others, so the amount effect can have a major effect on the isotopic composition of precipitation.

Differences in stable isotope compositions between summer and winter precipitation are seen all over the world. Differences observed in precipitation over Albuquerque are shown in Fig. 4.16. In addition to seasonal changes in temperature, the sources and travel paths (continentality effect) of water vapor in this region change as well. Air masses carrying winter precipitation originate over the Pacific Ocean and approach New Mexico from the west, while those carrying monsoonal rains in the summer originate in either the Gulf of Mexico or the Gulf of California and approach the area from the south and east. In St. Louis, there is a regular seasonal variation in the $\delta^{18}O$ values of precipitation. Criss (1999) proposed that this seasonality is related to the annual migration of the Gulf Stream, with different sources of precipitation being tapped in summer and winter. A number of additional minor complications can arise with seasonal change. (See Dansgaard, 1964, and Criss, 1999, for further details on seasonal effects.)

4.8 GROUNDWATER

Combined hydrogen and oxygen isotopic data provide us with a unique geochemical tool for evaluating the sources and flow paths of groundwaters. In most cases, the isotopic composition of groundwater represents the average isotopic composition of precipitation that fell relatively recently in the local recharge areas. In some areas, waters in deep aquifers can have an isotopic composition that is completely different from that of modern precipitation.

Sometimes this difference is a result of recharge regions being very distant from the collection site, but commonly such waters are quite old and were trapped in the aquifer when local climatic conditions were very different, as in the Pleistocene.

In areas of high topographic relief or those close to rivers fed from higher altitudes, groundwaters are often a mixture of local precipitation and distal (high-altitude) waters with much lower $\delta^{18}O$ values. These differences can be exploited nicely in environmental and hydrological studies. For example, groundwater pumping for domestic usage in the region of Sacramento, California, has lowered the head in the region and induced a flow of water into that system from the nearby Sacramento and American Rivers, whose sources lie high in the Sierra Nevada Mountains. Unperturbed groundwaters there have $\delta^{18}O$ values of about $-7‰$, whereas the $\delta^{18}O$ values of the two rivers are about $-11‰$. With such a sharp contrast in isotopic compositions, it is easy to monitor the extent of infiltration of the river waters into this groundwater system. A $\delta^{18}O$ contour map of the region shows a clear gradient in $\delta^{18}O$ values, from $-7‰$ for waters sampled far from the rivers to $-10.5‰$ for waters sampled closer to the rivers (Criss and Davisson, 1996).

Another example of tracing water sources is provided by waters discharging from the Grimsell gas pipeline tunnel in Switzerland. These waters are generally cold, but in the central part of the tunnel, waters issuing from a suite of late brittle faults have temperatures in excess of 24°C and have significantly higher flow rates than those of waters in other parts of the tunnel (where temperatures are only a few degrees above zero). The warm waters were heated to temperatures in excess of 100°C, a temperature calculated on the basis of their dissolved silica and cation contents. The thermal waters are clearly coming from a depth where they have been heated due to the high geothermal gradient. The hot waters could have arisen from metamorphic dewatering reactions or some other exotic source, but stable isotope analyses prove that they are purely meteoric in origin. The $\delta^{18}O$ and δD values of these waters plot on the MWL, but *lower* than those of adjacent cold waters. Although the hot waters ascend from ~5 km depth, the isotopic data indicate that they originate at altitudes higher than the source of the cold waters. The hot waters issuing in the Grimsell tunnel originate at several hundred meters above the recharge area of the cold waters and flow down the brittle fault systems to depths of about 5 km, where they are heated and rise due to their buoyancy.

Sources and flow paths of groundwaters can be traced on very large scales in major aquifer systems. A remarkable example is provided by groundwater that discharges from natural springs and artesian wells in central Missouri. Stable isotope analyses of all these waters lie on the MWL, so, clearly, the waters were not altered isotopically during their passage from recharge area to discharge area (Fig. 4.17). The freshwater springs have $\delta^{18}O$ values of about $-7‰$ and δD values of about $-50‰$, typical of modern precipitation in the region. Springs with higher salinities of up to 30 per mil have much lower $\delta^{18}O$ and δD values, down to about $-15‰$ and $-108‰$, respectively. These low-$\delta^{18}O$ waters most likely originated at high elevations in the Front Range of Colorado and traveled in the Western Plains aquifer system to the central lowlands of Missouri, a distance of more than 1000 km (Banner et al., 1989; Musgrove and Banner, 1993). The high salinity of these waters is attributed to their passage through the Permian salt deposits in Kansas without concomitant isotopic exchange with the enclosing rocks. Knowing that the Front Range was uplifted at about 65 Ma, flow rates in this system are calculated to be at least 0.015 m/y, a value that is compatible with model flow rates calculated on other bases. Stable isotope data can be very useful in understanding origins and flow patterns of large aquifer systems and in developing generalized models for such systems.

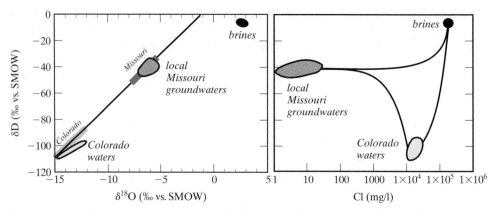

FIGURE 4.17: **Example of groundwater mixing from midcontinent of the United States.** Three distinct waters can be identified: local Missouri groundwaters, exotic waters from the Rocky Mountains with low δD values and moderate salinities, and brines with high salinities and high δD values. Other waters (not shown) plot in the mixing field between the three. After Musgrove and Banner (1993).

4.9 GEOTHERMAL SYSTEMS

Geothermal fields exist in volcanic regions all over the world. Some are exploited for geothermal energy and all hold the fascination of scientists and laymen alike. A well-defined body of magma has been identified under the system at Yellowstone National Park, and exotic chemical compositions characterize the waters of numerous systems in the world. It is understandable that many early workers believed that geothermal waters contained a significant component of primitive magmatic fluid whose components had never seen the surface of the Earth. Such a fluid is called *juvenile water*, and the quest to identify it occupied the research activities of many leading scientists in the first half of the 20th century. The origin of geothermal fluids was determined unambiguously by stable isotope measurements and remains one of the crowning achievements of the discipline. Friedman (1953) first recognized the dominance of meteoric water in these fluids by noting that δD values of certain hot springs were similar to those of local meteoric waters. A full understanding of the origin of the aqueous component of these fluids came through combined hydrogen and oxygen measurements by Craig (1963), who showed that δD values of geothermal waters are nearly identical to those of the local meteoric water, while the $\delta^{18}O$ values are usually shifted (the *^{18}O-shift*) to more positive values (Fig. 4.18). The important conclusion was drawn that geothermal waters all over the world are dominated by local meteoric waters. Juvenile waters, if present at all, cannot constitute more than about 1% of these fluids.

The modifications to the oxygen, but not hydrogen, isotope values of meteoric water can be understood in terms of hydrothermal interaction at depth. Volcanic terranes are riddled with cracks and fissures. Precipitation easily enters the ground through these cracks and descends to a significant depth, where it interacts with hot rocks. As a result of water–rock interactions, isotopes are exchanged and chemical constituents taken into solution. The evolved fluids ascend to the surface in convection systems that are numerous and varied in geothermal fields.

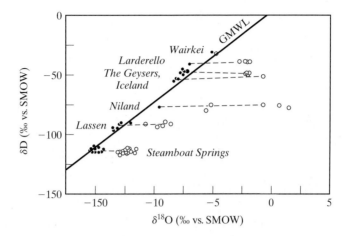

FIGURE 4.18: δD–$\delta^{18}O$ values for geothermal waters (open circles) and equivalent local meteoric waters (filled circles). The δD values of the geothermal waters are the same as those of the local meteoric waters, clearly identifying their origin. The $\delta^{18}O$ values of the geothermal waters, by contrast, are shifted higher, consistent with a meteoric source that has been modified by high-temperature interaction with the host rock. (After Craig, 1963.)

Oxygen in descending meteoric waters undergoes exchange reactions with crystalline igneous rocks at depth and with other rocks (often carbonates) that the fluid encounters on its journey back to the surface. Because rocks have very high $\delta^{18}O$ values relative to the negative $\delta^{18}O$ values of meteoric waters, high-temperature interaction will *raise* the $\delta^{18}O$ values of the waters and *lower* the $\delta^{18}O$ value of the rock. In contrast, hydrogen isotope compositions of neutral[8] geothermal waters are no different from those of local precipitation. This observation is explained by the simple fact that there is very little hydrogen in igneous rocks and none in carbonate rocks. Thus, there is simply no reservoir of hydrogen with which the water can exchange. (The concept of fluid–rock ratios and high-temperature isotopic exchange between rocks and fluids will be considered in more detail in Chapters 11 and 12.)

4.10 BASINAL BRINES AND FORMATION WATERS

Large sedimentary basins host old, highly saline waters called *basinal brines* or *formation waters* if associated with oil. Basinal brines are of particular interest because they are commonly encountered in regions hosting oil or economically important ore deposits. These fluids share a number of characteristics: (1) They are commonly found at depths between 500 and 3,700 m, (2) the reservoir rocks are of marine origin, and (3) salinities range from 5 to 30‰. Given the host rock and high salinities, it is natural to conclude that such fluids have a marine origin and are simply *connate* waters trapped during sedimentation. But such is not the case. The scientific literature is replete with studies of the origin of these important natural fluids, and stable isotope analyses have played a major role in understanding them. Isotopic analyses have been made of formation waters from California, the Gulf Coast, and the

---

[8]Some hydrothermal waters are acidic and δD values of such waters are normally more positive than those of local meteoric waters. This effect arises because of the isotopic properties of the H_3O^+ ion.

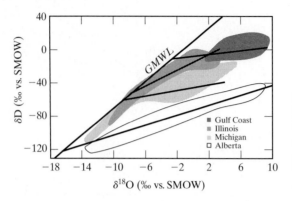

FIGURE 4.19: $\delta^{18}O$–δD values of basinal brines. The combined delta values track back to those of local meteoric water. The most saline brines are generally those with the highest combined $\delta^{18}O$–δD values. (After Clayton et al., 1966.)

Paris Basin and brines from the Illinois, Michigan, Delaware, and Alberta basins, among others. $\delta^{18}O$ and δD values of these fluids covary in a systematic way (Fig. 4.19). Typically, data for the lowest-salinity and lowest-temperature fluids plot near the LMWL, and there is a systematic, though scattered, increase in both $\delta^{18}O$ and δD values with increasing salinity and temperature. Because of significant scatter in the data initially published for brines from midlatitude basins, there was some doubt about the connection of such brines to locally derived meteoric water. The doubts were removed by the publication of data for brines from the high-latitude Alberta basin (Hitchon and Friedman, 1969). In Alberta, the light isotopic compositions of local meteoric waters provide a strong contrast with the isotopic compositions of marine components of the brines. At least 30% of the water in Alberta brines is meteoric water, and the other component(s) may be marine waters that were diagenetically altered.

The increase in $\delta^{18}O$ values and in the salinity of basinal brines with temperature is due primarily to the effect of temperature on the rate or extent of oxygen isotope and chemical exchange between water and enclosing rocks. Sedimentary rocks have very high $\delta^{18}O$ values, typically 20–30‰, and the main reservoir of oxygen in these systems resides in the rock. Therefore, for reasons already explained, the water becomes enriched in ^{18}O as a result of exchange reactions with the rocks. An increase in evaporation with increasing temperature may also play a role in this ^{18}O enrichment process.

The increase in δD values of basinal brines with increasing temperature is less well understood. Possible explanations include (1) exchange with hydrous minerals, because D concentrates in water relative to minerals, (2) exchange reactions with, or loss of, H-bearing gases such as H_2S, CH_4 and H_2, because D concentrates in water relative to these gases, and (3) membrane filtration, because isotopically lighter water is preferentially squeezed out of sediments when they are compacted (Coplen and Hanshaw, 1973).

In sum, the following explanations for the stable isotope systematics of basinal brines have been proposed:

1. With the passage of time, original connate waters were flushed away by meteoric waters that entered the system. In addition to this flushing process, water–rock interactions altered the chemical and isotopic compositions of the fluids (Clayton et al., 1966).
2. The original marine waters underwent evaporation and diagenetic alterations before being trapped and provide one component of a two-component system. The other component is locally derived meteoric water whose isotopic composition could vary with time (Hitchon and Friedman, 1969).

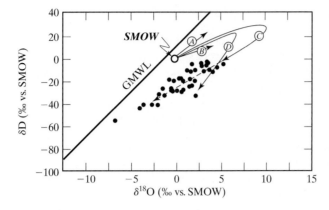

FIGURE 4.20: Trajectories of $\delta^{18}O$–δD values of evaporating seawater. (See Knauth and Beeunas, 1986, for details.)

3. Basinal brines are mixtures of meteoric water and evolved seawaters. One component of brine develops from the release of water of crystallization in gypsum.
4. Basinal brines developed early during extensive evaporation of seawater. Under certain conditions of humidity and chemical composition of the evaporating brine, the combined δD–$\delta^{18}O$ values define a *hook*, such that the data *fall back* on lines of various slopes characteristic of those measured for basinal brines (Fig. 4.20). Meteoric water could have entered the system with the passage of time (Knauth and Beeunas, 1986).
5. The present brines are mixtures of several components, one of which is meteoric water. The other components are relatively unaltered seawater and brines of different degrees of evolution from seawater (Fontes and Matray, 1993).

These explanations are basin specific, but certain features—most notably, the involvement of meteoric water and an evolved marine component—are common to brines in all basins. A proper understanding of the stable isotope systematics and evolution of basinal brines must include a careful consideration of the chemical compositions of the brines.

4.11 GLACIAL ICE

At various times in the history of our planet, including the present, glacial ice has been an important global reservoir of freshwater. Because glaciers form at high latitudes and high altitudes, they are isotopically unusually light. During periods of maximum glaciation in the past, the oceans were significantly enriched in ^{18}O and D because so much light water was locked up on land. Stable isotope measurements have always loomed in importance as a means of attacking classic problems in glaciology and paleoclimatology. In his classic paper on oxygen isotope variations in freshwaters, Dansgaard (1954) wrote, "Greenland Ice Cap..., in the opinion of this author, offers the possibility ... to determine climatic changes over a period of time of several hundred years of the past.... An investigation will be undertaken as soon as an opportunity exists." These words must surely register among the greatest understatements in the history of geochemistry. At that time, Dansgaard had no idea how wildly successful this idea would prove to be! The hope to evaluate climatic variations on a scale of several hundred years underestimated the success of this avenue of research by three orders of magnitude: Ice cores from Greenland and Antarctica provide a continuous record of precipitation in excess of 100,000 years (Johnsen et al., 1972).

4.11.1 Underlying Bases for Glacial Paleoclimatology

The basic premise of glacial paleoclimatology is that mean $\delta^{18}O$ and δD values of precipitation falling in a given season or longer are related to mean ambient temperatures. This correlation is especially robust at high latitudes, where condensation occurs near to the ground surface (Section 4.7.1).

Requisite conditions for the successful application of isotopic analyses of glacial ice to climate (temperature) reconstruction are the following:

1. Significant accumulation must occur in both winter and summer.
2. The pattern of precipitation, or trajectory of air masses, is the same year-round or is known to vary in a predictable manner.
3. Ablation (sublimation) of snow is minimal.
4. The topography at the site of accumulation remains constant. (The elevation does not change.)
5. A method for determining the age of the ice layers is available.

Stable isotope compositions of glacial ice in Antarctica and Greenland indeed vary with the seasons as expected: $\delta^{18}O$ (and δD) values of summer samples are higher than those of winter samples. Differences in $\delta^{18}O$ values between summer and winter ice, $\Delta\delta^{18}O_{summer-winter}$, range from greater than 10‰ for samples only decades old to 2 to 3‰ for samples that are several hundred years old (Fig. 4.21). Dansgaard determined a

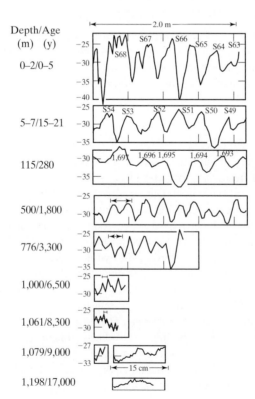

FIGURE 4.21: $\delta^{18}O$ values (SMOW) in firn and ice as a function of depth at Camp Century, Greenland, deep ice core. As the ice ages and becomes compressed, seasonal differences are reduced due to homogenization, and annual cycles become thinner. Y-axis is $\delta^{18}O$ in ‰ vs. SMOW. Years corresponding to summer values are given for first three curves. (After Johnsen et al., 1972.)

$\delta^{18}O$/temperature gradient for modern Antarctic ice of $d\delta^{18}O/dT = \sim 0.69\text{‰}/°C$, with higher $\delta^{18}O$ values corresponding to higher temperatures. In recent accumulations, summer ice is visibly different from winter ice because summer ice undergoes a certain amount of melting. With increasing age, the distinction between summer and winter layers becomes blurred. Ice recrystallizes and flattens, the layers become thinner, and optical distinctions between summer and winter ice disappear. Fortunately, the seasonal isotope variations are still recognizable. Diffusion has a small effect, such that differences between summer and winter layers are reduced (Johnsen et al., 1972).

4.11.2 Determining the Age of Glacial Ice

Assigning reliable ages to the ice poses a major challenge in isotopic studies of glacial ice. Tritium dating and ^{210}Pb dating are used frequently for samples of ice that are younger than about 100 years. For somewhat older samples, ^{14}C dating can be used, but this technique requires large samples of fracture-free ice. Annual layers can be counted by visual inspection, but for samples older than several hundred years, the task becomes increasingly difficult. For each year, the isotopic composition of about eight samples must be measured to establish clear seasonal variation. In addition, parts of the core may be missing and the annual layers become intractably thin.

An important method of dating ice is to link the age of ice layers containing volcanic ash or sulfurous aerosols to historic dates of the eruptions. Zielinski et al. (1994) identified 43 layers in the 8,000 years of the Greenland GISP 2 (Greenland Ice Sheet Project) core, whose ice had relatively high acidity (SO_4 in excess of 20 ppb). Stuiver et al. (1995) correlated over 80% of these layers with historic accounts of eruptions worldwide. Depending on distance to the volcano, there may be a lag of several years for sulfurous aerosols to reach Greenland, but the correlations made to date appear to be excellent. For many eruptions, there is a lowering of temperature of $\sim 1°C$ for several years *preceding* and *following* the eruption (Fig. 4.22).[9]

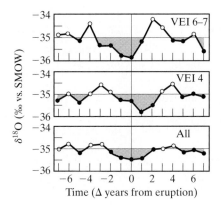

FIGURE 4.22: Mean stacked $\delta^{18}O$ values for ice cores immediately preceding and following volcanic eruptions. The top curve is for the largest volcanoes (Volcanic Eruption Index 6–7), the middle curve is for those with a moderate VEI index, and the bottom curve is for all eruptions. (After Stuiver et al., 1995.)

---

[9]Temperature changes *preceding* a volcanic eruption are logically explained by a lag in the arrival of aerosols to the extreme boreal regions.

4.11.3 Thinning of Ice Layers

With established differences between the isotopic compositions of summer and winter ice, it should be possible to combine layer thickness with isotopic composition to estimate accumulation rates of summer and winter ice. This approach has been used with some success, but layer thickness decreases with time due to flattening or thinning imposed by the load of ice above and recrystallization. Typical decreases in thickness with time are 0.35 m/y for ice less than 280 years old, 0.20 m/y for ice 1,800 years old, and 0.054 m/y for ice 8,300 years old. Dansgaard et al. (1969) proposed a mathematical model for evaluating the thinning of glacial ice as a function of age, assuming (1) a constant rate of accumulation, (2) constant thickness of the total ice sheet, (3) a constant flow rate with time, and (4) the same geographic position of deposition for each layer. In order to minimize uncertainties with these calculations, samples are taken from a glacial divide where there is little or no lateral flow and changes in elevation are minimal. Well-known sites in Greenland and Antarctica meet these criteria and have been investigated (Johnsen et al., 1972).

4.11.4 The Example of Camp Century, North Greenland

An example of the information that is retrievable from ice cores is seen in results from Camp Century, North Greenland (Dansgaard et al., 1971). The discussion that follows is divided into three parts, on the basis of distinct depths of the ice core.

0 to 283 meters The age of the ice over this core distance ranges from 0 to 780 B.P. (Fig. 4.23). Dansgaard et al. originally sampled the core at the decadal scale.[10] A number of cold and warm periods can be identified in the data shown in the figure:

1. A climatic optimum in 1930
2. The Little Ice Age (1600–1740)
3. A cold period in the 15th century that may have been responsible for the demise of the Viking settlements in Greenland
4. Especially cold years in 1352 and 1355 (recorded dates for severe Icelandic winters are 1350, 1351, and 1355)

296 to 1,150 meters The ages of the ice at these greater depths are not well determined, but most likely cover the last 15,000 years of ice formation. Among the significant details that can be extracted from the older core are the following (Fig. 4.24):

1. The postglacial climatic optimum from 4,100 to 8,000 years B.P. has consistently high $\delta^{18}O$ values.
2. The end of the Younger Dryas (the end of the Wisconsonian glaciation) is clearly visible.
3. There is a strong shift to lower $\delta^{18}O$ values in the late Pleistocene that may be in part related to changing local circulation patterns.

---

[10]Investigators now are able to sample at the annual scale going back in excess of 1,000 years (e.g., Stuiver et al., 1995).

Section 4.11: Glacial Ice 95

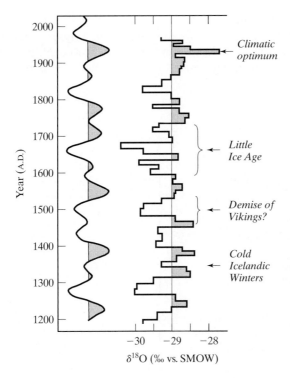

FIGURE 4.23: The upper 283 m of the ice core at Camp Century, Greenland. The shaded areas represent warmer times; the curve on the left is a Fourier transform fit to the data given at the right. The good fit to the measured data suggests that there are a number of periodicities related to variations in the radiation of the sun, such as sunspot activity that control temperature. The arrows refer to specific known cold and warm periods. (See text for details; after Dansgaard et al., 1971.)

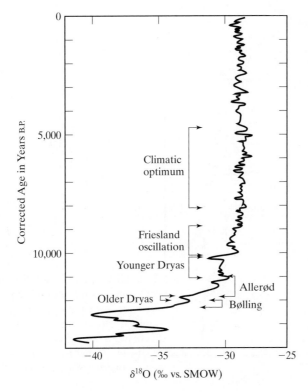

FIGURE 4.24: Most recent 14 thousand years of Camp Century ice core (296–1150 m). Note the dramatic shifts to lower $\delta^{18}O$ values in Pleistocene samples. The last millennium appears to be colder than the preceding one. The $\delta^{18}O$ values during the climatic optimum are higher than at any other time. The rapid oscillations in the Pleistocene are due to glacial events; the overall lowering of $\delta^{18}O$ values with increasing age may be in part related to changing local circulation patterns. (After Dansgaard et al., 1971.)

0 to 1,373 meters The annual layers of the core get significantly compressed with age (Fig. 4.25). Although it is difficult to know ages precisely, it appears that the core is over 126,000 years at the bottom (Dansgaard et al., 1982), a conclusion supported by comparison with other Greenland ice cores. Correlations with Antarctic ice cores are more

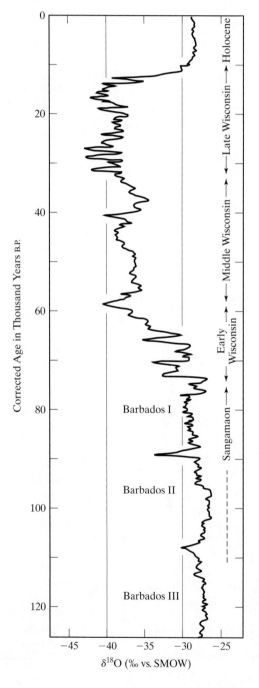

FIGURE 4.25: The full record of the Camp Century core, plotted against a corrected timescale. A rapid warming is seen at the end of the Pleistocene, as well as a more gradual cooling starting at the beginning of the Wisconsin. Numerous glacial events can be identified on the isotope curve (Dansgaard et al., 1971).

difficult due to the complex ocean circulation patterns around Antarctica. Several features are apparent:

1. The entire Wisconsonian glaciation is visible, from 73,000 to 13,000 B.P. The phenomenon can be divided into early, middle, and late events.
2. The Barbados I, II, and III high-sea-level stands are apparent in the lower parts of the diagram, resulting in the highest $\delta^{18}O$ values measured in the entire section.
3. There is a rapid and sudden change to significantly higher $\delta^{18}O$ values at the Pleistocene–Holocene boundary.

4.11.5 Example of the GRIP Summit Core: Flickering Climates

Perhaps the most significant development in stable isotope studies of glacial ice, and one that may bear on governmental policies regarding global warming, is the indication that changes in climate can be more rapid than ever thought possible (Alley et al., 2003). Members of the Greenland Ice Core Project (GRIP) observed remarkable changes in the $\delta^{18}O$ values of ice from the Eemian period (133,000 to 114,000 B.P.), an interglacial period that was previously thought to be of stable warmth (Johnsen et al., 1997). The GRIP data record

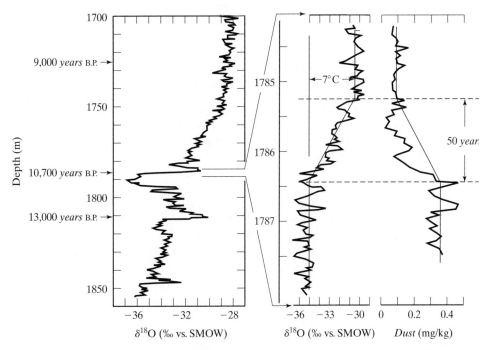

FIGURE 4.26: A rapid increase in $\delta^{18}O$ values and concomitant decrease in dust (warmer, less arid conditions) indicate a rapid change in both temperature and humidity during the Pleistocene–Holocene transition. A temperature change of ~7‰ in 50 years is seen from the $\delta^{18}O$ record of the South Greenland ice core at the Younger Dryas–Pre-Boreal transition. The changing dust concentration and deuterium excess parameter (not shown) over a 20-year period all point to rapid changes in climate, possibly related to changing ocean circulation patterns. After Dansgaard et al. (1989).

three warm substages in the Eemian that were about 7°C higher than the intervening cold stages. But, even more striking was the finding that temperatures dropped about 14°C over a period of only a decade or two at the end of the Eemian. If the interpretation of these data is correct, our present stable climate over the last 8,000 years should be considered an exception to the possible norm of rapid climate change on Earth.

An abrupt temperature change is also seen at the end of the Younger Dryas (Dansgaard et al., 1989). Rapid changes in humidity and in the accumulation of dust occurred in a period of about 20 years, while temperatures increased by 7°C in a 50-year period in South Greenland (Fig. 4.26). Such changes indicate the fragility of the surface temperature of our planet and the need to be concerned with the effects of our anthropogenic activity.

PROBLEM SET

1. **a.** Two rivers flow into a dammed lake. The flow rate through the dam is measured to be 1,057,000 m³/day. The engineers want to know the flow rate of the rivers. *Assuming no evaporation* and assuming constant flow rates and $\delta^{18}O$ values, compute the flow rates in each river if the $\delta^{18}O$ values of input rivers A and B and the dam output are −4.5, −12.7, and −9.2‰ respectively. Which river is sampling waters from higher altitudes.
   b. In summer, there is significant evaporation at the dam. Compute the amount of evaporation (and summer outflow rate) if the $\delta^{18}O$ value of the output is now −4.9‰ and the fractionation between water and water vapor is 9‰. (*Hint*: For this part of the problem, assume that the input rate and $\delta^{18}O$ values are identical to those of part a, but that the outflow rate is now different from part a).

2. Many of the isotopic processes occurring during precipitation can be understood from a logical evaluation of fractionation, temperature, and amount. Sometimes a variable may appear to control the delta values of precipitation in a particular direction when, in fact, the opposite is observed. Consider the positive correlation between temperature and delta value: Lower temperatures are proportional to lower delta values. However, the fractionation between water and water vapor *increases* with decreasing temperature, so precipitation should have a *higher* value. Why, then do T and δ have a positive correlation?

3. Table 4.1 gives the relative abundance and delta values for the different water reservoirs of the Earth. If all of the ice caps melt, what will the new $\delta^{18}O$ and δD values of the ocean be?

4. **a.** Plot the $\delta^{18}O$ values of a cloud mass and resulting precipitation as a function of F, given the following conditions: A total of 20% of water is lost to precipitation (F ranges from 1 to 0.8), $T = 5°C$, $\delta^{18}O_i$ (cloud) = −16‰. The fractionation between water and water vapor is given by the equation (Horita and Wesolowski, 1994)

$$1000 \ln \alpha_{\text{water–water vapor}} = a - \frac{b \times 10^3}{T} + \frac{c \times 10^6}{T^2} + \frac{d \times 10^8}{T^3},$$

where

$a = -7.69$ $b = 6.712$
$c = -1.667$ $d = 3.504$.

   b. In fact, the temperature must decrease during extended periods of precipitation in order to maintain the condition of condensation. Using equations (4.18) and (4.19) in the text, and assuming a regular change of temperature between 10 and 0°C, plot the $\delta^{18}O$ value of the precipitation and calculate the average $\delta^{18}O$ value of precipitation.

5. Craig defined SMOW in terms of NBS-1, a sample of Potomac River water, in the following way:

$$(^{18}O/^{16}O)_{SMOW} = 1.008(^{18}O/^{16}O)_{NBS-1}$$
$$(D/H)_{SMOW} = 1.050(D/H)_{NBS-1}.$$

   a. What is the $\delta^{18}O$ value of NBS-1 relative to SMOW and relative to PDB?
   b. What is the δD value of NBS-1 relative to SMOW?
   c. Does NBS-1 fall on the global meteoric water line? Does NBS-1 fall on the local meteoric water line (use the meteoric water line for North Carolina in Fig. 4.4 as your proxy).

6. Many sedimentological records indicate that far more windblown dust accumulates during glacial periods than at other times. Correlating this dust with humidity, would you expect the deuterium excess parameter d to be larger or smaller in glacial vs. interglacial periods? Think of two geological materials that would allow you to test your conclusions?

ANSWERS

1. The mass balance is given by $\delta^{18}O_{River\,A}(X_a) + \delta^{18}O_{River\,B}(1 - X_a) = \delta^{18}O_{outflow}$. Solving $-4.5(X_a) + -12.7(1 - X_a) = -9.2$, we obtain $X_a = 0.42$. Thus, 42% comes from River A, 68% from River B.

   For part (b), we have an input of $-9.2‰$. (Remember, all other parameters are constant from part (a), so a similar equation can be set up). The two sinks are the outflow ($\delta^{18}O = -4.9‰$, the δ value of the lake and vapor ($\delta^{18}O = -4.9 - 9 = -13.9.‰$) Answer: Evaporation accounts for 48% of the loss.

2. At low temperatures, the fractionation between liquid and vapor is large. Therefore the Rayleigh distillation effect becomes more important and overwhelms all other considerations.

3. The $\delta^{18}O$ and δD values of the two reservoirs (ocean and cryosphere) are given by the equation $\delta_{total} = \sum_i x_i \delta_i$. For oxygen, this yields $0.972 \times 0.0 + 0.0215 \times (-30) = -0.65‰$. For hydrogen, the δD value is $-4.95‰$.

4. a. The Rayleigh fractionation equation for this process is $\delta^{18}O_v = [\delta^{18}O_{v,i} + 1000]F^{(\alpha-1)} - 1000$ (equation 4.16). From equation and the given constants, the α value at 5°C is 1.01126. The $\delta^{18}O$ vapor value will be lighter than the coexisting liquid by $\delta^{18}O_l = \alpha(\delta^{18}O_v + 1000) - 1000$. The $\delta^{18}O_l$ value is plotted in the figure.

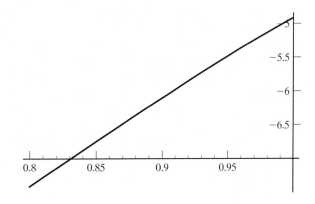

5. From the definition of α, $\alpha = \dfrac{1000 + \delta_a}{1000 + \delta_b}$, for $\alpha = 1.008$ and $\delta^{18}O_a = 0.0‰$ (i.e., SMOW), $\delta^{18}O_{NBS-1} = -7.937$ vs. SMOW and 22.728 vs. PDB. Likewise, the δD value of NBS-1 is -47.62 vs. SMOW. If a sample lies on the meteoric water line, the following equation is valid: $\delta D = 8 \times \delta^{18}O + 10$. For a $\delta^{18}O$ value of -7.939, the δD value for a sample falling on the meteoric water line is $-53.7‰$. Thus, NBS-1 plots well above the global meteoric water line. For the North Carolina meteoric water line, $\delta D = 6.3\delta^{18}O + 2.9$, so the calculated δD value is $-47.1‰$ in excellent agreement with the measured value.

REFERENCES

Aldaz, L., and Deutsch, S. (1967). On a relationship between air temperature and oxygen isotope ratio of snow and firn in the South Pole region. *Earth and Planetary Science Letters* **3**, 267–274.

Alley, R. B., Marotzka, J., Nordhaus, W. D., Ovrepeck, J. T., Peteet, D. M., Pielke, R. A., Jr., Pierrehumbert, R. T., Rhines, P. B., Stocker, T. F., Talley, L. D., and Wallace, J. M. (2003). Abrupt climate change. *Science* **299**, 2005–2010.

Banner, J. L., Wasserburg, G. J., Dobson, P. F., Carpenter, A. B., and Moore, C. H. (1989). Isotopic and trace element constraints on the origin and evolution of saline groundwaters from central Missouri. *Geochimica et Cosmochimica Acta* **53**, 383–398.

Chamberlain, C. P., Poage, M. A., Craw, D., and Reynolds, R. C. (1998). Topographic development of the Southern Alps recorded by the isotopic composition of authigenic clay minerals, South Island, New Zealand. *Chemical Geology* **155**, 279–294.

Clayton, R. N., Friedman, I., Graf, D. L., Mayeda, T. K., Meents, W. F., and Shimp, N. F. (1966). The origin of saline formation waters, I: Isotopic composition. *Journal of Geophysical Research* **71**, 3869–3882.

Coplen, T. B., and Hanshaw, B. C. (1973). Ultrafiltration by a compacted clay membrane—I. Oxygen and hydrogen isotopic fractionation. *Geochimica et Cosmochimica Acta* **37**, 2295–2310.

Coplen, T. B., Hopple, J. A., Böhlke, J. K., Peiser, H. S., Rieder, S. E., Krouse, H. R., Rosman, K. J. R., Ding, T., Vocke, R. D. J., Révész, K. M., Lamberty, A., Taylor, P., and DeBièvre, P. (2002). *Compilation of minimum and maxium isotope ratios of selected elements in naturally occurring terrestrial materials and reagents.* Reston, VA: United States Geological Survey.

Craig, H. (1961) Isotopic variations in meteoric waters. *Science* **133**, 1702–1703.

Craig, H. (1963) The isotopic geochemistry of water and carbon in geothermal areas. In *Nuclear geology in geothermal areas* (ed. E. Tongiorgi), pp. 17–53. Pisa, Italy: Consiglio Nazionale della Ricerche, Laboratorio de Geologia Nucleare.

Craig, H., and Gordon, L. I. (1965). Deuterium and oxygen-18 variations in the ocean and the marine atmosphere. *Stable Isotopes in Oceanographic Studies and Paleotemperatures,* 9–130.

Criss, R. E. (1999). *Prinicples of stable isotope distribution.* New York: Oxford University Press.

Criss, R. E., and Davisson, M. L. (1996). Isotopic imaging of surface water/groundwater interactions, Sacramento Valley, California. *Journal of Hydrology* **178**, 205–222.

Dansgaard, W. (1954). The O^{18}-abundance in fresh water. *Geochimica et Cosmochimica Acta* **6**, 241–260.

Dansgaard, W. (1964). Stable isotopes in precipitation. *Tellus* **16**, 436–468.

Dansgaard, W., Clausen, H. B., Gundestrup, N., Hammer, C. U., Johnsen, S. F., Kristinsdottir, P. M., and Reeh, N. (1982). A new Greenland deep ice core. *Science* **218**, 1273–1277.

Dansgaard, W., Johnsen, S. J., Clausen, H. B., and Langway, C. C., Jr. (1971). Climatic record revealed by the camp century ice core. In *The Late Cenozoic Glacial Ages* (ed. K. K. Turekian), pp. 37–56. New Haven: Yale University Press.

Dansgaard, W., Johnsen, S. J., Moller, J., and Langway, C. C., Jr. (1969). One thousand centuries of climatic record from Camp Century on the Greenland ice sheet. *Science* **166,** 377–381.

Dansgaard, W., White, J. W. C., and Johnsen S. J. (1989). The abrupt termination of the Younger Dryas climate event. *Nature* **339,** 532–534.

Eiler, J. M., and Schauble, E. (2004). $^{18}O^{13}C^{16}O$ in Earth's atmosphere. *Geochimica et Cosmochimica Acta* **68,** 4767–4777.

Epstein, S., and Mayeda, T. K. (1953). Variation of ^{18}O content of waters from natural sources. *Geochimica et Cosmochimica Acta* **4,** 213–224.

Fontes, J. C., and Gonfiantini, R. (1967). Comportement isotopique au cours de l'évaporation de deux bassins Sahariens. *Earth and Planetary Sciences* **3,** 258–266.

Fontes, J. C., and Matray, J. M. (1993). Geochemistry and origin of formation brines from the Paris Basin, France; 2, Saline solutions associated with oil fields. *Chemical Geology* **109,** 177–200.

Fricke, H. C., and O'Neil, J. R. (1999). The correlation between $^{18}O/^{16}O$ ratios of meteoric water and surface temperature; its use in investigating terrestrial climate change over geologic time. *Earth and Planetary Science Letters* **170,** 181–196.

Friedman, I. (1953). Deuterium content of natural water and other substances. *Geochimica et Cosmochimica Acta* **4,** 89–103.

Friedman, I., Machta, L., and Soller, R. (1962). Water vapor exchange between a water droplet and its environment. *Journal of Geophysical Research* **67,** 2761–2770.

Gat, J. R. (1996). Oxygen and hydrogen isotopes in the hydrologic cycle. *Annual Review of Earth and Planetary Sciences* **24,** 225–262.

Gat, J. R., and Carmie, I. (1970). Evolution of the isotopic composition of atmospheric waters in the Mediterranean Sea area. *Journal of Geophysical Research* **86,** 3039–3048.

Harmon, R. S., and Schwarcz, H. P. (1981). Changes of ^2H and ^{18}O enrichment of meteoric water and Pleistocene glaciation. *Nature* **290,** 125–128.

Hitchon, B., and Friedman, I. (1969). Geochemistry and origin of formation waters in the western Canada sedimentary basin: I. Stable isotopes of hydrogen and oxygen. *Geochimica et Cosmochimica Acta* **33,** 1321–1349.

Horita, J., and Wesolowski, D. J. (1994). Liquid–vapor fractionation of oxygen and hydrogen isotopes of water from the freezing to the critical temperature. *Geochimica et Cosmochimica Acta* **58,** 3425–3437.

Johnsen, S. J., Clausen, H. B., Dansgaard, W., Gundestrup, N. S., Hammer, C. U., Andersen, U., Andersen, K. K., Hvidberg, C. S., Dahl-Jensen, D., Steffensen, J. P., Shoji, H., Sveinbjornsdottir, A. E., White, J., Jouzel, J., and Fisher, D. (1997). The $\delta^{18}O$ record along the Greenland Ice Core Project deep ice core and the problem of possible Eemian climatic instability. *Journal of Geophysical Research, C, Oceans* **102,** 26397–26410.

Johnsen, S. J., Dansgaard, W., Clausen, H. B., and Langway, C. C., Jr. (1972). Oxygen isotope profiles through the Antarctic and Greenland ice sheets. *Nature* **235,** 429–434.

Jouzel, J., Lorius, C., Petit, J. R., Genthon, C., Barkov, N. I., Kotlyakov, V. M., and Petrov, V. M. (1987). Vostok ice core; a continuous isotope temperature record over the last climatic cycle (160,000 years). *Nature* **329,** 403–408.

Jouzel, J., and Merlivat, L. (1984). Deuterium and oxygen 18 in precipitation; modeling of the isotopic effects during snow formation. *Journal of Geophysical Research. D. Atmospheres* **89,** 11749–11757.

Kakiuchi, M., and Matsuo, S. (1979). Direct measurements of D/H and $^{18}O/^{16}O$ fractionation factors between vapor and liquid water in the temperature range from 10° to 40°C. *Geochemical Journal* **13,** 307–311.

Kendall, C., and Coplen, T. B. (2001). Distribution of oxygen-18 and deuterium in river waters across the United states. *Hydrological Processes* **15,** 1363–1393.

Knauth, L. P., and Beeunas, M. A. (1986). Isotope geochemistry of fluid inclusions in Permian halite with implications for the isotopic history of ocean water and the origin of saline formation waters. *Geochimica et Cosmochimica Acta* **50,** 419–433.

Lloyd, R. M. (1966). Oxygen isotope enrichment of sea water by evaporation. *Geochimica et Cosmochimica Acta* **30,** 801–814.

Majoube, M. (1971). Fractionnement en oxygene 18 et en deuterium entre l-eau et sa vapeur. *Journal de Chimie et Physique* **68,** 1423–1436.

Meehan, T. D., Giermakowski, J. T., and Cryan, P. M. (2004). GIS-based model of stable hydrogen isotope ratios in North American growing-season precipitation for use in animal movement studies. *Isotopes in Environmental and Health Studies* **40,** 291–300.

Merlivat, L., and Jouzel, J. (1979). Global climatic interpretation of the deuterium–oxygen 18 relationship for precipitation. *Journal of Geophysical Research* **84,** 5029–5033.

Musgrove, M., and Banner, J. L. (1993). Regional ground-water mixing and the origin of saline fluids; midcontinent, United States. *Science* **259,** 1877–1882.

Poage, M. A., and Chamberlain, C. P. (2001). Empirical relationships between elevation and the stable isotope composition of precipitation and surface waters; considerations for studies of paleoelevation change. *American Journal of Science* **301,** 1–15.

Rozanski, K., Araguas, A. L., and Gonfiantini, R. (1992). Relation-between long-term trends of oxygen-18 isotope composition of precipitation and climate. *Science* **258,** 981–985.

Sharp, Z. D., Masson, H., and Lucchini, R. (2005). Stable isotope geochemistry and formation mechanisms of quartz veins; extreme paleoaltitudes of the Central Alps in the Neogene. *American Journal of Science* **305,** 187–219.

Stuiver, M., Grootes, P. M., and Braziunas, T. F. (1995). The GISP2 $\delta^{18}O$ climate record of the past 16,500 years and the role of the sun, ocean and volcanoes. *Quaternary Research* **44,** 341–354.

Zielinski, G. A., Mayewski, P. A., Meeker, L. D., Whitlow, S., Twickler, M. S., Morrison, M., Meese, D. A., Gow, A. J., and Alley, R. B. (1994). Record of volcanism since 7000 B.C. from the GISP 2 Greenland ice core and implications for the volcano–climate system. *Science* **264,** 948–952.

CHAPTER 5

THE OCEANS

5.1 OVERVIEW

The $\delta^{18}O$ values of *surface* marine waters vary by several per mil, particularly in coastal marine waters at high latitudes, where isotopically light glacial and stream waters feed into the ocean. The large variations are due to evaporation, the influx of freshwater, and melting or freezing of ice. In contrast, the oxygen isotope compositions of *deep* ocean waters are nearly constant, varying by only about 1‰. Salinities are also nearly constant, at 34 to 35‰.[1] Nonetheless, the subtle variations in both salinity and isotopic composition of these waters can be used to provide valuable information on mass accumulations and circulation patterns in the deep oceans.

The earliest study of oxygen isotope variations in ocean waters was made by Epstein and Mayeda (1953), whose main interest at that time was focused on the extent of oxygen isotope variability of ocean water as it pertained to determining paleotemperature estimates from $\delta^{18}O$ values of carbonates. (See Chapter 6.) These two researchers quickly realized that the failure to consider possible variations in $\delta^{18}O$ values of surface waters could lead to a disturbing uncertainty of up to about 10°C in temperature measurements made with the carbonate–water oxygen isotope thermometer. Further uncertainty is introduced when possible and probable temporal variations in the oxygen isotope composition of the oceans are taken into consideration.

Has the oxygen isotope composition of ocean water changed in any significant way over geologic time? From simple mass-balance considerations involving the amount and isotopic composition of water now locked up on land in glaciers, we *know* that $\delta^{18}O$ values of ocean waters were about 1‰ lower in ice-free times than they are at present. But were variations in the past larger than this? On the basis of oxygen isotope analyses of marine sediments deposited at different times in the geologic past, $\delta^{18}O$ values as low as −15‰ have been

---

[1]Salinity is often reported in per mil notation, where 34‰ salinity is equivalent to 3.4 wt % dissolved NaCl. Salinity should not be confused with isotopic composition.

proposed for open ocean waters! A change of this magnitude in the isotopic composition of the vast quantity of water in the oceans would mirror an enormous change in some major (plate tectonic?) process operating on Earth, but the evidence presented for such a change is highly controversial. The question of major oxygen isotope variability in ancient oceans remains one of the most important unresolved problems in earth science. We will return to this intriguing subject later in the chapter.

5.2 OXYGEN ISOTOPE VARIATIONS IN MODERN OCEANS

5.2.1 Salinity–$\delta^{18}O$ Relations in Shallow Marine Waters

Epstein and Mayeda (1953) first recognized the almost linear correlation between salinity (S) and the $\delta^{18}O$ value of ocean waters, and the anticipated sympathetic correlation between salinity and δD was documented a few years later by Friedman, Schoen, and Harris (1961). The correlation between salinity and stable isotope ratios of ocean waters is explained by two simple processes: evaporation and the addition of freshwater. Evaporation from the ocean surface preferentially removes light isotopologues from the water and increases its salinity, so values of both $\delta^{18}O$ and S of surface waters increase with increasing degree of evaporation. Upon the introduction of freshwater from melting ice, river systems, or rain, the salinity of the affected parcel of ocean water is lowered. The $\delta^{18}O$ and δD values of this water are also lowered because meteoric waters have lower stable isotope ratios than ocean water has. The magnitude of the isotopic effect is related to the water source. $\delta^{18}O$ values of glacial waters are very low (a typical value is $-30‰$), while those of rivers in temperate zones are much higher (e.g., $-4.9‰$ for the Mississippi River). Thus, for a given dilution factor, high-latitude glacial waters will shift $\delta^{18}O$ values to a much greater extent than low-latitude river waters will for the same shift in salinity. In an ice-free world, $\delta^{18}O$ would change by about $0.2‰$ per unit change in salinity (although this will vary in accordance with the $\delta^{18}O$ value of the freshwater), so that a $5‰$ fluctuation in salinity would be accompanied by a $1.0‰$ fluctuation in the O^{18}/O^{16} ratio of the oceans.

The carbonate–water paleothermometer is valid only when the $\delta^{18}O$ value of the water from which the carbonates precipitated is known. Large variations in $\delta^{18}O$ values of surface waters resulting from evaporation and the influx of meteoric waters could bear significantly on ocean paleotemperature estimates made from analyses of carbonate from the shells of planktonic organisms. The rather loose relation observed between salinity and $\delta^{18}O$ values is called the *salinity effect* and implies that we can approximate one parameter from knowledge of the other. That is, if the paleosalinity can be determined independently, a correction for the departure of $\delta^{18}O$ from values of normal marine waters can be made (Railsback et al., 1989). A surprising number of authors have misused the term *salinity effect*, assuming that salinity itself affects the oxygen isotope fractionation between carbonates and ocean water, which is not the case. There is indeed a small effect on the fractionation between water in highly saline solutions and other phases (O'Neil and Truesdell, 1991), but at the relatively low salinities of ocean water, the effect on oxygen isotope fractionation is negligible. *The salinity effect employed in paleotemperature studies is not related to salinity fractionation effects.*

Craig and Gordon (1965) measured isotopic compositions of both vapor and surface waters along north–south and east–west transects across the Atlantic Ocean. These data, shown in Figure 4.8, reflect the nonequilibrium nature of the evaporation process and also point to relative humidity as a dominant control on vapor compositions. As one travels in a northerly direction in the Atlantic Ocean, there is little change in $\delta^{18}O$ values of liquid water, but $\delta^{18}O$ values of vapor in equilibrium with these waters decreases with increasing latitude, due to decreasing sea-surface temperatures and concomitant increases in the magnitude of the isotopic fractionation between liquid and vapor. Because of kinetic isotope effects, however, the measured $\delta^{18}O$ values of the vapor are about 4.5‰ more negative than the values expected at equilibrium. During the process of evaporation, light molecules preferentially diffuse through the near-surface layers of liquid water and into the vapor phase, with little or no back exchange. The lowest $\delta^{18}O_{vapor}$ values are found in the region of the trade-wind belts, at 18–26° of latitude, where evaporation is highest and relative humidity is lowest. Variations in the $\delta^{18}O$ value of water vapor are minimal in an east–west direction, as expected.

The various factors that contribute to changes in S and $\delta^{18}O$ are summarized in Fig. 5.1. Intense evaporation in the region of the trade winds causes both $\delta^{18}O$ and S values to increase, whereas intense precipitation at equatorial latitudes lowers both $\delta^{18}O$ and S values. At high latitudes, meltwater entering the ocean has very low $\delta^{18}O$ values, so that oxygen isotope ratios are more strongly affected than salinity. In contrast, the formation of sea ice, a particularly important process in the Weddell sea off Antarctica, affects salinity more strongly than oxygen isotope ratios. This is because the oxygen isotope fractionation between ice and seawater is small (O'Neil, 1968), but essentially no salt is incorporated into the ice as it forms. The salinity of the surrounding water thus increases significantly, while its $\delta^{18}O$ value decreases, but only slightly. Assuming that $\Delta^{18}O_{ice-water} = 2‰$, a quantity of ice formation that changes the salinity by 2‰ will change the $\delta^{18}O$ value by only 0.1‰. Knowledge of the stable isotope systematics shown in Fig. 5.1 bears importantly on our understanding of ocean circulation patterns.

5.2.2 Salinity–$\delta^{18}O$ Relations in Deep Ocean Waters

Craig and Gordon (1965) developed deep ocean circulation models on the basis of combined $\delta^{18}O$–S data. Because the total variation in the isotopic composition of deep ocean water is only about 0.3‰, they needed extremely high precision for their analyses. To meet this challenge, Craig and Gordon developed a procedure to greatly increase the precision of their $\delta^{18}O_{water}$ determinations. This procedure is worth detailing here because, although many researchers may quote extremely high precision, the protocol developed by Craig and Gordon illustrates just how hard it is to achieve such a goal in reality.

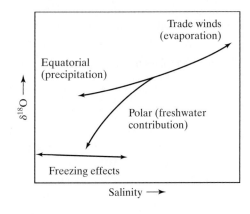

FIGURE 5.1: Schematic illustration of various effects on the isotopic composition and salinity of ocean surface waters.

Craig and Gordon analyzed relatively large samples of water to limit the effects of vapor loss and contamination. In addition, they equilibrated their water samples with CO_2 that had previously been isotopically equilibrated with water of "normal" marine composition. In this way, changes in the $\delta^{18}O$ value of the equilibrating CO_2 gas would be very small during exchange with their samples, thus minimizing changes in the isotopic composition of the water sample during the equilibration process. The mass spectrometer working standard was CO_2 that had also been equilibrated with water of marine composition, and both reference and sample gas were introduced through the *sample* inlet side, thereby eliminating any fractionation effects that might arise during gas transfer to the mass spectrometer through different inlets. By adopting this protocol, Craig and Gordon were able to increase their precision to better than ±0.02‰.

The isotopic compositions and salinities of near-surface waters vary considerably with latitude. But with increasing depth, there is a convergence at or below ~1,000 m to values that are remarkably consistent for large, distinct ocean basins. Values of $\delta^{18}O$ and S for 10 samples of water that spanned 40° of latitude were identical to within 1σ standard deviations of 0.04 and 0.03‰, respectively. $\delta^{18}O-S$ values for deep ocean basins are given in Table 5.1. Although subtle, the distinctions are significant and can be explained completely by specific circulation patterns.

The $\delta^{18}O-S$ values of North Atlantic Deep Waters (NADW) plot directly on the linear array defined by North Atlantic surface waters (Fig. 5.2). The deep waters are thus consistent with downwelling of cold surface waters in the Norwegian sea, in accord with independently developed circulation models. In the South Atlantic region near the Weddell Sea (59–65°S), the slope $d\delta^{18}O/dS$ is close to zero (Fig. 5.3), consistent with the earlier explained relations associated with the formation of sea ice (Fig. 5.1). Salinities range from 33.4 to 34.7 per mil, while $\delta^{18}O$ values are constant at 0.4–0.5‰ for all but the most saline waters. The $\delta^{18}O-S$ values for the Antarctic Bottom Waters (AABW) are distinct from those of the NADW because the mechanism of their formation differs from that occurring in the North Atlantic. North Atlantic waters achieve sufficient density to sink by cooling and evaporation (increase in salinity) of Gulf Stream waters. In contrast, waters in the Weddell sea are so cold that significant freezing occurs, increasing salinity and density, but not changing $\delta^{18}O$ values. On the basis of salinity and isotopic data, two distinct bodies of downwelling waters are clearly identified (Fig. 5.3).

In contrast to the compositions of North Atlantic waters, data for the Pacific Deep Waters (PDW) do not lie on the $\delta^{18}O-S$ line for surface Pacific waters, implying an *exotic* source for the PDW. In other words, PDW *cannot* be generated by the downwelling of Pacific Ocean waters (Fig. 5.4), and the same applies to deep waters in the Indian Ocean. In addition, neither of the two Atlantic downwelling sources (NADW or AABW) matches the combined $\delta^{18}O-S$ values of the PDW. Instead, the PDW data lie intermediate to the NADW and AABW data and are explained by simple mixing of these two deep water sources (Fig. 5.3). Such mixing is

TABLE 5.1: Salinity and $\delta^{18}O$ values of principal deep-water ocean masses.

	Salinity (‰)	$\delta^{18}O$ ‰
North Atlantic Deep Water (NADW)	34.93	+0.12
Antarctic Bottom Waters (AABW)	34.65	−0.45
Indian Deep Waters (IDW)	34.71	−0.18
Pacific Deep Waters (PDW)	34.70	−0.17*

*Pacific samples near Antarctica have a slightly lower $\delta^{18}O$ value of −0.21‰.
(From Craig and Gordon, 1965.)

Section 5.2: Oxygen Isotope Variations in Modern Oceans

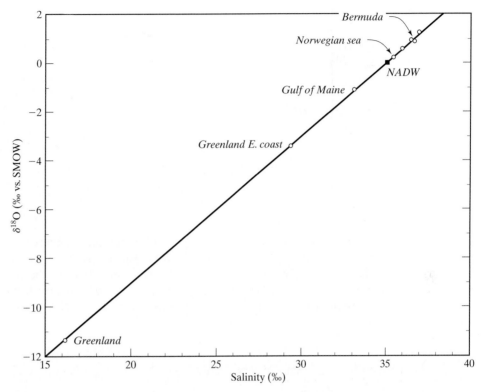

FIGURE 5.2: $\delta^{18}O$–S **values for Atlantic surface waters and North Atlantic Deep Water (NADW, black square).** The NADW $\delta^{18}O$–S values fall on the mixing trend for surface waters. (After Craig and Gordon, 1965, with data from Epstein and Mayeda, 1953.)

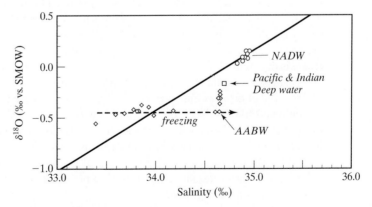

FIGURE 5.3: $\delta^{18}O$ **vs. salinity for Atlantic waters and Antarctic Bottom Waters (AABW).** The near-horizontal trend for deep waters in the Weddell Sea indicates a freezing-related trajectory. The AABW appear to have their source in the downwelling waters of the Weddell Sea. The Pacific and Indian Deep Waters can be explained by a simple mixing relation between the NADW and AABW. (After Craig and Gordon, 1965.)

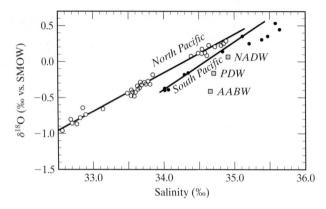

FIGURE 5.4: $\delta^{18}O–S$ data for North and South Pacific surface waters and various deep waters. Pacific Deep Waters (*PDW*) cannot be generated by any combination of Pacific surface waters, but can be generated by mixing Antarctic Bottom Waters (*AABW*) with North Atlantic Deep Waters (*NADW*). (After Craig and Gordon, 1965.)

consistent with our current understanding of global ocean circulation patterns. Appreciable ocean downwelling occurs only in the North Atlantic and the Antarctic. The North Atlantic data reflect the downwelling source in the Norwegian Sea, and the Weddell Sea data record the local cold-water downwelling that is associated with the production of ice. These two downwelling waters move eastward and mix, eventually traversing both the Pacific and Indian Oceans.

5.3 DEPTH PROFILES IN MODERN OCEANS: $\delta^{18}O(O_2)_{aq}$ AND $\delta^{13}C(\Sigma CO_2)$

Atmospheric oxygen all over the globe has a constant $\delta^{18}O$ value of 23.8 ± 0.3‰ (Kroopnick and Craig, 1972).[2] The oxygen isotope fractionation between O_2 dissolved in seawater and O_2 in the atmosphere is 0.6‰ at 25°C (note the direction of this fractionation; $\delta^{18}O$ of O_2 dissolved in seawater is 0.6‰ higher than in air) and is reflected in measured $\delta^{18}O$ values of O_2 in shallow seawater. Because of the huge generation of O_2 in the atmosphere from photosynthesis, and the relatively low amounts of dissolved O_2 in the oceans, the $\delta^{18}O$ value of dissolved O_2 in the shallow marine environment is controlled by atmospheric oxygen. With increasing depth, the abundance of dissolved oxygen in the ocean decreases from a value of ~210 μmoles/kg to a minimum of 150 μmoles/kg at about 1 km depth (Fig. 5.5). This decrease is due to the consumption of dissolved oxygen by the oxidation of dead organic matter. With increasing depth, the O_2 concentration increases again to 270 μmoles/kg, due to the O_2 contribution of downwelling waters. $\delta^{18}O$ values of dissolved oxygen mirror the oxygen concentration profile. O_2 dissolved in surface waters has $\delta^{18}O$ values in excess of 24‰ as a result of equilibrium exchange with O_2 in the atmosphere. $\delta^{18}O$ values increase with depth because light oxygen is preferentially consumed in the oxidation of organic matter. At a depth of ~1 km, dissolved oxygen attains maximum $\delta^{18}O$ values in excess of 30‰. $\delta^{18}O$ values then decrease to a near-constant value of 26‰, a process related to the continual contribution of fresh O_2 in downwelling waters.

---

[2]Kroopnick and Craig published a value of 23.5‰ using a $\alpha(CO_2\text{-water}) = 1.0409$ at 25°C. Using the currently accepted value of 1.0412 for this fractionation, the $\delta^{18}O$ value of atmospheric O_2 is 23.8‰.

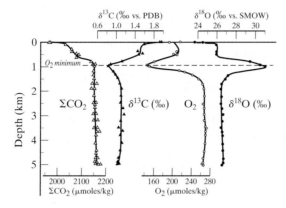

FIGURE 5.5: Profiles of dissolved $\Sigma(CO_2)$ and its $\delta^{13}C$ values (left) and dissolved O_2 and its $\delta^{18}O$ values (right). The profile changes are due to the oxidation of organic matter and, at depth, to mixing with deep waters. Corrected for $\alpha_{CO_2-H_2O}$ fractionation of 1.0412. (After Kroopnick et al., 1972.)

The $\delta^{13}C$ curve for total dissolved inorganic carbon (ΣCO_2) is antipodal to the $\delta^{18}O$ curve for dissolved O_2. The abundance of ΣCO_2, which is primarily HCO_3^-, increases steadily with depth as organic matter is oxidized. The concentrations increase from ~2,000 µmoles/kg near the surface to ~2,160 µmoles/kg at a depth of 1 km, after which the concentration remains almost constant. The $\delta^{13}C$ values of ΣCO_2 decrease from values near 2‰ at the surface to 0.7‰ at a depth of ~1 km and then rise back up to an almost constant value of about 1‰ near the middle of the deep waters and below (Fig. 5.5). In the absence of biological activity, $\delta^{13}C$ of HCO_3^- in surface waters would be about −1‰, a value appropriate to equilibrium exchange with atmospheric CO_2. But isotopically light HCO_3^- is preferentially used by organisms in the formation of their soft tissues, driving the $\delta^{13}C$ values of any HCO_3^- remaining in the upper (photic) zone to 2‰, or even higher in regions of intense biological productivity. The return to more negative $\delta^{13}C$ values at depth is caused by the addition of light carbon to these waters from the oxidation of organic matter. The formation of organic matter in the productive region of the shallow waters and its systematic removal by oxidation at depth make up the *biological carbon pump* (Box 7.1).

5.4 ISOTOPIC COMPOSITIONS OF ANCIENT OCEANS

How frequently and by how much has the isotopic composition of the global oceans changed with time? Why does present-day seawater have the isotopic composition that it does? What exchange mechanisms or buffering capacity exist in the oceans to cause $^{18}O/^{16}O$ and D/H ratios to change or remain the same with time? These questions are important for a number of reasons, the most compelling being that the reservoirs of oxygen and hydrogen in the oceans are so large that only a major geologic event could cause a significant change in their isotopic ratios.

5.4.1 Primitive Oceans

If our oceans are dominated by water that has degassed from primitive Earth, then the $\delta^{18}O$ and δD values of the primitive oceans would have been around 7 ± 1‰ and -60 ± 20‰, respectively, constrained by equilibrium exchange between water and the mantle at high temperatures (Silverman, 1951). Some vapor initially present over the hot oceans was undoubtedly dissociated to the elements by ionizing radiation, and a fraction of the newly formed H_2 could have escaped the gravitational pull of Earth. Light hydrogen must have left our planet preferentially, and this process could account in some small part for the high D/H ratio of modern oceans relative to presumed D/H ratios of the mantle.[3] (Hydrogen in the mantle is represented by OH groups in hydrous minerals and molecular water in nominally anhydrous minerals. The reason that the escape process can account for only a very small part of the increase in D/H ratio of the oceans is a matter of material balance: An impossibly high fraction of the hydrogen present in the initial degassed water would need to have escaped to the exosphere to raise the D/H ratio to the value of the modern oceans.) As primitive Earth cooled, there were interactions between ocean water and hot basalt. Any hydrous minerals formed in this early high-temperature stage of ocean water development would have raised δD of the primitive waters to a small extent, but the effect of this process on $\delta^{18}O$ would have been negligible.

In contrast, marine *sediments* and alteration minerals formed by low-temperature weathering of terrestrial rocks have D/H ratios that are considerably lower than the D/H ratio of ocean water, so the formation of hydrous minerals at low temperatures is a highly efficient way to raise the D/H ratio of the world's oceans. The effect is somewhat lessened by the fact that marine sediments are subducted into the mantle and some fraction of the water entrained within them is returned to the surface in volcanic emanations.

In contrast to hydrogen isotope ratios, oxygen isotope ratios of the ocean will be lowered significantly by the formation of marine sediments and minerals from the weathering of igneous rocks. Oxygen isotope fractionations between sedimentary materials (carbonates, silica, phosphates, clays) and water are so large (20–30‰) at low temperatures, that locking up ^{18}O in weathering minerals and in marine sediments should have played a major role in lowering the $\delta^{18}O$ value of the early oceans from high initial values to those near 0‰. The degree to which $^{18}O/^{16}O$ ratios were lowered and D/H ratios were raised by weathering and sedimentary processes depends once again on material-balance considerations and how much sedimentary material was reworked over time.

The hydrogen isotope history of primitive oceans is seldom addressed and, for the most part, remains unresolved. Limited hydrogen isotope analyses of rocks that formed in ancient oceans (greenstones and ophiolites) are similar to those of modern equivalents (Sheppard and Epstein, 1970; Yui et al., 1990; Lécuyer et al., 1996), suggesting that D/H values of the ocean have remained fairly constant throughout geologic time, except during periods of glaciation, when light water is locked up in continental ice sheets. Modeling of hydrogen fluxes into and out of the ocean led Lécuyer et al. (1996) to conclude that the δD of the Archean ocean may have been several tens per mil lower than the modern value, a suggestion that is compatible with values inferred from analyses of δD of fluid inclusions in younger, but still ancient, Silurian and Permian evaporites (Knauth and Roberts, 1991).

The oxygen isotope history of ancient oceans, by contrast, has received considerable attention, via a plethora of approaches. In the early years of stable isotope geochemistry, much attention was devoted to the carbonate–water isotope thermometer, and the lack of knowledge regarding the $\delta^{18}O$ values of ancient oceans was a major concern. The oxygen isotope composition of ancient oceans has relevance not only to ocean paleotemperatures, but also to understanding ocean hydrothermal systems, models of ocean growth, and plate tectonic histories, as well as to the reconstruction of environmental conditions in the past.

---

[3] Rare-gas isotope geochemistry confirms a large loss of volatile elements early in Earth's history.

Perry and Tan (1972) proposed that the Archean oceans were volumetrically much smaller than modern ones. Under this size constraint, material-balance arguments would favor significant lowering of the $\delta^{18}O$ values of early oceans by the intense low-temperature alteration of oceanic basalt. No rock record exists to confirm or deny this suggestion for very early ocean genesis. Samples exist well back into the Archean, however, and they provide clues to the genesis and isotopic history of the ancient oceans. (See Chapter 8 for more details.)

5.4.2 Secular Changes in $\delta^{18}O$ of Marine Sediments

Marine sediments have been used to address the question of the oxygen isotope compositions of ancient oceans. The $\delta^{18}O$ values of cherts, iron formations, phosphorites, and limestones provide information on how $\delta^{18}O$ values of ancient oceans may have differed from those of the modern oceans. In every suite of samples examined to date, measured $\delta^{18}O$ values of marine sediments decrease with increasing age (Fig. 5.6).

Using oxygen isotope ratios of marine sediments to reconstruct conditions in ancient oceans is plagued with uncertainty. Consider only that the $\delta^{18}O$ value of a chemical sediment in the ocean is a function of both temperature and the $\delta^{18}O$ value of the water. To complicate matters even further, the sediments examined may not have retained their original isotopic ratios through time. There are three ways to explain the observed secular trends for marine sediments:

1. $\delta^{18}O$ values of the oceans were more negative in the past.
2. Temperatures of ancient oceans were higher in the past.
3. Sediments become lighter with time through diagenetic reactions.

If explanation 1 is correct, mammoth and, for the most part, unidirectional changes occurred in the way the Earth operates, specifically in plate tectonic processes, including the cessation of activity for very long periods. Explanation 2 is attractive because temperatures were certainly higher in the very old oceans. Some data, however, require marine organisms to have thrived at temperatures as high as about 70°C, a temperature most biologists would find unacceptable.[4] Most workers in the field favor (at least in part) explanation 3, but it, too,

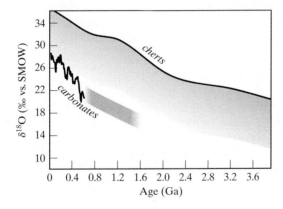

FIGURE 5.6: Schematic variation in the $\delta^{18}O$ values of cherts and carbonates. (See Chapters 6 and 8 for sources, in addition to Anderson and Arthur, 1983.) The lower $\delta^{18}O$ values of ancient sediments may be due to lower $\delta^{18}O_{ocean}$ values in the past, higher ocean temperatures in the past, or more intense diagenesis in old samples.

---

[4]However, the recent discovery of thermophillic bacteria thriving in near-boiling waters lessens the force of this argument.

is fraught with uncertainty. Most importantly, some specific sedimentary minerals are grossly more resistant to oxygen isotope exchange with diagenetic fluids than are other such minerals. As a consequence, rates of diagenetic alteration are expected to be grossly different as well. Nonetheless, the same secular trend is seen in the oxygen isotope compositions of all sedimentary rocks. It is well to point out that there are rare, but important, exceptions to these systematics documented in the literature. That is, there are reported cases of ancient sedimentary minerals whose oxygen isotope compositions are similar to those of modern equivalents.

Some combination of all three processes might explain the trends, but assigning the relative contribution of each process is not possible with the data at hand. Understanding these secular trends *and* the exceptions bears importantly on our understanding of how the Earth worked in the past and poses an exciting challenge to earth scientists.

5.5 SEAWATER–BASALT INTERACTIONS: BUFFERING THE $\delta^{18}O$ VALUE OF THE OCEAN

Muehlenbachs and Clayton (1976) forever changed the "ocean composition" landscape with their landmark paper on interactions between seawater and the oceanic crust. These authors measured oxygen isotope compositions of samples dredged from the ocean floor, as well as samples from Deep Sea Drilling Project (DSDP) boreholes. Confirming early measurements of Garlick and Dymond (1970), they reported that $\delta^{18}O$ values of ocean crust covered a broad range, from <4‰ to >25‰ (SMOW). This wide range of $\delta^{18}O$ values was explained by seawater alteration of oceanic crust at both low and high temperatures (discussed in Sections 5.5.1 and 5.5.2). Combining the two alteration regimes, Muehlenbachs and Clayton devised a model of ocean-crust interaction that effectively buffers the $\delta^{18}O$ value of the ocean to its present steady-state value near 0‰. Refinements of this model (Gregory, 1991) indicate that drastic and *long-term* (many tens of millions of years) changes in plate tectonic activity would be required to cause significant changes of several per mil in the $^{18}O/^{16}O$ ratio of the ocean from its present value of 0‰.

Pristine Mid Ocean Ridge Basalt (MORB) has a uniform $\delta^{18}O$ value of 5.7‰. In fact, so constant is this value that any sample of fresh MORB glass could be used as an oxygen isotope reference standard. Oceanic basalts with lower and higher values than this contain hydrous alteration phases. Muehlenbachs and Clayton recognized two types of alteration that were later confirmed by high-spatial-resolution studies of DSDP cores.

5.5.1 Low-Temperature Alteration

The vast majority of oceanic basalts are altered to some extent at low temperatures and have $\delta^{18}O$ values that are more positive than 5.7‰. Only the youngest glassy MORB preserves its primary $\delta^{18}O$ value of 5.7‰. Most alteration of this type occurs between 0 and 15°C, but some occurs at temperatures as high as about 50°C. $\delta^{18}O$ values of basalts that were intensely altered at the lowest temperatures near 0°C can have values as high as 25‰, because equilibrium fractionations between clay minerals and water are very large at low temperatures and, consequently, values of $\Delta^{18}O$ (oceanic crust water) increase with increasing degree of alteration. The positive correlation between $\delta^{18}O$ values and H_2O^+ contents of the rock illustrates this point (Fig. 5.7). The high $\delta^{18}O$ values are unambiguously explained by the presence of hydrous phases that form during the low-temperature

Section 5.5: Seawater–Basalt Interactions: Buffering the $\delta^{18}O$ Value of the Ocean

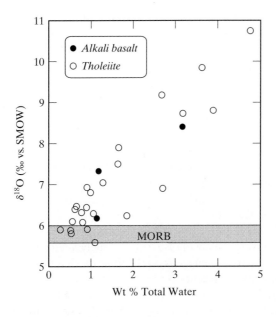

FIGURE 5.7: Variation in the $\delta^{18}O$ values and wt % water in ocean basalts. The increase in both parameters is related to low-temperature interaction with seawater. (After Muehlenbachs and Clayton, 1972.)

alteration of anhydrous basalt on the ocean floor. In most cases, measurements are made of mixtures, hereafter written as *basalt*, of pristine MORB and alteration phases. In the initial stages of alteration, the alteration phases may be only incipient precursors to smectite minerals and are not visible under the microscope. But even minor degrees of alteration can be easily determined by a measurement of H_2O^+.

Low-temperature alteration starts almost immediately after extrusion of the basalt and is a very heterogeneous process that normally affects about 10% of the rock to depths of about 600 meters in the crust. After several million years, the basalts reach a relatively constant $\delta^{18}O$ value of 8–10‰. Lack of further change with time indicates effective cessation of circulation of low-temperature seawater through the rocks. The low-temperature interaction between basalt and seawater causes an *increase* in the $\delta^{18}O$ value of the rock and a corresponding *decrease* in the $\delta^{18}O$ value of the water, the magnitudes of which are determined by the ratios of water to rock, or W/R ratio. Stakes and O'Neil (1982) suggested that these low-temperature systems are water dominated, with W/R > 50/1.

5.5.2 High-Temperature Alteration

Low-temperature alteration represents one half of the picture. Heat flow measurements at midocean ridges show that newly formed crust is rapidly cooled by extensive penetration of vast quantities of water into the rock. The geophysical predictions were confirmed with the discovery of black smokers—ephemeral plumes of superheated water near midocean ridges. Petrographic evidence for high-temperature alteration is provided by greenschist-facies rocks that are found on the seafloor. $\delta^{18}O$ values of these *submarine greenstones*, or *spilites*, are lower (down to ~1.8‰) than those of unaltered MORB, in contrast to what is observed for the vast majority of seafloor basalts, which are altered at much lower temperatures. These low $\delta^{18}O$ values are consistent with high-temperature reactions and isotopic exchange between basalt and seawater, wherein $\Delta^{18}O$(*basalt*–water) values are very small. At 300°C, equilibrium oxygen isotope fractionations between *basalt* and water are less than the 5.7‰ difference between unaltered MORB and ocean water. Therefore, seawater–*basalt* interactions at high temperatures *decrease* the $\delta^{18}O$ value of the *basalt* and *increase* the $\delta^{18}O$ value of the water. These high-temperature systems extend to crustal depths in excess of 1 km and are rock dominated, with W/R ratios ranging from about 50/1 to values as low as about 1/1.

5.5.3 Evidence from Drill Core Material

Preliminary conclusions about the oxygen isotope systematics of seawater–*basalt* interactions in the modern oceanic crust were made on the basis of analyses of samples dredged from the ocean floor. Further work has shown that alteration processes in the crust are extremely heterogeneous and can be seen on millimeter scales in tiny veinlets and on kilometer scales along extensive fracture zones. Our understanding of the vertical distribution of $\delta^{18}O$ values in seafloor basalts was greatly improved by the recovery of deep sea drill cores. Unfortunately, the cores are often incomplete, especially lacking extremely altered material. Accepting these constraints, Alt et al. (1986) measured an oxygen isotope depth profile of samples from DSDP borehole 504B. They presented evidence for low-temperature alteration in the upper 600 meters of core, followed by an abrupt change in mineralogy and isotopic composition that is consistent with high-temperature metamorphism (Fig. 5.8). The sharp division between styles of alteration at this site can be explained by a model whereby hot seawater migrated up through the relatively impermeable dike section and cold seawater descended through the more permeable upper layers.

5.5.4 Evidence from Obducted Material

Obducted equivalents of oceanic crust, the ophiolites, pose no problem of sampling and offer a way of studying the problem of seawater–crust interaction in ancient materials. Because the nature of both high- and low-temperature alteration on the seafloor has not varied through time, the patterns and $\delta^{18}O$ values of seafloor alteration should be the same in ancient ophiolites as they are in the modern crust, and indeed, this is the case (Muehlenbachs, 1986).[5] The ophiolitic sequence from the 3.5-Ga Onverwacht Group has $\delta^{18}O$ values whose ranges are indistinguishable from those of modern ocean samples (Hoffman et al., 1986), and similar patterns are seen in other ancient ophiolites and hydrothermal deposits (e.g., Holmden and Muehlenbachs, 1993). The most complete and best-exposed section of obducted oceanic crust is provided by the Samail ophiolite in Oman. Gregory and Taylor (1981) found that $\delta^{18}O$ variations in this ophiolite are consistent with the seafloor interactions proposed by Muehlenbachs and Clayton (Fig. 5.9). Pillow basalts and the diabase dike complex have $\delta^{18}O$ values that are higher than 5.7‰, and gabbros deeper down in the section have $\delta^{18}O$ values that are lower than 5.7‰. The amount of ^{18}O depletion by high-temperature alteration appears to be balanced by the amount of ^{18}O enrichment by low-temperature alteration in the Samail rocks, providing strong evidence for the buffering capacity of these alteration processes.

---

[5]Ophiolites are not exactly analogous to midocean-ridge spreading centers, but they are all that is available.

Section 5.5: Seawater–Basalt Interactions: Buffering the $\delta^{18}O$ Value of the Ocean 115

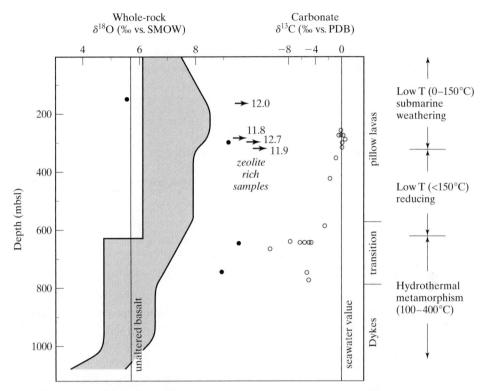

FIGURE 5.8: $\delta^{18}O$ **values of DSDP Hole 504B.** There is an abrupt change from $\delta^{18}O$ values higher than those of typical MORB to values lower than those of MORB at 624 meters below sea level (mbsl), a depth that marks an interface between hot upwelling waters and cold downwelling waters. $\delta^{13}C$ values of carbonate in the rocks (open circles) below 624 m have a mantle signature. Anomalous $\delta^{18}O$ values of basalt are plotted as solid circles or arrows. (After Alt et al., 1986.)

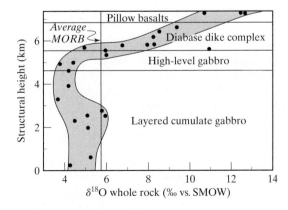

FIGURE 5.9: Vertical section of the Samail ophiolite and measured $\delta^{18}O$ values of different units. The lower, "high-temperature," section has low $\delta^{18}O$ values, while the upper, "low-temperature," section has high $\delta^{18}O$ values, consistent with the model of Muehlenbachs and Clayton. (After Gregory and Taylor, 1981.)

5.6 BUFFERING THE $^{18}O/^{16}O$ RATIO OF OCEAN WATER

The two competing types of seawater–basalt interactions known to occur in the oceanic crust are not coincidental and are natural consequences of basaltic magmatism occurring on the seafloor. These interactions clearly bear on the oxygen isotope budget of the ocean, and there is strong evidence that, in concert with continental weathering, they buffer the $^{18}O/^{16}O$ ratio of the ocean to a steady-state value.

5.6.1 Summing the Processes Affecting the $^{18}O/^{16}O$ Ratio of Seawater

There are five natural processes that affect the oxygen isotope composition of seawater:

1. Continental weathering—lowers $\delta^{18}O_{ocean}$
2. Seafloor *weathering* (a misnomer meaning low-temperature alteration)—lowers $\delta^{18}O_{ocean}$
3. Hydrothermal alteration of ocean basalt (high-temperature alteration)—raises $\delta^{18}O_{ocean}$
4. Water cycling associated with subduction and magmatism—lowers and raises $\delta^{18}O_{ocean}$
5. Glaciation—raises $\delta^{18}O_{ocean}$

Glaciation is different in kind from the other processes: The removal of light water vapor to continental ice sheets results in unidirectional (positive) changes of 1–2 per mil in the $\delta^{18}O$ value of seawater and occurs on a relatively short time scale of 10^3 years.

The first four processes enumerated constitute the backbone of the buffering hypothesis. These different processes operate in such a way that $\delta^{18}O$ of the oceans is maintained at a value near 0‰. In fact, the buffering model is quite robust because the $\delta^{18}O$ value of MORB is fixed and the temperatures of both low- and high-temperature alteration are also fixed within narrow limits and cannot change with time. If the $\delta^{18}O$ value of seawater could be shifted to either higher or lower values by some unknown process, basalt–seawater interactions would force it back to a value near 0‰ in ~250 million years. The buffering rate is therefore on the order of ~100 Ma.

5.6.2 Model Calculations

Gregory (1991) presented the following material-balance equation relating the rate of change $\delta^{18}O$ of seawater to exchange processes between the lithosphere and hydrosphere:

$$d\delta W/dt = \Sigma k_i(\delta W + \Delta_i - \delta_i^o). \tag{5.1}$$

In this equation, δW is the instantaneous $\delta^{18}O$ value of seawater, δ_i^o are the initial $\delta^{18}O$ values of the different rock reservoirs (i), Δ_i are bulk fractionation factors between the i reservoirs and water, and k_i are the rate constants in units of $1/t$.

From the definition of Δ_i, it follows that $\delta_i^o - \Delta_i$ is the $\delta^{18}O$ value of seawater (δW) when steady-state conditions are reached with the ith reservoir and

$$\delta W_{steady-state} = [\Sigma k_i(\delta_i^o - \Delta_i)]/\Sigma k_i. \tag{5.2}$$

In both simple and more refined models developed to address the fluxes of ^{18}O in and out of the ocean, relative rates of continental weathering and seafloor spreading are key parameters. There are obvious constraints on these rates, one being that the rate of continental weathering must be less than that of continental growth. Rates of seafloor spreading have very little effect on the overall buffering process, as long as the average global spreading rate is greater than 1 km²/yr. Present global spreading rates are 3–4 km²/yr and were probably higher in the past. If seafloor spreading rates were doubled, the time required for the $\delta^{18}O$ of the ocean to return to zero from a perturbed value would be on the order of 60 million years. $\delta^{18}O$ values of the oceans could change only if seafloor spreading effectively ceased and continental weathering continued. The rate of change would be on the scale of billions of years, however. If seafloor spreading really did stop, there would be a reduction in global tectonic activity as well, and the effect of continental weathering would diminish accordingly.

5.6.3 Unresolved Controversy

From the very onset of stable isotope geochemistry, there have been strong proponents for and against the existence of major changes in oxygen isotope ratios of the ocean over geologic time. At present, the arguments are centered on evidence for changes in $\delta^{18}O$ of only about 3–6 per mil over relatively short geologic periods of tens of millions of years. Most of the strong evidence comes from oxygen isotope variations in carefully selected brachiopod shells and inorganic cements deposited in Devonian times. (See Chapter 6.) The arguments are cogent, and yet model calculations indicate that such changes are unlikely. Biogenic phosphate is much more resistant to diagenetic changes of oxygen isotope ratios, and isotopic analyses of conodonts and fossil fish remains may provide penetrating evidence on this important question, one way or the other.

PROBLEM SET

1. Extensive evaporation takes place in the tropics, and yet the isotopic composition of the tropical oceans is relatively constant. Why?
2. Denitrification is the process whereby NO_3^- and NO_2^- are converted to N_2O and, ultimately, to N_2 gas by anaerobic bacteria. N_2 gas strongly favors the light isotope of nitrogen. If denitrification is most intense in the upper water column, and the resulting N_2 gas is lost to the atmosphere, what should the $\delta^{15}N$ profile of dissolved NO_3^- look like?
3. Figure 5.1 shows a schematic of the effects of various processes on the $\delta^{18}O$ and salinity of ocean waters. Explain why the slopes of each curve are different, Justify your explanation.
4. Why do the Pacific Deep Waters fall below the salinity/$\delta^{18}O$ line for surface Pacific waters?
5. Why do the Antarctic bottom waters plot at higher salinity than those defined by the trend for Atlantic surface waters?
6. The concentration and δ values for both oxygen and carbon in depth profiles of the ocean are anti-correlated (Fig. 5.5). Why?
7. Discuss the oxygen isotope composition of the ocean through time, indicating how it might have changed and what evidence is used both for and against your hypothesis. Include the following topics in your discussion: measured $\delta^{18}O$ values of ancient sediments, the ocean buffering model of Muehlenbachs and Clayton, and ideas about early ocean formation.

ANSWERS

1. Although there is significant evaporation, most of the water is returned to the ocean by precipitation, so that the entire evaporation–precipitation cycle is a closed system.
2. See Chapter 9 for further discussion.

3. Four trajectories are shown in Fig. 5.1: Equatorial precipitation, trade winds evaporation, polar contribution and freezing. 1. The equatorial precipitation will lower salinity, but only slightly, as the dilution is countered by precipitation. The curve is relatively flat because the source of water is locally derived evaporation of the ocean, so that the $\delta^{18}O$ value of precipitation is similar to that of evaporation; 2. Trade wind evaporation will raise both $\delta^{18}O$ and salinity values because the vapor that is removed has a lower $\delta^{18}O$ value than its source (the ocean) and a portion of the vapor is removed by transport to continental settings. 3. Polar freshwater contribution has a strong slope, because the $\delta^{18}O$ value of the infiltrating freshwater has very low $\delta^{18}O$ values. 4. Freezing has a large effect on the salinity because salt is not incorporated into ice. The $\delta^{18}O$ changes little because the fractionation between ice and water is small. The negative slope in Fig. 5.1 is explained because $\Delta^{18}O_{ice-water}$ is positive.

4. The Pacific deep waters do not originate at the surface of the Pacific Ocean. Instead, they are a mixture of the North Atlantic Deep Waters and the Antarctic Bottom Waters. See Fig. 5.3 and accompanying discussion.

5. Antarctic waters freeze. Ice formation concentrates dissolved salt in the remaining water, driving up salinity. However, the isotopic fractionation between water and ice is small, so that the isotopic composition of the remaining water changes only slightly.

6. The $\delta^{18}O$ values increase with depth because oxidation of organic matter preferentially removes the heavy isotope. In doing so, the concentration of oxygen decreases with depth. In contrast, $\delta^{13}C$ values decrease with depth because there is a removal of light isotopes at the surface by incorporation into organic material and a rerelease at depth by oxidation of the light organic material, the latter causing an increase in $\delta^{13}C$ values. The concentration of dissolved carbon increases with depth because of the biological pump, discussed further in Chapter 7.

REFERENCES

Alt, J. C., Muehlenbachs, K., and Honnorez, J. (1986). An oxygen isotopic profile through the upper kilometer of the oceanic crust, DSDP Hole 504B. *Earth and Planetary Science Letters* **80**, 217–229.

Anderson, T. F., and Arthur, M. A. (1983). Stable isotopes of oxygen and carbon and their application to sedimentologic and paleoenvironmental problems. In *Stable Isotopes in Sedimentary Geology*, Vol. 10 (ed. M. A. Arthur, T. F. Anderson, I. R. Kaplan, J. Veizer, and L. S. Land), pp. 1–151. Columbia, SC: SEPM Short Course.

Craig, H., and Gordon, L. I. (1965). Deuterium and oxygen-18 variations in the ocean and the marine atmosphere. *Stable Isotopes in Oceanographic Studies and Paleotemperatures*. Pisa, Italy: Consiglio Nazionale delle Ricerche, 9–130.

Epstein, S., and Mayeda, T. K. (1953). Variation of ^{18}O content of waters from natural sources. *Geochimica et Cosmochimica Acta* **4**, 213–224.

Friedman, I., Schoen, B. S., and Harris, J. L. (1961). The deuterium concentration in Arctic sea ice. *Journal of Geophysical Research* **66**, 1861–1864.

Garlick, G. D., and Dymond, J. R. (1970). Oxygen isotope exchange between volcanic materials and ocean water. *Geological Society of America Bulletin* **81**, 2137–2141.

Gregory, R. T. (1991). Oxygen isotope history of seawater revisited; timescales for boundary event changes in the oxygen isotope composition of seawater. In *Stable Isotope Geochemistry: A Tribute to Samuel Epstein*, Vol. 3 (ed. H. P. Taylor, Jr., J. R. O'Neil, and I. R. Kaplan), pp. 65–76. San Antonio: Geochemical Society.

Gregory, R. T., and Taylor, H. P., Jr. (1981). An oxygen isotope profile in a section of Cretaceous oceanic crust, Samail ophiolite, Oman: Evidence for ^{18}O buffering of the oceans by deep (>5 km) seawater–hydrothermal circulation at mid-ocean ridges. *Journal of Geophysical Research* **86**, 2737–2755.

Hoffman, S. E., Wilson, M., and Stakes, D. S. (1986). Inferred oxygen isotope profile of Archaean oceanic crust, Onverwacht Group, South Africa. *Nature* **321**, 55–58.

Holmden, C., and Muehlenbachs, K. (1993). The $^{18}O/^{16}O$ ratio of 2-billion-year-old seawater inferred from ancient oceanic crust. *Science* **259**, 1733–1736.

Knauth, L. P., and Roberts, S. K. (1991). The hydrogen and oxygen isotope history of the Silurian–Permian hydrosphere as determined by direct measurement of fossil water. In *Stable Isotope Geochemistry: A Tribute to Samuel Epstein*, Vol. 3 (ed. H. P. Taylor, Jr., J. R. O'Neil, and I. R. Kaplan), pp. 91–104. University Park, PA: Geochemical Society.

Kroopnick, P., and Craig, H. (1972). Atmospheric oxygen: Isotopic composition and solubility fractionation. *Science* **175**, 54–55.

Kroopnick, P., Weiss, R. F., and Craig, H. (1972). Total CO_2, ^{13}C, and dissolved oxygen—^{18}O at GEOSECS II in the North Atlantic. *Earth and Planetary Science Letters* **16**, 103–110.

Lécuyer, C., Gruau, G., Frueh-Green, G. L., and Picard, C. (1996). Hydrogen isotope composition of early Proterozoic seawater. *Geology* **24**, 291–294.

Muehlenbachs, K. (1986). Alteration of the oceanic crust and the ^{18}O history of seawater. In *Stable Isotopes in High Temperature Geological Processes*, Vol. 16 (ed. J. W. Valley, H. P. Taylor, Jr., and J. R. O'Neil), pp. 425–444. Chelsea, MI: Mineralogical Society of America.

Muehlenbachs, K., and Clayton, R. N. (1972). Oxygen isotope geochemistry of submarine greenstones. *Canadian Journal of Earth Sciences* **9**, 471–478.

Muehlenbachs, K., and Clayton, R. N. (1976). Oxygen isotope composition of the oceanic crust and its bearing on seawater. *Journal of Geophysical Research.* **81**, 4365–4369.

O'Neil, J. R. (1968). Hydrogen and oxygen isotope fractionation between ice and water. *Journal of Physical Chemistry* **72**, 3683–3684.

O'Neil, J. R., and Truesdell, A. H. (1991). Oxygen isotope fractionation studies of solute–water interactions. In *Stable Isotope Geochemistry: A Tribute to Samuel Epstein*, Vol. 3 (ed. H. P. J. Taylor, J. R. O'Neil, and I. R. Kaplan), pp. 17–25. San Antonio: Lancaster Press, Inc.

Perry, E. C., Jr., and Tan, F. C. (1972). Significance of oxygen and carbon isotope variations in Early Precambrian cherts and carbonate rocks of Southern Africa. *Geological Society of America Bulletin* **83**, 647–664.

Railsback, L. B., Anderson, T. F., Ackerly, S. C., and Cisne, J. L. (1989). Paleoceanographic modeling of temperature–salinity profiles from stable isotopic data. *Paleoceanography* **4**, 585–591.

Sheppard, S. M. F., and Epstein S. (1970). D/H and $^{18}O/^{16}O$ ratios of minerals of possible mantle or lower crustal origin. *Earth and Planetary Science Letters* **9**, 232–239.

Silverman, S. R. (1951). The isotope geology of oxygen. *Geochimica et Cosmochimica Acta* **2**, 26–42.

Stakes, D. S., and O'Neil, J. R. (1982). Mineralogy and stable isotope geochemistry of hydrothermally altered oceanic rocks. *Earth and Planetary Science Letters* **57**, 285–304.

Yui, T. F., Yeh, H. W., and Lee, C. W. (1990). A stable isotope study of serpentinization in the Fengtien Ophiolite, Taiwan. *Geochimica et Cosmochimica Acta* **54**, 1417–1426.

CHAPTER 6

BIOGENIC CARBONATES: OXYGEN

Carbonates are one of the most studied phases in stable isotope geochemistry. They are found at all but the oldest chronological intervals, and the information from oxygen and carbon isotopes can be used to infer paleotemperatures, paleoproductivity, circulation patterns, water depth, and more. Oxygen isotope ratios of marine carbonates most often provide information about water temperature and ice volume, while carbon isotopes provide information about biological productivity and other biological conditions. Because of the vastness of the field and the different information obtained from each isotope, they are presented in separate chapters (6 and 7), although some overlap is unavoidable.

6.1 INTRODUCTION

The potential of biogenic carbonates as an indicator of paleoclimate played a pivotal role in the development of stable isotope geochemistry. The discipline as we know it today was spawned by Harold Urey, who recognized the possibility of determining temperatures of ancient oceans by the measurement of preserved oxygen isotope ratios of biogenic carbonate deposited by marine organisms. But for meaningful application to paleoceanography, these temperatures had to be determined to within ±0.5°C, which in turn required increasing the precision of mass spectrometric analyses of $^{18}O/^{16}O$ ratios by an order of magnitude. Accomplishing this feat of engineering not only made possible the development of the oxygen isotope paleotemperature scale, but opened many doors of discovery by allowing precise measurements to be made of small variations in stable isotope ratios of several light elements.

On his 1947 lecture tour sponsored annually by the Royal Society of London, Harold Urey was speaking about the physical fractionation of stable isotopes between ideal gases and simple aqueous solutions. He finished his lecture at ETH Zürich and accepted a question from Paul Niggli, the renowned Alpine geologist. Niggli asked whether the fractionation between carbonates and water might be large enough and sensitive enough to temperature variations so that they could be used for reconstructing ancient marine temperatures. The story goes that Urey thought a second, said that he didn't know, but that it seemed reasonable. Later calculations led him to believe that there was promise in this avenue of research. But numerous hurdles presented themselves before he would be able to apply the paleoclimatic technique. Specifically, Urey needed (1) a more precise mass spectrometer, (2) a method of reproducibly converting carbonates to a gaseous phase that he could analyze in the mass spectrometer, and (3) a means of quantifying the fractionation between carbonate and water as a function of temperature. Putting together an incredible team of young scientists, including Samuel Epstein, Charles McKinney, John McCrea, Harold Lowenstam, and Harmon Craig, he was able to work out the details in record time and clear the hurdles necessary to bring the paleotemperature technique to fruition.

The basic idea for the carbonate paleothermometer is as follows: The fractionation between calcite and water is a function of temperature, so the difference in the $\delta^{18}O$ values of calcite and water can be used to determine the temperatures of the ocean at the time the carbonate formed. In order for this method to work, however, a number of potential problems need to be addressed. In addition to the analytical problems mentioned in the previous paragraph, there must be an understanding of whether the calcite formed in isotopic equilibrium or whether the biogenic "vital effect" (discussed in Section 6.4.2) caused the calcite to precipitate out of isotopic equilibrium. Next, whether the $\delta^{18}O$ value of the calcite has or has not been preserved since its formation must be addressed. Finally, the $\delta^{18}O$ value of the ocean at the time the calcite formed must be estimated; in almost all cases, the $\delta^{18}O$ value of the ocean is not measured directly. All of these problems, noted by Urey in 1948, are addressed in the sections that follow.

6.2 THE PHOSPHORIC ACID METHOD

6.2.1 A Major Breakthrough

The quest for reliable methods to pretreat shells and perform precise oxygen isotope analyses of their constituent carbonates was described by Urey as "the toughest chemical problem I ever faced." The development of the phosphoric acid method of carbonate analysis by his doctoral student John McCrea (1950) was a seminal chapter in the history of stable isotope geochemistry. This method, modified only slightly in the ensuing years, involves converting carbonate to CO_2 gas by reaction with phosphoric acid. The CO_2 is then analyzed in the mass spectrometer. The technique remains the protocol for analyzing carbonates in stable isotope laboratories all over the world, and McCrea's methods paper is almost certainly the most quoted work in the stable isotope literature. Carbonate analysis today is routine, but there were major obstacles to overcoming this chemical challenge in 1949.

McCrea initially tried liberating CO_2 from carbonates by thermal decomposition:

$$CaCO_3 + \text{heat} \longrightarrow CaO + CO_2. \tag{6.1}$$

Despite good chemical yields (i.e., the reaction went to completion), the extracted CO_2 had scattered $\delta^{18}O$ values, far outside the required reproducibility of ±0.1‰, and the approach was abandoned. We know now that, under proper conditions with good temperature control, it is possible to obtain reliable isotopic data by means of thermal decomposition (Sharp et al., 2003). Nonetheless, such methods are rarely used routinely.

McCrea next turned to acid decomposition:

$$2H^+ + CaCO_3 \longrightarrow Ca^{2+} + H_2O + CO_2. \tag{6.2}$$

The procedure involves reacting the carbonate with an acid in an evacuated vessel and then purifying, collecting, and finally analyzing the CO_2 gas as a measure of the $\delta^{18}O$ value of the original carbonate. The only common acids whose vapor pressures of water and other compounds are low enough to use in a vacuum system are concentrated H_2SO_4 and H_3PO_4. It soon became apparent that 100% H_3PO_4 was the acid of choice. The early workers were concerned about contamination from organic matter present in commercial acids, because fragments of organic molecules made in the source of the mass spectrometer could have masses in the 43–47 range that would interfere with determinations of 46/44 and 45/44 ratios. Thus, the acid recipe (see Box 6.1) finally adopted by the Chicago group ensured that no contamination arose from the specially prepared acid, a light green, very viscous, and unstable liquid that can spontaneously solidify at any time. Most likely, some of the procedures developed by the early workers are unnecessary or have become obsolete by the commercial availability of pure H_3PO_4, but most laboratories take the approach "If it ain't broken, don't fix it," so the acid ritual survives. Some laboratories have found that commercially available 85% phosphoric acid can be vacuum distilled to obtain a 100% phosphoric acid, which apparently works just fine.

The phosphoric acid and calcite are reacted at a constant temperature. Most calcium in solution is present as the calcium phosphate ion pairs $(CaPO_4)^-$ and $CaHPO_4$. The water that is produced is taken up by excess P_2O_5 to form H_3PO_4. During extraction of the CO_2, a very

---

BOX 6.1 Recipe for preparation of 100% H_3PO_4 for stable isotope analysis of carbonates

1. Pour 2.5 L (one bottle) of commercial 85% phosphoric acid into a large Pyrex® beaker that is placed on a hot plate in a fume hood.
2. Slowly stir in about 1.4 kg of analytical grade P_2O_5. The ensuing reaction is exothermic and should be done with care.
3. Add a spatula-tip quantity of CrO_3. The solution turns pale yellow.
4. Raise the temperature to 200°C slowly and heat for about 7 hours. The solution turns green.
5. Raise the temperature to 220°C and heat 4–5 hours.
6. Stop heating when the density at 25°C = 1.9, the density of 100% H_3PO_4.
7. When the acid cools, stir in 3mL of H_2O_2.
8. Store the acid in brown glass bottles with rubber seals. 500 mL is a convenient size.
9. Age the H_3PO_4 for about one month before using.

Notes

1. The density must be more than 1.8. $[H_3PO_4]$ > 100% is desirable, as excess P_2O_5 readily takes up water. Solutions with $[H_3PO_4]$ > 103%, however, are more prone to crystallize, and they inhibit the diffusion of CO_2 out of the acid into the headspace.
2. The acid is highly corrosive and will destroy the markings on glass thermometers. Use a glass sleeve or some other means of protecting these markings.
3. The acid turns green as a result of the reduction of Cr(VI) to Cr(III) by organic matter present in the commercial acid.
4. H_2O_2, a reducing agent in this case, is added to reduce residual Cr(VI) to Cr(III).
5. Avoid exposing the acid to air for extended periods, as it is hygroscopic.
6. No one understands what happens during the aging period, but aging seems to be necessary. At least, it can't hurt!

small amount of some other volatile compounds, mostly water vapor, is also liberated and is separated from the CO_2 by the judicious use of cryogenic traps.

6.2.2 Acid Fractionation Factors

In both thermal and acid decompositions of carbonates, the liberated CO_2 contains *all* the carbon, but only two-thirds of the oxygen, in the carbonate (equation (6.2)). As a consequence, the $\delta^{13}C$ value of the evolved CO_2 and that of the parent carbonates are identical, but the oxygen isotope ratios are different due to a fractionation between the CO_2 gas and oxygen from the carbonate that remains dissolved in the acid. $^{18}O/^{16}O$ ratios are always higher in liberated CO_2 than in the original carbonate (a phenomenon related to the stronger C=O double bonds in CO_2 gas). The magnitude of the oxygen isotope fractionation is probably controlled by both kinetic and equilibrium effects. A so-called *acid fractionation factor* must be applied to the CO_2 analysis to obtain the oxygen isotope ratio of the carbonate. The oxygen isotope fractionation between evolved CO_2 and a particular carbonate is given by

$$\alpha_{CO_2-\text{carbonate}} = \frac{1000 + \delta^{18}O_{CO_2}}{1000 + \delta^{18}O_{\text{carbonate}}} \qquad (6.3)$$

and is constant at a given temperature of reaction. The α value becomes smaller with increasing temperature. Because of the temperature effect, H_3PO_4–carbonate reactions must be run at a constant temperature. For many years, these reactions were carried out at 25°C, but temperatures as high as 90°C are commonly used today, both to ensure that reactions are complete in the relatively short times used in automated systems and to reduce the solubility of CO_2 in the acid, allowing for smaller samples to be analyzed. It makes no difference what temperature is used, because the method is calibrated to international reference standards reacted at the same temperature as the samples. As long as temperatures are kept constant and an α value is determined on the basis of accepted IAEA standards, measured (and corrected) $\delta^{18}O$ values of samples will be consistent with IAEA-accepted scales (SMOW or PDB). (See problem 7 for sample calculation).

For many years, the values of acid fractionation factors were unknown, because the $\delta^{18}O$ values of the carbonate themselves were unknown. All that could be measured was the $\delta^{18}O$ value of the evolved CO_2 gas. Later, Sharma and Clayton (1965) and others finally measured the $\delta^{18}O$ value of the *total* carbonate by the method of fluorination.[1] Once the baseline $\delta^{18}O_{\text{carbonate}}$ value was determined, it became a trivial exercise to determine the α value at any temperature.[2] Fractionation factors have values on the order of 1.01000 (10 per mil) and are different for different carbonates (Table 6.1). They can even vary among different specimens of the same mineral. Complicating matters, measurements of these fractionations taken in different laboratories do not always agree. Fortunately, most problems are eliminated when laboratories calibrate their analyses to internationally agreed upon standards. Complications appear only in establishing absolute $^{18}O/^{16}O$ ratios of international reference standards or in relating analyses of carbonates to those of waters, silicates, and oxides, phases analyzed using a different procedure.

---

[1] The fluorination reaction is approximated by $2CaCO_3 + 6F_2 \longrightarrow 2CaF_2 + 2CF_4 + 3O_2$. All of the O_2 is measured for its $\delta^{18}O$ value—hence the "total carbonate" value.

[2] With a known $\delta^{18}O$ value of the carbonate (from fluorination) and a measured $\delta^{18}O$ value of evolved CO_2 gas from phosphoric acid digestion, the α value could be determined by plugging these values into equation (6.3).

TABLE 6.1: Acid fractionation factors.

Mineral	Temperature (°C)	Fractionation factor α	Reference
Calcite, $CaCO_3$	25	1.01025	1
	25	1.01049	4
"sealed vessel"	50	1.009311	2
"common acid bath"	50	1.009002	2
aragonite, $CaCO_3$	25	1.01034	1
	25	1.01107	4
dolomite, $CaMg(CO_3)_2$	25	1.01109	1
	50–100	$4.23 + 6.65 \times 10^5/T^2$	3
siderite, $FeCO_3$	50–150	$3.85 + 6.84 \times 10^5/T^2$	3
	25	1.01017	6
	50	1.01016	6
ankerite, $CaFe(CO_3)_2$	50–150	$4.15 + 6.68 \times 10^5/T^2$	3
magnesite, $MgCO_3$	50	1.01160	5
strontianite, $SrCO_3$	25	1.01049	1
witherite, $BaCO_3$	25	1.01097	1
	25	1.01063	4
smithsonite, $ZnCO_3$	25	1.01130	1
otavite, $CdCO_3$	25	1.01145	1
	25	1.01124–1.01369	4
rhodocrosite, $MnCO_3$	25	1.01012	1
cerussite, $PbCO_3$	25	1.01013	1

1, Sharma and Clayton, 1965; 2, Swart et al., 1991; 3, Rosenbaum and Sheppard, 1986; 4, Kim and O'Neil, 1997; 5, Perry and Tan, 1972; 6, Carothers et al., 1988.

6.2.3 Applicability

The H_3PO_4 method is one of the most robust used in stable isotope geochemistry and is applicable to the analysis of all carbonates. Some carbonate minerals, like magnesite and smithsonite, *require* relatively high reaction temperatures because they react so slowly. Using even the most basic extraction line, one can measure $\delta^{18}O$ and $\delta^{13}C$ values of samples weighing about 1 mg or more to a precision of better than 0.1 and 0.05‰, respectively. With sophisticated extraction systems connected directly to the mass spectrometer (*online systems*), samples as small as tens to hundreds of micrograms can be analyzed routinely to the same precision. A few authors have reported that grain size and carbonate–acid ratio can significantly influence the isotopic analyses (Wachter and Hayes, 1985; Barrera and Savin, 1987; Al-Aasm et al., 1990; Swart et al., 1991), but these effects are not seen in all laboratories and are effectively eliminated when the reactions are carried out at relatively high temperatures. Contamination by organic matter, chlorine- and sulfur-bearing compounds, or inclusions of other carbonates pose more serious problems (Charef and Sheppard, 1984). Organic matter is a particular concern for modern samples and should be removed before analysis. This can be done a number of ways, including roasting the sample in a stream of helium, treating it with mild oxidizing agents, and exposing it to an oxygen plasma. The small amount of organic matter present in fossil carbonate generally has no effect on the measured $\delta^{13}C$ and $\delta^{18}O$ values of the carbonate, but the samples are routinely sent through a pretreatment step in any case. One can always analyze treated and untreated splits of a given sample to ascertain whether pretreatment is necessary. Recently,

6.3 THE OXYGEN ISOTOPE PALEOTEMPERATURE SCALE

Armed with an improved isotope ratio mass spectrometer and the phosphoric acid method of carbonate analysis, the Chicago group faced the challenge of calibrating a temperature scale based on the temperature-sensitive oxygen isotope fractionation between biogenic carbonate and ocean water. They had neither any established reference standards at that time nor knowledge of the fractionation factors for the analytical methods they employed: acid–carbonate reaction for carbonates (McCrea, 1950), and CO_2–H_2O equilibration for waters (Cohn and Urey, 1938). Undaunted, Samuel Epstein addressed the problem by making a simple empirical calibration. He collected shells and water from cold- and warm-water environments. He also cultured shells in laboratory aquariums. He then measured the difference between the oxygen isotope composition of CO_2 liberated from the carbonate by phosphoric acid at 25°C and that of CO_2 equilibrated with the ambient H_2O at 25°C. Fortuitously, this difference for normal marine calcite and ocean water is very small at 25°C. The point is illustrated in Fig. 6.1, which use a $\delta^{18}O$ value of water = 0‰ vs SMOW as an example.[3] The $\alpha(CO_2$–$H_2O)$ value at 25°C is 1.0412 (O'Neil et al., 1975), corresponding to a $\delta^{18}O$ value of CO_2 equilibrated with water at 25°C of 41.2‰ on the SMOW scale. The standard PDB calcite is 30.91‰ heavier than SMOW (Coplen et al., 1983). Finally, the $\alpha(CO_2$–calcite) value for acid fractionation is 1.01025 at 25°C (corresponding to a $\Delta^{18}O_{CO_2-cc}$ of 10.57 for $\delta^{18}O_{cc}$ = 30.91 or α_{CO_2-SMOW} = 1.03091 × 1.01025 = 1.04148), so that the CO_2 liberated from PDB calcite has a $\delta^{18}O$ value of 41.48‰. In other words, the difference between the $\delta^{18}O$ value of CO_2 equilibrated with SMOW and that liberated by the phosphoric acid digestion of PDB is only 0.28‰.[4] Craig (1965) reported a value of 0.22‰ for this difference. The discrepancy is due to an earlier α(PDB–SMOW) = 1.03086, compared with the now-accepted value of 1.03091.

To calibrate the temperature dependence of the isotopic fractionation between biogenic carbonate and water, Epstein made analyses of shell material from attached or sedentary organisms, including mussels, brachiopods, red and black abalone, and limpets, that were living in the cool waters off Puget Sound (lowest T = 7.4°C), in the temperate waters of Monterey Bay, and in warm waters along the coast of Baja California (highest T = 20°C). Two higher temperature calibration points at 29 and 31°C were obtained from analyses of calcite regenerated by a cultured snail and a bivalve (*Pinna* sp.) to repair holes that were drilled into their shells.[5] In addition to the ambient waters, oxygen isotope analyses of the natural and cultured carbonate samples provided data adequate to establish an empirical calibration for

---

[3]The difference between the $\delta^{18}O$ value of CO_2 and H_2O varies slightly in relation to the $\delta^{18}O$ value of the phases relative to the standard, although the α value does not. For example, the $\alpha(CO_2$–$H_2O)$ value at 25°C is 1.0412. If the $\delta^{18}O$ value of the water is 0, then the coexisting CO_2 is 41.2‰ heavier. However, if the $\delta^{18}O$ value of the water is 20‰, for instance, then the $\delta^{18}O$ value of the CO_2 is 62.02, 42.0‰ heavier. The α value is the same in both cases.

[4]Interestingly, the temperatures obtained from the Pee Dee Belemnite (PDB standard) correspond to a temperature of 15.8°C, assuming a $\delta^{18}O$ value of the ocean of 0 on the SMOW scale.

[5]These warm-water organisms were cultured in a tank in Bermuda.

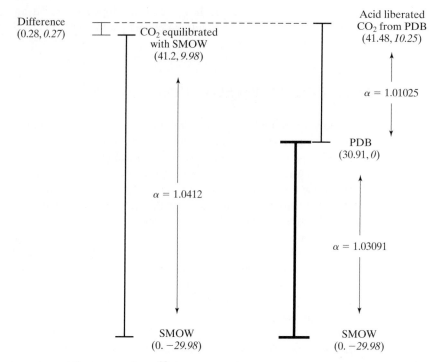

FIGURE 6.1: Difference in the $\delta^{18}O$ value of acid-liberated CO_2 from Pee Dee Belemnite (PDB) and that equilibrated with SMOW. Delta values are shown on the SMOW scale (normal) and PDB scale (italic). In either case, the difference between the CO_2 gases measured in the mass spectrometer is only on the order of 0.2‰. The water–CO_2 equilibration and CO_2 released from carbonates are constant. The fractionation between carbonate and water (shown as a thick line) is a function of the carbonate formation temperature. The α value of 1.03091 corresponds to an equilibration temperature of 15.8°C.

the fractionation between biogenic calcite and water over the range of temperatures found in modern oceans. After publication of the paleotemperature equation in 1951, it was recognized that the helium roasting procedure used to remove organic matter from the shells had introduced extraneous oxygen into the system. Epstein et al. (1953) corrected the procedure and published the following revised equation, which became the classic paleotemperature equation:

$$t(°C) = 16.9 - 4.2(\delta_c - \delta_w) + 0.13(\delta_c - \delta_w)^2. \tag{6.4}$$

In this equation, δ_c is the $\delta^{18}O$ value of CO_2 liberated from reaction between the carbonate and phosphoric acid at 25°C, and δ_w is the $\delta^{18}O$ value of CO_2 equilibrated with water at 25°C. Epstein's data and fit are shown in Fig. 6.2. There is no theoretical basis for the form of equation (6.4); it is simply a best fit of the data to a second-order polynomial. Over the 0–30°C range of modern ocean waters, $\delta^{18}O$ (PDB) values of marine carbonates range from about +3 to −3‰. The first measurements of paleotemperatures using this method are shown in Fig. 6.3.

Both δ_c and δ_w are the values relative to the same working standard of the mass spectrometer. CO_2 from PDB was the working standard used in the early days at the University of Chicago. Note that water analyses normalized to the SMOW scale and carbonates

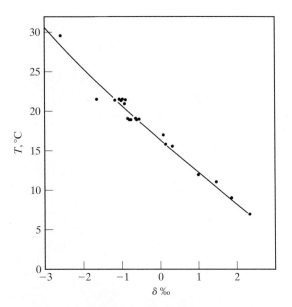

FIGURE 6.2: Plot of Epstein et al.'s (1953) corrected data set for determining the fractionation factor for carbonate–water. The x-axis (δ‰) refers to the ($\delta_c - \delta_w$) of equation (6.4), while $T(°C)$ is the measured water temperature. Reprinted with permission from the Geological Society of America.

normalized to the PDB scale *cannot be used* in the classic paleotemperature equation, but can be rewritten in the following form, appropriate for δ_c and δ_w values on the PDB and SMOW scales, respectively:

$$t(°C) = 15.75 - 4.3(\delta^{18}O_{c-PDB} - \delta^{18}O_{w-SMOW}) + 0.14(\delta^{18}O_{c-PDB} - \delta^{18}O_{w-SMOW})^2 \quad (6.5)$$

Keep in mind that δ_c on the PDB scale is about 30 per mil lower than δ_c on the SMOW scale. In the first case, equation (6.4), one uses the isotopic composition of CO_2 released from the carbonate by acid decomposition; in the second case, equation (6.5), the isotopic composition of total oxygen in the solid carbonate is reported. To emphasize this important point that is frequently misunderstood, recall that $\delta^{18}O$ of the PDB carbonate standard is 0‰ on the PDB scale and 30.91‰ on the SMOW scale (equation (2.21)). This difference seldom poses a problem in the practical world, because the PDB scale for oxygen isotope analyses is restricted to analyses of carbonates only. The SMOW scale is used to report oxygen isotope analyses of every other substance, including water.

O'Neil et al. (1969) measured the equilibrium oxygen isotope fractionation between inorganic calcite and water from 0 to 500°C using precipitation methods at low temperatures and recrystallization methods at high temperatures. The data were then fit to the following equation,[6] whose form has a basis in statistical mechanics:

$$1000 \ln \alpha_{\text{calcite-water}} = \frac{2.78 \times 10^6}{T^2} - 2.89. \quad (6.6)$$

---

[6]The additive term originally published was -3.39, which was later corrected to -2.89 after it was recognized that an error was made in a mass spectrometer correction factor.

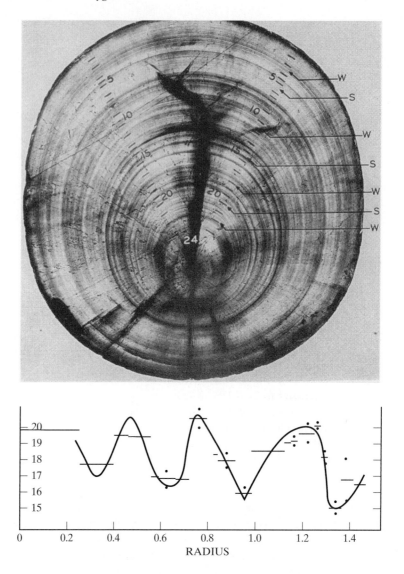

FIGURE 6.3: Cross section (top) of 150-million-year-old belemnite showing growth rings. The graph (bottom) illustrates isotopic temperatures obtained from oxygen isotope analyses of concentric layers of the skeleton. The regular variations reflect seasonal variations in growth temperature and indicate that the animal was born in the summer and died four years later in the spring. Reprinted from Urey et al. (1951) with permission from the Geological Society of America.

In equation (6.6), T is in kelvins. As good fortune dictates, equations (6.4) and (6.6) are indistinguishable from one another within the limits of experimental uncertainty, lending credence to the idea that there is no significant difference in the isotopic properties of biogenic and inorganic calcite. Recall that the $\delta^{18}O$ values for calcite and water in equation (6.6) must be on the same scale, either SMOW or PDB. Kim and O'Neil (1997) revised both the α factor for calcite–phosphoric acid fractionation (Table 6.1) and the calcite–water fractionation factor. The normalized paleotemperature equations (6.5) and (6.6) are compared in Fig. 6.4.

Overall, the *precision* of an oxygen isotope temperature can be as high as ±0.5°C, but given the uncertainties explained in the following sections, the *accuracy* is probably no better than ±3°C for all but the most recent samples. Relative temperature differences are nonetheless highly accurate, and reliable estimates of temperature *change* are desired in studies of palaeoclimate.

Many organisms deposit aragonite in their shells, so it is important to know whether different polymorphs of $CaCO_3$ have significantly different oxygen isotope properties. From limited experimental data, Tarutani et al. (1969) determined an inorganic aragonite–calcite fractionation of 0.6‰ at 25°C. While this fractionation is relatively small, as is expected for two such similar minerals, it is significant in terms of paleotemperature determinations. That is, the paleotemperature equation developed for calcite is not appropriate for shells made of aragonite. Grossman and Ku (1986) determined the aragonite–water fractionation between living organisms and seawater over a temperature range of 4–20°C (Fig. 6.4). Their equation, presented in terms of the $\delta^{18}O$ of aragonite on the PDB scale and the $\delta^{18}O$ of water on the SMOW scale, is

$$T(°C) = 19.7 - 4.34(\delta^{18}O_{aragonite} - \delta^{18}O_{water}) \tag{6.7}$$

Combining the equations of Epstein et al. and Grossman and Ku yields an aragonite–calcite fractionation that is somewhat larger than the inorganic fractionation determined by Tarutani et al. at 25°C. Note that the temperature coefficients of the aragonite–water and calcite–water fractionations are so similar that the possibility of using the aragonite–calcite fractionation as a thermometer is excluded.

Tarutani et al. (1969) precipitated Mg calcites in the laboratory and reported that the oxygen isotope fractionation between Mg–calcite and water increases by 0.06‰ per mole percent $MgCO_3$ at 25°C. Mg/Ca ratios vary considerably in skeletal carbonates, and corrections for this effect should be made in paleotemperature determinations.

6.4 FACTORS AFFECTING OXYGEN ISOTOPE PALEOTEMPERATURES

The carbonate paleotemperature equation has three variables: $\delta^{18}O_{carbonate}$, $\delta^{18}O_{water}$, and temperature (t). We estimate t from the measured $\delta^{18}O_{carbonate}$ value and an assumed $\delta^{18}O_{water}$

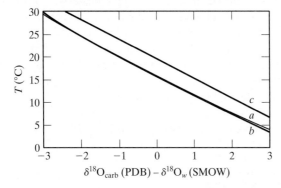

FIGURE 6.4: Comparison of the calcite–water fractionation curves of Epstein et al. (1953; curve a) normalized to the PDB and SMOW standards (equation (6.5)) and O'Neil et al. (1969; equation (6.6), curve b). Also shown is the aragonite–water fractionation curve of Grossman and Ku (1986; curve c from equation (6.7)).

value. The validity of the estimate depends on a number of factors, already recognized by Urey early on (1948):

1. *The $\delta^{18}O_{water}$ value at the time of calcite growth.* The $\delta^{18}O$ values of the ocean have undoubtedly changed in the past, due to glacial–interglacial periods. Over the long term, the $\delta^{18}O$ values of the oceans appear to be buffered by hydrothermal interaction with the seafloor (Chapter 5), but the level of fluctuation is not known. Isolated basins or shallow seas could be perturbed from the normal marine value by the evaporation or influx of freshwater. We know that ancient carbonates have low $\delta^{18}O$ values, supporting (but in no way proving) that the $\delta^{18}O$ value of the ancient ocean was lower than it is today.
2. *The degree to which the $\delta^{18}O$ values of carbonates have been altered.* Even low-temperature diagenesis can alter the $\delta^{18}O$ value of a carbonate, due to its ease of dissolution in freshwaters. Great care is taken to avoid the effects of diagenesis, but no foolproof method exists to prove that no diagenesis has taken place. The low $\delta^{18}O$ values of ancient carbonates are equally explained by diagenesis as by changing ocean composition.
3. *The degree to which the carbonates precipitated in equilibrium with water.* It is known that some organisms (e.g., corals) secrete carbonate that is not in isotopic equilibrium with water. This so-called vital effect must be also be considered.

6.4.1 Variations in $\delta^{18}O$ of Ocean Water in Space and Time

We have seen that variations in oxygen isotope compositions of surface waters in modern oceans arise from both evaporation and the influx of freshwater. These processes must be taken into consideration in working with the carbonates formed in the near-surface environment, with planktic organisms[7] especially in samples from shallow epicontinental seas or restricted marine basins where an influx of freshwater and evaporation could have caused large isotopic shifts. The problem can be at least partly addressed because oxygen isotope values and salinity are generally correlated (Chapter 5). For fossil materials, therefore, it is sometimes possible to estimate the $\delta^{18}O$ value of the ocean if an independent estimate of salinity can be made from salinity-dependent cation ratios in the carbonate (Carpenter and Lohmann, 1992), from Sr/Ca ratios (Beck et al., 1992; DeVilliers et al., 1994), or by some other means.

The Quaternary period presents a unique problem, in that we have to deal with fluctuating $\delta^{18}O_{ocean}$ values related to glacial–interglacial cycles. On a timescale of tens of thousands of years, the $\delta^{18}O$ value of the entire ocean mass changed when isotopically light water was transferred from the ocean to continental ice sheets. Numerous isotopic studies have shown that, during periods of advance and retreat of continental glaciers, $\delta^{18}O$ values of marine carbonate changed repeatedly and in a regular manner. Are the isotopic shifts due to a changing ocean temperature or the changing composition of the ocean related to the growth of ice sheets? When the temperature of seawater decreases, the fractionation between carbonate and water increases (lower temperature, larger fractionation), so that organisms should precipitate carbonate with higher $\delta^{18}O$ values. But when temperatures decrease, ice caps grow, removing light water from the ocean and increasing the $\delta^{18}O_{ocean}$ value. Both effects—lowering the temperature and raising the $\delta^{18}O$ value of the ocean—will cause the

---

[7]The words *planktic* (surface dwelling) and *benthic* (bottom dwelling) are also spelled *planktonic* and *benthonic*, particularly in the older literature.

$\delta^{18}O$ values of carbonates to increase. We cannot tell a priori whether high $\delta^{18}O$ values in glacial times are due to lower ocean temperatures or larger ice caps. Emiliani (1955; 1966) attempted to deconvolute this problem by analyzing carbonates that were precipitated in warm-water regions of the Caribbean and equatorial Atlantic (Fig. 6.5). He reasoned that temperatures would be relatively constant in the Central Atlantic, so that changes in the secular isotope record would be related to changes in the $\delta^{18}O$ values of the oceans and not temperature. He therefore interpreted the regular variations in oxygen isotope ratios of the carbonates as a record of changing ice volumes.

Emiliani's conclusion was reaffirmed in a later series of works by Shackleton and Opdyke (1973; 1976) using $\delta^{18}O$ values of benthic and planktic foraminifera from the western tropical Pacific. Following Emiliani's (1955; 1966) reasoning, they assumed that deep waters, generated at high latitudes and buffered by the presence of ice, have relatively constant temperatures. Variations in the $\delta^{18}O$ values of deep (benthic) foraminifera should therefore track the isotopic composition of the ocean. They found that deep (benthic) and shallow (planktic) foraminera had the same secular isotopic patterns, offset only by a constant amount related to their *relative* temperature differences. The magnitude of the isotopic variations were therefore related to changes in the ocean's isotopic composition and not temperature. This view is now the consensus of active workers in the field.

The usefulness of measuring coeval benthic and planktic foraminifera is illustrated in a comparative chemostratigraphic study of the central and intermediate Pacific Ocean over a much longer interval (Fig. 6.6). The similarity and gradual increase in $\delta^{18}O$ values of benthic and planktic foraminifera from intermediate latitudes over the past 80 million years shows that temperatures have decreased during this mostly ice-free time. In contrast, planktic foraminifera from the Central Pacific have a nearly constant $\delta^{18}O$ value, suggesting that the surface temperatures at the more tropical latitudes have remained constant, as have the $\delta^{18}O$ values of the oceans.

6.4.2 Vital Effects

Epstein's early calibration of the paleotemperature scale was made using molluscs. Fortunately, molluscs—especially belemnites and brachiopods—*tend* to precipitate their carbonate shells in oxygen isotope equilibrium with ambient waters (Lowenstam, 1961; see Carpenter and

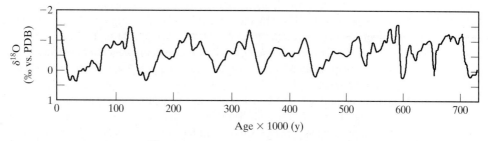

FIGURE 6.5: Variation in $\delta^{18}O$ values vs. age for foraminifera from Caribbean cores (Emiliani, 1966). General observations include a cyclic glacial–interglacial trend with a period of ~100,000 years. The sawtooth pattern suggests that there is a rapid change from glacial (high $\delta^{18}O$ values) to interglacial cycles and a more gradual return to glacial conditions. Note that the $\delta^{18}O$ values on the y-axis are reversed, decreasing upwards as age decreases to the left.

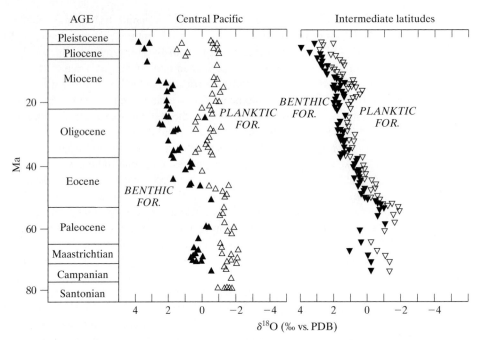

FIGURE 6.6: $\delta^{18}O$ values of benthic and planktic foraminifera for central and intermediate latitudes. The deep and shallow water data parallel each other at the intermediate latitudes, suggesting that temperatures at the surface and at depth were the same (or offset by a constant amount) and that temperatures generally decreased throughout the Tertiary. (Changing $\delta^{18}O$ values of the ocean could also explain the change, but there's no reason to expect such changes, and the planktic foraminifera data from the Central Pacific nullify this possibility.) Benthic foraminifera data for the Central Pacific parallel the intermediate-latitude data throughout the column, consistent with deeper water temperature being controlled by downwelling cold waters from higher latitudes. In contrast, the equatorial planktic foraminifera data are relatively constant throughout, suggesting constant surface temperatures *and* constant $\delta^{18}O_{ocean}$ values. (Modified from Anderson and Arthur, 1983.)

Lohmann, 1995, for additional details). Some organisms secrete shells out of equilibrium with ambient water, leading to the so-called *vital effect*. For purposes of thermometry, it is critical to identify those organisms (e.g., corals) whose life processes *always* introduce an oxygen isotope offset by the vital effect and to recognize the conditions under which vital effects operate only *sometimes* for other organisms (e.g., foraminifera).

The commonly used planktic foraminifera often, but not always, secrete their tests in oxygen isotope equilibrium. Divergence from equilibrium is related to environmental factors, including the intensity of sunlight, temperature stress, nutrient supplies, and the like. Thus, planktic foraminifera can secrete carbonate out of equilibrium with ocean water, especially at tropical temperatures. Most benthic foraminifera, by contrast, live in a more uniform environment and are isotopically much better behaved. In fact, the extremely uniform temporal variations in data for certain benthic foraminfera (and corrected planktic data) allow precise correlations to be made between cores that are thousands of kilometers apart (Prell et al., 1986).

Not only do coccoliths deposit calcium carbonate out of oxygen isotope equilibrium with environmental waters, but the magnitude of the vital effect for this class of organisms varies irregularly with both temperature and taxa. Echinoderms, corals, red algae, and certain benthic foraminifera notoriously precipitate their carbonate out of equilibrium with ambient waters (Fig. 6.7).

Over the course of 50 years or more, we have learned which species to use in oxygen isotope studies of paleoclimate and also to make sensible corrections to isotopic analyses that are predictably offset by a vital effect. Biogeochemists have even used disequilibrium depositions of carbonate to study details of life processes of modern and extinct marine organisms. (Proposed explanations of the vital effect, particularly for carbon, are discussed in more detail in Section 7.4.3.)

6.4.3 Diagenesis

Original isotopic ratios must be preserved in carbonate shells for meaningful studies of paleotemperature to be conducted. Reactions between biogenic carbonate and diagenetic fluids can easily erase the original isotopic record if fluid/carbonate ratios are large. During carbonate diagenesis, little or no *direct* isotopic exchange takes place between solid carbonates and aqueous fluids, because the rates of solid-state diffusion of carbon and oxygen in carbonates at low temperatures are negligible (O'Neil, 1977). Carbon and oxygen isotope ratios of biogenic carbonate can be changed by two diagenetic processes: (1) the addition of new carbonate by cementation and (2) the dissolution of unstable carbonate and the reprecipitation of a stable mineral, normally low-magnesium calcite.

Cementation is the most common diagenetic process leading to a change in the isotopic composition of a marine carbonate. Isotopic measurements of a cemented biogenic carbonate reflect mixtures of original (unaltered) and new carbonate, rather than of original carbonate

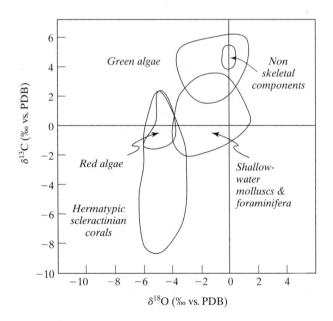

FIGURE 6.7: Oxygen and carbon isotope offsets of selected modern organisms due to the vital effect. (After Anderson and Arthur, 1983.)

alone. Cementation of marine carbonates commonly occurs where sediments are exposed to high-energy conditions. Because carbonate cements are abiotic, there can be no vital effect, and they are likely to be deposited in both carbon and oxygen isotope equilibrium with their parent fluids. The pore waters in equilibrium with the earliest-formed cements often are the same as marine water, so that these early cements will be in isotopic equilibrium with ocean water. Unlike many biogenic carbonates, which are often unstable aragonite or high-Mg calcite (see below), cements are often thermodynamically stable low-Mg calcite (although this varies between ice age and greenhouse conditions, due to changing ocean chemistry) and therefore are not prone to recrystallization at some later time. Clearly, then, cements have a high potential for providing information relating to the original isotopic composition of seawater and/or temperature.

Not all cements give primary information. The cementing fluid may be locally confined and not in rapid communication with the major aqueous oceanic reservoir. In such cases, the breakdown of organic matter can lead to the formation of cements with very low $\delta^{13}C$ values. Secondary cements are often coarse grained, but reflect equilibration with meteoric water. Carbonate cements are frequently large enough to sample cleanly, but infilling cements, particularly in shells of small organisms, can be analyzed separately only by using special in situ methods such as laser ablation or ion probe techniques.

Solution and reprecipitation is a combined process that is thermodynamically driven towards a lower free-energy state. For example, calcite is stable relative to aragonite, and low-Mg calcite is stable relative to high-Mg calcite. During the thermodynamically driven chemical reactions to more stable phases, isotopic exchange can take place as well. Recrystallization can occur on a very fine scale (e.g., replacement reactions) such that original textures are retained despite chemical and isotopic changes. *Recrystallization* and *neomorphism* are terms used to describe the process of solution and redeposition and are strictly used to describe the isochemical process of grain coarsening.

The magnitude of the change in isotopic compositions of carbon and oxygen in biogenic carbonate that undergoes recrystallization depends on four factors: (1) temperature, (2) the isotopic compositions of H_2O and HCO_3^- in the fluid, (3) the fluid/solid ratio, and (4) the susceptibility of the carbonate to recrystallization to a more stable phase. As an end-member case, consider a biogenic marine carbonate that recrystallizes with a tiny amount of pore water of marine composition at a temperature close to the original deposition temperature. Under these conditions, recrystallization will not significantly alter the isotopic ratios of the original carbonate, and indeed, such recrystallization occurs in marine sediments directly after deposition. At the other extreme, isotopic ratios of the same biogenic carbonate would change dramatically if it underwent neomorphism bathed in a diagenetic fluid containing a component of freshwater (of low $\delta^{18}O$) that carried soil-derived bicarbonate (of low $\delta^{13}C$). In general, open-system diagenesis (high water/rock ratios) leads to loss of primary isotopic information, but if diagenetic trends in isotopic ratios are regular, it may be possible to extrapolate back to original compositions.

The *diagenetic potential* of a mineral in a given system can be described as the tendency of that mineral to undergo reaction with a given diagenetic fluid. The greater the departure from chemical (not isotopic) equilibrium between mineral and fluid, the greater is the diagenetic potential. Several factors control the diagenetic potential of biogenic calcium carbonate, and the strongest of these is chemical composition—specifically, the Mg/Ca ratio. Low-magnesium calcite is *stable* and thus relatively insoluble and nonreactive in diagenetic fluids, whereas high-magnesium calcite is *metastable* relative to pure calcite and thus is more soluble and reactive. Crystal size is also an important parameter, because small particles have relatively high surface areas and can lower the free energy of the system by dissolving and recrystallizing to larger grains. The polymorphic form of $CaCO_3$ is another factor that controls diagenetic potential. Aragonite and vaterite[8] are metastable in surface environments and are highly prone to dissolution and reprecipitation to the more stable calcite, particularly when exposed to fluids with a freshwater component.

Freshwater or seawater containing a meteoric component is undersaturated with respect to marine carbonates, and this disequilibrium promotes dissolution. Fresh rainwater is a particularly corrosive agent to carbonates, as it is slightly acidic as well. Marine carbonates exposed to such fluids, either in shallow coastal waters or on land, undergo *meteoric diagenesis*. Aragonite and high-magnesium calcite are common metastable constituents of biogenic carbonate and react with diagenetic fluids to form more stable low-magnesium calcite with a different isotopic composition.

Meteoric diagenesis almost always lowers both carbon and oxygen isotope ratios of carbonates because $\delta^{18}O$ values of meteoric waters are normally lower than those of seawater and $\delta^{13}C$ of soil bicarbonate is lower than that of seawater bicarbonate.[9] In arid regions, where evaporation is intense, meteoric waters can have positive $\delta^{18}O$ values such that meteoric diagenesis of marine carbonate could shift $\delta^{18}O$ values to higher values.

Alteration patterns on $\delta^{18}O$–$\delta^{13}C$ diagrams have characteristic shapes depending on the magnitude of the various diagenetic parameters. At the onset of diagenesis in a given region, a tiny amount of water enters the system and dissolves some carbonate. The fluid/rock ratio is perforce very low at this stage, and the system is rock-dominated. The bicarbonate in solution generated by the dissolution of solid carbonate will exchange oxygen isotopes with the water, mix with the soil bicarbonate already present in the fluid, and reprecipitate as a cement or replacement carbonate. Under the rock-dominated conditions that prevail initially, newly precipitated carbonate will have isotopic ratios that are similar to those of the original carbonate. With ever-increasing water/rock (*W/R*) ratios, both $\delta^{18}O$ and $\delta^{13}C$ values of the carbonate will decrease, but they will not change at the same rate. Oxygen is a major component of water, while dissolved carbon is only a trace component. Therefore, early diagenesis will affect the carbonate isotope ratios of oxygen far more than those of carbon (Brand and Veizer, 1981; Lohmann, 1988; Marshall, 1992). In effect, the *W/R* ratios for oxygen are higher than those for carbon for the same amount of water. As diagenesis proceeds and fluid/rock ratios increase, $\delta^{18}O$ values of successively deposited carbonate (cement) become increasingly more negative, while $\delta^{13}C$ values remain nearly constant. With the passage of time, $\delta^{18}O$ values reach a final limiting value that is controlled by the isotopic compositions of the diagenetic fluid and effective water/rock ratios. With still-increasing fluid/rock ratios, $\delta^{13}C$'s of ensuing cements become more negative, once again approaching a final, limiting value defined by the $\delta^{13}C$ value of the infiltrating fluid. In the context of a time sequence, a rotated J pattern develops on a $\delta^{13}C$–$\delta^{18}O$ diagram (Fig. 6.8). The data points on the upper right limb of the curve represent unaltered material. In combination with careful petrographic examination and chemical analysis, these stable isotope patterns can provide a detailed diagenetic history in a given carbonate terrane.

---

[8]Measurements are made only of calcite and aragonite. Vaterite, a rare, naturally occurring polymorph of $CaCO_3$, has been studied only under laboratory-controlled conditions.

[9]The $\delta^{13}C$ value of atmospheric CO_2 is -6 to $-7‰$ and is in near equilibrium with marine-dissolved inorganic carbon. (See Chapter 7.)

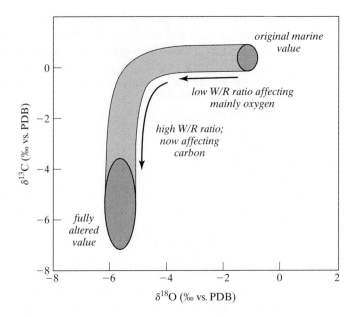

FIGURE 6.8: Illustration of changes in $\delta^{18}O$ and $\delta^{13}C$ that accompany the alteration of a marine carbonate by meteoric water. Initial dissolution and reprecipitation by interaction with pore water fluids with a small meteoric component will shift carbonates to lower $\delta^{18}O$ values but leave $\delta^{13}C$ values relatively unchanged because there is so little carbon in a water-rich fluid. Only after F/R ratios become very large, will $\delta^{13}C$ values decrease, due to exchange with light dissolved C in solution. The verticle line defined by increasing F/R interaction is sometimes called the 'meteoric calcite line' (Lohmann, 1988), which defines the $\delta^{18}O$ value of meteoric water responsible for precipitation of late calcite cements.

During diagenesis, a number of chemical changes occur, and these changes can be used to identify altered and primitive portions of the carbonates. Most commonly, the concentrations of the trace elements Mn (promotes cathodoluminescence) and Fe (diminishes cathodoluminescence) of the cement increase during diagenesis under reducing conditions, and concentrations of Sr and Mg decrease. $^{87}Sr/^{86}Sr$ ratios can increase or decrease, depending on the source of strontium in the local meteoric water. All these changes are specific to the conditions of diagenesis and sources of fluids, so no one geochemical tracer is completely diagnostic.

A number of strategies can be employed to circumvent and even exploit the effects of diagenesis, particularly for older material. Thick, nonluminescent portions of brachiopods and marine cements are likely to have preserved their original mineralogy, as well as chemical and isotopic compositions (Figure 6.9). Certain portions of brachiopods *can*, and frequently *do*, precipitate stable, low-magnesium calcite in equilibrium with seawater, and in addition, this carbonate is relatively massive and coarse grained. Analyses of nonluminescent portions of these shells can be used to determine original isotopic composition, and analyses of altered portions are used to study diagenesis.

Another strategy involves specifically *searching out* the metastable phases. Aragonite is a metastable polymorph of $CaCO_3$ at surface conditions. It is easily altered to the more stable calcite. Therefore, if aragonite can be found, its very preservation implies that diagensis has been minimal. Accordingly, we have two seemingly diametrically opposed philosophies at

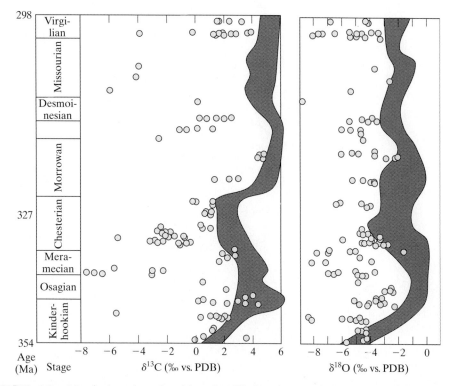

FIGURE 6.9: North American brachiopods (filled region) and coexisting cements (circles) from Carboniferous sediments. The $\delta^{18}O$ and $\delta^{13}C$ values of cements are lower than those for coexisting brachiopods, indicating diagenesis with meteoric water. Thick non-luminescent samples tend to be the least affected by diagenesis. (Modified from Mii et al., 1999.)

our disposal. One is to search for metastable material, such as aragonite. The reasoning is simply that if it had been altered, it would have recrystallized as stable low-Mg calcite. The other approach is to find the most stable material—samples that were precipitated as stable low-Mg calcite. Because it is already stable, it is less likely to undergo recrystallization. Obviously, this strategy is valid only if we can be sure that the carbonate originally was low-Mg calcite. A simplified schematic of acceptable samples and strategies to use in studies of the stable isotope composition of marine biogenic carbonate is given in Figure 6.10.

For samples of Cenozoic age, both diagenesis and ambiguities in the isotopic composition of ocean water are negligible. Consequently, numerous isotopic studies have been made of pristine fossils of Cenozoic age in undisturbed cores from the deep sea. In Mesozoic or older sediments, isotopic measurements are limited mostly to shelf deposits, where preservation is generally poor (and the oxygen isotope composition of the water is suspect).

Early workers active in oxygen isotope paleothermometry established guidelines for assessing diagenesis of their carbonates, and these guidelines are still valid today. If any of the following is true, the carbonate is more likely (though not certain) to have retained its original $\delta^{18}O$ value:

1. The skeletal material or cement is made of unstable minerals, such as aragonite or high-Mg calcite. These minerals would not survive exposure to diagenetic fluids. Their survival indicates that interaction with diagenetic fluids has been minimal.

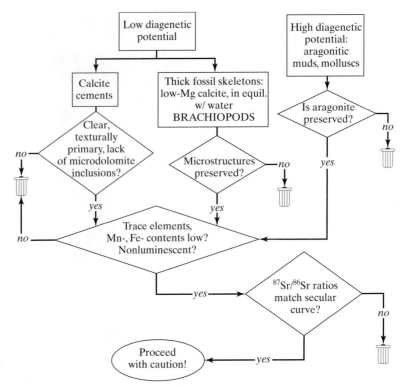

FIGURE 6.10: Schematic flowchart of common procedures for identifying diagenesis. (For further information, see Carpenter et al., 1991; Marshall, 1992; and Grossman, 1994.)

2. The skeletal material *secreted by the organism* is stable low-Mg calcite. Stable minerals have a low diagenetic potential.
3. There are seasonal variations in isotopic ratios along the direction of growth. Recrystallization would obliterate these signals. (This point has been challenged.)
4. The material has the highest $\delta^{18}O$ value in the population. Diagenesis normally lowers the $\delta^{18}O$ value.
5. The material is not luminescent. Diagenesis often introduces cathodoluminescent Mn to neoformed carbonate.

6.4.4 Ecology of the Organism

Carbon and oxygen isotope compositions of shells reflect local conditions of productivity and temperature *at the time of deposition*. Some organisms spend one part of their lives in one environment and other parts of their lives in different environments that can be, for example, darker, more saline, or colder. Even among the same species, larger and more robust individuals tend to build their shells in deeper, colder waters and therefore have higher $^{18}O/^{16}O$ ratios than their more fragile counterparts. Researchers must be aware of the ecology of the organisms they use if they are to interpret the stable isotope data properly. It was precisely for these reasons that the Chicago group analyzed shells of attached or sessile organisms to establish the paleotemperature scale.

6.5 APPLICATIONS OF OXYGEN ISOTOPE PALEOTHERMOMETRY

Taking into account all the factors that can affect $\delta^{18}O$ values of biogenic carbonate, including the application of appropriate correction factors, enables scientists to address many important issues of oceanography and paleoclimatology, using the method of oxygen isotope paleothermometry. The principles employed are generally the same, so only a few examples of applications will be given here.

6.5.1 The Quaternary

The Quaternary record is very well preserved, in terms of both a complete detailed stratigraphy and a minimal amount of diagenesis. The glacial periods ("ice-house" conditions) in the Quaternary pose a complication that is nearly unique in chemostratigraphic reconstruction. Most other times in history are free of extensive glacial ice, and the $\delta^{18}O$ values of the oceans can be considered to be constant in the short term. In the Quaternary period, by contrast, the $\delta^{18}O$ values of marine carbonates are affected by changes in the isotopic composition of the ocean as much as by changes in temperature, as was already discussed in Section 6.4.1.

Arguably the most important discovery made through oxygen isotope analyses of biogenic carbonate was the delineation of important details of Pleistocene glaciation. The oxygen isotope curve originally presented by Emiliani (1966) and modified in Emiliani et al. (1978) is shown in Fig. 6.6. Emiliani worked on shells of pelagic foraminifera from the perennially warm waters of the Caribbean in order to avoid the problem of temperature variations as the cause of changes in $\delta^{18}O$ of the shells. Several important features of glaciation over the last 700,000 years are apparent from Fig. 6.5:

1. The patterns are saw toothed, implying that glaciation is a slow process and that deglaciation occurs much more rapidly.
2. A periodicity of ~100,000 years in the patterns can be reasonably linked to one of the Milankovich periodicities in orbital forcing,
3. There are many more glacial–interglacial stages in the Pleistocene than was previously thought.

An extremely useful method of correlating stratigraphic sections can be made from the regularities observed in the oxygen isotope record in Quaternary foraminiferal tests from all over the world's oceans. The uniformity in the record stems from two facts: (1) The effect measured is primarily a change in $\delta^{18}O$ of the oceans that resulted from changes in ice volumes on land, and (2) the mixing time of the oceans is very short ($\sim 10^3$ years). Since the early work of Emiliani, synchronous *oxygen isotope stages* have been recognized by many workers. Odd numbers are assigned to warmer, interglacial times, even numbers to colder, glacial times. There are five recognized oxygen isotope stages in the isotopic record of the last 130,000 years, and substages are recognized as well, particularly in the well-studied stage 5.

Age assignments are critical in stratigraphic work and are frequently the subject of considerable debate. A novel approach to age assignment is to assume that astronomically driven changes in climate (orbital forcing) are responsible for the waning and growth of ice sheets. Using this approach to tune several oxygen isotope records, Imbrie et al. (1984) established a reference chronostratigraphy for the late Quaternary called the SPECMAP

(Spectral Mapping Project). The SPECMAP composite chronology is frequently used to adjust oxygen isotope records when no reliable ages are available. Indeed, it is now common parlance to speak of events that occurred in a particular stage as, for example, oxygen isotope stage 2 or 5.

6.5.2 The Paleogene and Neogene (Cenozoic)

From a historical point of view, our state of knowledge of oceanographic features in the Paleogene–Neogene was greatly advanced by oxygen isotope analyses of benthic and planktic foraminifera made in the early 1970s. At that time, sampling techniques were primitive by standards acceptable today, but clear patterns were evident in the data obtained. As a point of reference, the large differences in $\delta^{18}O$ between planktic and benthic shells forming in modern oceans from low latitudes reflect the large differences in temperature between surface and bottom waters. Bottom-water temperatures are established by the sinking of cold, saline waters in the Antarctic and North Atlantic Oceans. But ice volumes and ocean circulation patterns change with time, and these changes are reflected in the oxygen isotope data. Few such changes are as dramatic as those which occurred in Paleogene–Neogene time.

Data on planktic and benthic foraminiferal shells separated from cores collected at several sites in the North Pacific Ocean are shown in Fig. 6.6. Differences in $\delta^{18}O$ between planktic and benthic shells were relatively small in the early Tertiary, reflecting relatively small differences in surface- and bottom-water temperatures at that time. The cooling at the late Eocene to Oligocene and the divergence of planktic and benthic foraminifera in Central Pacific samples suggest the beginning of cold downwelling waters originating at high latitudes. The dramatic cooling in the middle Miocene is related to the circumpolar Antarctic circulation and the formation of Antarctic ice sheets (Savin et al., 1975).

The huge number of analyses of Paleogene–Neogene ocean core samples results in an extremely high resolution record for the past 50–60 million years. Zachos et al. (2001) present a detailed compilation of stable isotope variations from the Paleogene to the present (Fig. 6.11). Variations on the scale of 10^4 to 10^5 y are related to orbital parameters, while longer scale, irreversible variations are related to tectonic processes. Spectral fitting of the data show a strong periodicity at 100 ky and 41 ky for samples 0 to 4 Ma, with a loss of intensity of the 100-ky band in older samples.

6.5.3 Older Samples

Well-preserved deep-sea cores for Cretaceous samples exist, so the foraminifera-based record extends back that far. Most Mesozoic and all Paleozoic samples are limited to shelf deposits. These lithologies have often undergone diagenesis, and great care must be taken to retrieve unaltered samples. In addition, the $\delta^{18}O$ value of the water in a shallow shelf setting may have been affected by a large meteoric contribution, lowering the apparent "seawater" value. Workers in the field have been careful to alleviate these problems by analyzing diagenetic-resistant material and by correlating salinity and $\delta^{18}O$ values of seawater by means of a temperature–salinity–density model or salinity-dependent cation ratios in the carbonate. As the age of samples increases, the uncertainties regarding diagenesis and $\delta^{18}O$ values of the ocean increase as well. Precambrian carbonates invariably have low $\delta^{18}O$ values, which can be correlated with one or more of three variables: diagenesis, low marine $\delta^{18}O$ values, or warm ocean temperatures. Other stable isotope sedimentary proxies support the low $\delta^{18}O$ values of carbonates (cherts, iron formations), but the significance of the low values is still debated.

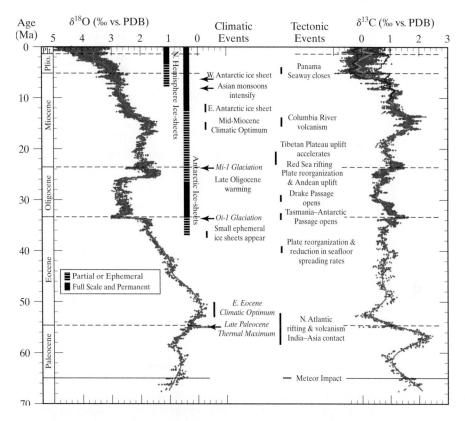

FIGURE 6.11: High-resolution $\delta^{18}O$ curve compiled from 40 ocean core samples. Most of the data are from the benthic foraminifera taxa *Cibicidoides* and *Nuttallides*, corrected for vital effect. Absolute ages are corrected to the paleomagnetic timescale. Major events, such as the Eocene climatic optimum and the timing and intensity of ice-sheet formation, are correlated with $\delta^{18}O$ values. (Reprinted from Zachos et al., 2001, with permission from AAAS.)

Although extraction of the $\delta^{18}O$ value of unaltered carbonates is complicated by the effects of diagenesis, and the relationship between the $\delta^{18}O$ values of carbonates and the ocean temperature is complicated by uncertainties about the oxygen isotope composition of the ocean through time, there have been a number of attempts to determine a secular curve for $\delta^{18}O$ values of marine carbonates. Figure 6.12 shows the extraordinary compilation of Veizer et al. (1999) for low-magnesium calcite shells, mainly brachiopods and belemnites. High $\delta^{18}O$ values correspond to times of glaciation and generally cold conditions, and there is a diminution in $\delta^{18}O$ values of Ordovician and older samples.

6.6 APPLICATION TO CONTINENTAL CARBONATES

Carbonates form in equilibrium with meteoric water in a number of different environments. Climatic information can be retrieved from these samples, but interpretation of the data is often complicated by uncertainties in the $\delta^{18}O$ values of the water forming the carbonate. The

142   Chapter 6   Biogenic Carbonates: Oxygen

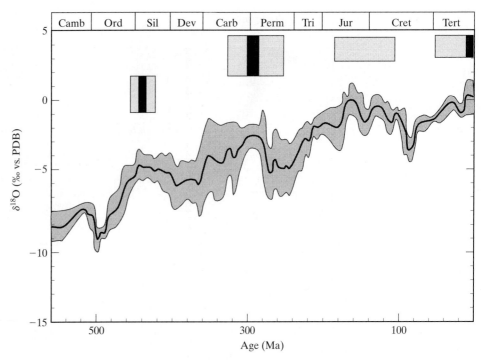

FIGURE 6.12: Variations in the oxygen isotope ratio of shell carbonates. 1σ uncertainties are shown as the shaded region around the central line. Cold periods are indicated by the shaded boxes above the curve, with ice ages illustrated with filled black boxes. There is a general increase in $\delta^{18}O$ values with decreasing age that may be due to changing $\delta^{18}O$ values of the oceans or more intense diagenesis in the older samples. Note that the cold periods generally correspond to higher $\delta^{18}O$ values, as expected. (Modified from Veizer et al., 1999.)

$\delta^{18}O$ values of terrestrial (or continental) carbonates often are used to estimate the $\delta^{18}O$ value of the meteoric water, as opposed to the temperature of formation. Samples analyzed include speleothems, lake sediments, vein calcite, travertines, and soil carbonates.

Interpretations of $\delta^{18}O$ values from terrestrial carbonates differ from those in the marine setting (Grootes, 1993). In Quaternary marine samples, high $\delta^{18}O$ values are caused either by increasing $\delta^{18}O$ values of the ocean due to glacial deposition on continents or by lower temperatures (or by a combination of the two). In contrast, the $\delta^{18}O$ values of water from lakes are primarily a function of the $\delta^{18}O$ value of meteoric water. (See problem 6.) In the marine setting, low temperatures increase the $\delta^{18}O$ value of carbonates. In the continental setting, cold causes the $\delta^{18}O$ values of meteoric water to decrease (see Section 4.7.1), which in turn lowers the $\delta^{18}O$ value of the precipitating carbonate. The effect is essentially the reverse of that seen in the marine environment!

As an example, McKenzie and Hollander (1993) measured the oxygen isotope profile in a sedimentary sequence from a varved lake in Switzerland. There is a regular decrease in the $\delta^{18}O$ values of the lacustrine chalk from $-8‰$ prior to 1887 to values of close to $-11‰$ in modern sediments. McKenzie and Hollander attribute the lower $\delta^{18}O$ values in the young sediments to changes in atmospheric circulation in Central Europe. Pre-1887 meteoric waters were sourced primarily from cold prevailing northwesterly winds with low $\delta^{18}O$ values. The moisture source changed to one dominated by warmer westerly-to-southwesterly winds in modern times.

Cave deposits provide another important archive for terrestrial paleoclimate. Deep caves with poor air circulation have nearly constant year-round temperatures. Drip waters entering a cave are often at or near carbonate saturation. When the waters enter the cave proper, CO_2 is lost, causing cabonate precipitation. If CO_2 loss is slow, equilibrium between the dissolved species HCO_3^-, CO_2^{2-}, and CO_2 is maintained, and carbonates precipitate in carbon and oxygen isotope equilibrium with formation waters (e.g., Schwarcz, 1986). Combined with accurate mass spectrometric uranium-series dating of the carbonates, detailed oxygen and carbon isotope records of terrestrial climates can be obtained. The oxygen isotope shifts at 5900 ybp and 3600 ybp in a stalagmite from Cold Water Cave, northeast Iowa, in the United States, illustrate an example of using terrestrial records to identify climate shifts on land (Fig. 6.13). The rapid shifts are linked to changes from forest to prairie and then prairie to savannah. The rapid climate changes are related to changing weather patterns in the Midwestern United States.

Certainly one of the most remarkable records comes from Devils Hole, Nevada, United States (Winograd et al., 1992), where a continuous 500,000 year record of vein calcite precipitation is preserved. The oxygen isotope changes mimic precisely those seen in the Vostok ice core in Antarctica from $\delta^{18}O$ isotope ratios in ice. The Devils Hole record extends the climate record hundreds of thousands of years beyond the limits of the ice cores, however. Combining data from Devils Hole with precise mass spectrometric uranimum-series dating, the authors concluded that the Pleistocene glacial cycles are not tied to Milankovitch cycles, a view that was quickly challenged (Imbrie et al., 1993).

The foregoing discussion is really only half the picture. The information from carbon isotopes is complementary to that from oxygen isotopes. Together, they provide far more information than either isotope alone. Chapter 7 continues with the carbon isotope story.

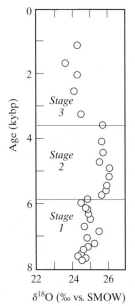

FIGURE 6.13: $\delta^{18}O$ value of speleothem calcite from Cold Water Cave, Iowa, United States. A warming event of 3°C is seen at 5,900 ybp, with a later 4°C cooling at 3,600 ybp. The climate changes can be related to well-known vegetation changes that have occurred in the area. (After Dorale et al., 1992.)

PROBLEM SET

1. Show that the aragonite–calcite fractionation of 0.6 per mil at 25°C determined by Tarutani et al. (1969) is roughly compatible with a combination of the aragonite–water equation of Grossman and Ku (1986) and the calcite–water equation of Epstein et al. (1953). (*Hint*: Start by choosing some reasonable values of δ_c and δ_w, and substitute them into the Epstein equation.)

2. A carbonate and a water sample are collected. The water is equilibrated with CO_2 at 25°C and the calcite is reacted with 100% phosphoric acid at 25°C, to extract pure CO_2. Both are measured on a mass spectrometer. The $\delta^{18}O$ value of the reference gas on the PDB scale is -19.64‰ voltages in the following table were obtained:
   a. What is the temperature of calcite formation obtained from Epstein's equation (6.4).
   b. What are the $\delta^{18}O$ values of the calcite and water on the PDB and SMOW scales, respectively?
   c. What is the temperature of calcite formation obtained from O'Neil's equation (6.6)? Is the answer the same as in part (a)?
   d. Would this carbonate sample be more likely to come from Barbados or the North Atlantic?

Water	Mass 44 (mV)	Mass 45 (mV)	Mass 46 (mV)
Std.	4032.853	4870.4308	5750.977
Sample	4116.429	4964.9088	5986.587
Calcite			
Std.	2891.286	3488.431	4122.685
Sample	2323.186	2818.992	3370.188

3. a. Delta values of marine biogenic carbonate almost always decrease during diagenesis. Under what circumstances could $\delta^{18}O$ increase during diagenesis?
   b. A Cretaceous shell comprises high-magnesium calcite. What is the likelihood that it retained its original isotopic composition through time?
   c. Is it possible for an organism to deposit its shell in oxygen isotope equilibrium and carbon isotope disequilibrium with ocean water? How about the reverse situation? Explain.

4. The $\delta^{18}O$ values of marine carbonates are controlled by temperature and the $\delta^{18}O$ value of the ocean. Separating these two factors during a glacial–interglacial period is difficult. Discuss all factors that control the $\delta^{18}O$ values of carbonates in lakes. Recognize that the $\delta^{18}O$ value of meteoric water is related to temperature, etc.

5. Convert Epstein's equation (6.4) to a form in which the $\delta^{18}O$ value of carbonate is related to PDB and the $\delta^{18}O$ value of the water is related to SMOW. Here is some useful information: The α value for SMOW–PDB is 1.03091; the α value for calcite acid fractionation ($\alpha_{CO_2-calcite} = 1.01025$); the α value for CO_2–water equilibration at 25°C is 1.0412; the combined fractionation from two factors with α_1 and α_2 is given by $\alpha_3 = \alpha_1 \times \alpha_2$.

6. For a carbonate precipitated in a lake, the $\delta^{18}O$ value of the lake water will be controlled by the annual average weighted temperature. The $\delta^{18}O$ value of the carbonate will be controlled by the $\delta^{18}O$ value of the lake water and temperature. These two effects control the $\delta^{18}O$ value of carbonate in an opposite manner. Which factor is more important? Make a semiquantitative assessment of the relative importance of both factors.

7. Your constant-temperature water bath operates only at 70°C. In order to determine the fractionation factor during phosphoric acid digestion, you react a sample of NBS-19 (which has a $\delta^{18}O$ value of -2.2‰ vs. PDB) with 100% phosphoric acid at 70°C and get a $\delta^{18}O$ value for the liberated CO_2 gas of 38.23 (vs. SMOW). What is the acid fractionation factor for calcite at 70°C? What is the fractionation factor for carbon (e.g., $\alpha_{CO_2 \text{ gas–calcite}}$) at 70°C.

ANSWERS

1. With reasonable values of $\delta_c = -2$ and $\delta_w = 0$, or $\Delta(\text{calcite}-\text{water}) = -2$, the Epstein equation yields a temperature of 25.8°C. Substituting this temperature into the Hayes and Grossman equation gives $\Delta(\text{aragonite}-\text{water}) = -1.41$. Comparing the two Δ values yields $\Delta(\text{aragonite}-\text{calcite}) = 0.6$.

2. **a.** From the definition $\delta = \left(\dfrac{R_x - R_{\text{std}}}{R_{\text{std}}}\right) \times 1{,}000$, we have

$$\delta = \left(\dfrac{\dfrac{V(46sa)}{V(44sa)} - \dfrac{V(46st)}{V(44st)}}{\dfrac{V(46st)}{V(44st)}}\right) \times 1000 = \left(\dfrac{\dfrac{5986.587}{4116.429} - \dfrac{5750.977}{4032.853}}{\dfrac{5750.977}{4032.853}}\right) 1000 = 19.84\text{‰ for the}$$

water vs. the reference gas. Similarly, for the carbonate, the $\delta^{18}O$ value is 17.38‰. Plugging these values into equation (6.4), $t(°C) = 16.9 - 4.2\,(\delta_c - \delta_w) + 0.13\,(\delta_c - \delta_w)^2$ gives 28°C.

   b. To convert from once scale to another, we use equation (2.3):
   $\delta_{X-B} = \delta_{X-A} + \delta_{A-B} + 0.001\delta_{X-A}\delta_{A-B}$. So, on the PDB scale, $\delta^{18}O_{\text{carbonate}} = -19.64 + 17.38 + 0.001(-19.64)(17.38) = -2.60$‰ vs. PDB. To convert to the SMOW scale, we use the equation $\delta^{18}O_{\text{SMOW}} = 1.03091(\delta^{18}O_{\text{PDB}}) + 30.91 = 28.22$‰ vs. SMOW.

4. The oxygen isotope composition of the water is controlled by that of the inflowing meteoric water (streams, groundwater, or rain) and the degree of evaporation. The isotopic composition of the water is controlled by temperature, distance from the source and changing sources. The $\delta^{18}O$ value of the carbonate is also controlled by temperature, as the fractionation between water and carbonate changes with temperature.

6. The $\delta^{18}O$ value of meteoric water has a temperature coefficient of 0.69‰/(°C) (from Chapter 4). The fractionation between carbonate and water has a temperature coefficient of 0.24‰/°C (= 1/4.2, from equation (6.4)). Therefore, temperature will change the $\delta^{18}O$ value of meteoric water almost three times as much as will the carbonate–water fractionation. Also, carbonates often form only at specific temperature intervals, making water an even stronger control on the $\delta^{18}O$ value of the precipitating carbonate.

7. $\alpha = \dfrac{1{,}000 + \delta_{(CO_2)}}{1{,}000 + \delta_{cc}}$, but both must be on the same scale. Convert -2.2 vs. PDB to SMOW, and you get $\delta^{18}O_{\text{SMOW}} = 1.03091(-2.2) + 30.91 = 28.64$, so $\alpha = \dfrac{1{,}038.23}{1{,}028.64} = 1.00932$. If we now convert to PDB, the sample gas is 7.10‰ (PDB), and $\alpha = \dfrac{1{,}007.10}{997.8} = 1.00932$. For carbon, $\alpha = 1$, because the $\delta^{13}C$ value of the CO_2 is identical to that of the carbonate.

REFERENCES

Al-Aasm, I. S., Taylor, B. E., and South, B. (1990). Stable isotope analysis of multiple carbonate samples using selective acid extraction. *Chemical Geology* **80**, 119–125.

Anderson, T. F., and Arthur, M. A. (1983). Stable isotopes of oxygen and carbon and their application to sedimentologic and paleoenvironmental problems. In *Stable Isotopes in Sedimentary Geology*, Vol. 10 (ed. M. A. Arthur, T. F. Anderson, I. R. Kaplan, J. Veizer, and L. S. Land), pp. 1–151. Columbia, SC: SEPM Short Course.

Barrera, E., and Savin, S. M. (1987). Effect of sample preparation on the δ^{18}O-value of fine-grained calcite. *Chemical Geology* **66**, 301–305.

Beck, J. W., Edwards, R. L., Ito, E., Taylor, F. W., Recy, J., Rougerie, F., Joannot, P., and Henin, C. (1992). Sea-surface temperature from coral skeletal strontium/calcium ratios. *Science* **257**, 644–647.

Brand, U., and Veizer, J. (1981). Chemical diagenesis of a multicomponent carbonate system—2: Stable isotopes. *Journal of Sedimentary Petrology* **51**, 987–997.

Carothers, W. W., Adami, L. H., and Rosenbauer, R. J. (1988). Experimental oxygen isotope fractionation between siderite–water and phosphoric acid liberated CO_2–siderite. *Geochimica et Cosmochimica Acta* **52**, 2445–2450.

Carpenter, S. J., and Lohmann, K. C. (1992). Sr/Mg ratios of modern marine calcite: Empirical indicators of ocean chemistry and precipitation rate. *Geochimica et Cosmochimica Acta* **56**, 1837–1849.

Carpenter, S. J., and Lohmann, K. C. (1995). δ^{18}O and δ^{13}C values of modern brachiopod shells. *Geochimica et Cosmochimica Acta* **59**, 3749–3764.

Carpenter, S. J., Lohmann, K. C., Holden, P., Walter, L. M., Huston, T. J., and Halliday, A. N. (1991). δ^{18}O values, ^{87}Sr/^{86}Sr and Sr/Mg ratios of Late Devonian abiotic marine calcite: Implications for the composition of ancient seawater. *Geochimica et Cosmochimica Acta* **55**, 1991–2010.

Charef, A., and Sheppard, S. M. F. (1984). Carbon and oxygen isotope analysis of calcite or dolomite associated with organic matter. *Chemical Geology* **46**, 325–333.

Cohn, M., and Urey, H. C. (1938). Oxygen exchange reactions of organic compounds and water. *Journal of the American Chemical Society* **60**, 679–687.

Coplen, T. B., Kendall, C., and Hopple, J. (1983). Comparison of stable isotope reference samples. *Nature* **302**, 236–238.

Craig, H. (1965). The measurement of oxygen isotope paleotemperatures. In *Stable Isotopes in Oceanographic Studies and Paleotemperatures* (ed. E. Tongiorgi), pp. 161–182. Pisa, Italy: Consiglio Nazionale delle Ricerche, Laboratorio de Geologia Nucleare.

DeVilliers, S., Shen, G. T., and Nelson, B. K. (1994). The Sr/Ca–temperature relationship in corralline aragonite: Influence of variability in $(Sr/Ca)_{seawater}$ and skeletal growth parameters. *Geochimica et Cosmochimica Acta* **58**, 197–208.

Dorale, J. A., González, L. A., Reagan, M. K., Pickett, D. A., Murrell, M. T., and Baker, R. G. (1992). A high-resolution record of Holocene climate change in speleothem calcite from Cold Water Cave, Northeast Iowa. *Science* **258**, 1626–1630.

Emiliani, C. (1955). Pleistocene temperatures. *Journal of Geology* **63**, 538–578.

Emiliani, C. (1966). Paleotemperature analysis of Caribbean cores P6304-8 and P6304-9 and a generalized temperature curve for the past 425,000 years. *Journal of Geology* **74**, 109–126.

Emiliani, C., Hudson, J. H., Shinn, E. A., and George, R. Y. (1978). Oxygen and carbon isotopic growth record in a reef coral from the Florida Keys and a deep-sea coral from Blake Plateau. *Science* **202**, 627–629.

Epstein, S., Buchsbaum, R., Lowenstam, H. A., and Urey, H. C. (1953). Revised carbonate–water isotopic temperature scale. *Geological Society of America Bulletin* **64**, 1315–1326.

Grootes, P. M. (1993). Interpreting continental oxygen isotope records. In *Climate Change in Continental Isotopic Records*, Vol. 78 (ed. P. K. Swart, K. C. Lohmann, J. A. McKenzie, and S. Savin), pp. 37–46. Washington, DC: American Geophysical Union.

Grossman, E. L. (1994). The carbon and oxygen isotope record during the evolution of Pangea; Carboniferous to Triassic. In *Pangea: Paleoclimate, Tectonics, and Sedimentation during Accretion, Zenith, and Breakup of a Supercontinent*, Vol. 288 (ed. G. D. Klein), pp. 207–228. Geological Society of America, Special Paper.

Grossman, E. L., and Ku, T. L. (1986). Oxygen and carbon isotope fractionation in biogenic aragonite; temperature effects. *Chemical Geology* **59**, 59–74.

Imbrie, J., Hays, J. D., Martinson, D. G., McIntyre, A., Mix, M., Morley, J. J., Pisias, N. G., Prell, W., and Shackleton, N. J. (1984). The orbital theory of Pleistocene climate: Support from a revised chronology of the marine $\delta^{18}O$ record. In *Milankovitch and Climate* (ed. A. Berger, J. D. Hays, G. Kukla, and B. Salzman), pp. 269–305. Dordrecht, The Netherlands: Reidel.

Imbrie, J., Mix, A. C., and Martinson, D. G. (1993). Milankovitch theory viewed from Devils Hole. *Nature* **363**, 531–533.

Kim, S.-T., and O'Neil, J. R. (1997). Equilibrium and nonequilibrium oxygen isotope effects in synthetic carbonates. *Geochimica et Cosmochimica Acta* **61**, 3461–3475.

Lohmann, K. C. (1988). Geochemical patterns of meteoric diagenetic systems and their application to studies of paleokarst. In *Paleokarst* (ed. N. P. James and P. W. Choquette), pp. 50–80. Berlin: Springer-Verlag.

Lowenstam, H. A. (1961). Mineralogy, O^{18}/O^{16} ratios, and strontium and magnesium contents of recent and fossil brachiopods and their bearing on the history of the oceans. *Journal of Geology* **69**, 241–260.

Marshall, J. D. (1992). Climatic and oceanographic isotopic signals from the carbonate rock record and their preservation. *Geological Magazine* **129**, 143–160.

McCrea, J. M. (1950). On the isotopic chemistry of carbonates and a paleotemperature scale. *Journal of Chemical Physics* **18**, 849–857.

McKenzie, J. A., and Hollander, D. J. (1993). Oxygen-isotope record in recent carbonate sediments from Lake Greifen, Switzerland (1750–1986); application of continental isotopic indicator for evaluation of changes in climate and atmospheric circulation patterns. In *Climate Change in Continental Isotopic Records*, Vol. 78 (ed. P. K. Swart, K. C. Lohmann, J. A. McKenzie, and S. Savin), pp. 101–111. Washington, DC: American Geophysical Union.

Mii, H.-S., Grossman, E. L., and Yancey, T. E. (1999). Carboniferous isotope stratigraphies of North America: Implications for Carboniferous paleoceanography and Mississippian glaciation. *Geological Society of America Bulletin* **111**, 960–973.

O'Neil, J. R. (1977). Stable isotopes in mineralogy. *Physics and Chemistry of Minerals* **2**, 105–123.

O'Neil, J. R., Adami, L. H., and Epstein, S. (1975). Revised value for the O^{18} fractionation between CO_2 and H_2O at 25°C. *Journal of Research of the U. S. Geological Survey* **3**, 623–624.

O'Neil, J. R., Clayton, R. N., and Mayeda, T. K. (1969). Oxygen isotope fractionation in divalent metal carbonates. *Journal of Chemical Physics* **51**, 5547–5558.

Perry, E. C., Jr., and Tan, F. C. (1972). Significance of oxygen and carbon isotope variations in Early Precambrian cherts and carbonate rocks of Southern Africa. *Geological Society of America Bulletin* **83**, 647–664.

Prell, W. L., Imbrie, J., Martinson, D. G., Morley, J. J., Pisias, N. G., Shackleton, N. J., and Streeter, H. F. (1986). Graphic correlation of oxygen isotope stratigraphy application to the late Quaternary. *Paleoceanography* **1**, 137–162.

Rosenbaum, J., and Sheppard, S. M. F. (1986). An isotopic study of siderites, dolomites and ankerites at high temperatures. *Geochimica et Cosmochimica Acta* **50**, 1147–1150.

Savin, S. M., Douglas, R. G., and Stehli, F. G. (1975). Tertiary marine paleotemperatures. *Geological Society of America Bulletin* **86**, 1499–1510.

Schwarcz, H. (1986). Geochronology and isotopic geochemistry of speleothems. In *Handbook of Environmental Isotope Geochemistry*, Vol. 2 (ed. P. Fritz and J. C. Fontes), pp. 271–303. Amsterdam: Elsevier.

Shackleton, N. J., and Opdyke, N. D. (1973). Oxygen isotope and palaeomagnetic stratigraphy of Equatorial Pacific Core V28-238: Oxygen isotope temperatures and ice volumes on a 10^5 and 10^6 year scale. *Quaternary Research* **3**, 39–55.

Shackleton, N. J., and Opdyke, N. D. (1976). Oxygen-isotope and paleomagnetic stratigraphy of Pacific Core V28-239, late Pliocene to latest Pleistocene. *Geological Society of America Memoir* **145**, 449–464.

Sharma, T., and Clayton, R. N. (1965). Measurement of 0–18/0–16 ratios of total oxygen of carbonates. *Geochimica et Cosmochimica Acta* **29**, 1347–1353.

Sharp, Z. D., Papike, J. J., and Durakiewicz, T. (2003). The effect of thermal decarbonation on stable isotope compositions of carbonates. *American Mineralogist* **88**, 87–92.

Swart, P. K., Burns, S. J., and Leder, J. J. (1991). Fractionation of the stable isotopes of oxygen and carbon in carbon dioxide during the reaction of calcite with phosphoric acid as a function of temperature and technique. *Chemical Geology* **86**, 89–96.

Tarutani, T., Clayton, R. N., and Mayeda, T. K. (1969). The effect of polymorphism and magnesium substitution on oxygen isotope fractionation between calcium carbonate and water. *Geochimica et Cosmochimica Acta* **33**, 987–996.

Urey, H. C., Epstein, S., McKinney, C., and McCrea, J. (1948). Method for measurement of paleotemperatures. *Bulletin of the Geological Society of America (abstract)* **59**, 1359–1360.

Urey, H. C., Epstein, S., and McKinney, C. R. (1951). Measurement of paleotemperatures and temperatures of the Upper Cretaceous of England, Denmark, and the southeastern United States. *Geological Society of America Bulletin* **62**, 399–416.

Veizer, J., Ala, D., Azmy, K., Bruckschen, P., Buhl, D., Bruhn, F., Carden, G. A. F., Diener, A., Ebneth, S., Godderis, Y., Jasper, T., Korte, C., Pawellek, F., Podlaha, O. G., and Strauss, H. (1999). $^{87}Sr/^{86}Sr$, $\delta^{13}C$ and $\delta^{18}O$ evolution of Phanerozoic seawater. *Chemical Geology* **161**, 59–88.

Wachter, E. A., and Hayes, J. M. (1985). Exchange of oxygen isotopes in carbon dioxide–phosphoric acid systems. *Chemical Geology* **52**, 365–374.

Winograd, I. J., Coplen, T. B., Landwehr, J. M., Riggs, A. C., Ludwig, K. R., Szabo, B. J., Kolesar, P. T., and Revesz, K. M. (1992). Continuous 500,000-year climate record from vein calcite in Devils Hole, Nevada. *Science* **258**, 255–260.

Zachos, J., Pagani, M., Sloan, L., Thomas, E., and Billups, K. (2001). Trends, rhythms, and aberrations in global climate 65 Ma to Present. *Science* **292**, 686–693.

CHAPTER 7

CARBON IN THE LOW-TEMPERATURE ENVIRONMENT

7.1 INTRODUCTION

Carbon dioxide has been called "the most important substance in the biosphere" (Revelle, 1985). It makes life on Earth possible and warms our planet to the habitable condition such that H_2O is in the liquid state. Carbon is the foundation of the world's major energy source, and as fossil fuels continue to be burned at an alarming rate of 30 trillion metric tons of CO_2 emitted per year, we have come to recognize the serious effects of anthropogenic-induced global warming. The carbon cycle is a complex system of feedback mechanisms at many levels and has been the object of intense geologic research. Stable carbon isotopes have played a critical role in this research, constraining amounts and fluxes into and out of the various carbon reservoirs. Other applications of stable carbon isotopes include variations in temperature and productivity in the past, photosynthetic pathways, diets, metabolic pathways, evidence for early life on earth, and variations in greenhouse gas abundances through time. In this chapter, we will review the carbon cycle, discuss plant photosynthesis and carbonate formation as they relate to carbon isotopes, and provide several examples of the application of carbon isotope geochemistry to geological processes.

Carbon exists in oxidized, elemental, and reduced forms. Oxidized forms include CO_2 and carbonates, elemental forms include graphite and diamond, and reduced forms include methane and organic matter. Other examples of each form exist as well. As is the case for almost all compounds, the heavy isotope ^{13}C is concentrated in the more oxidized forms.[1] The metabolic reduction of carbon (i.e., the formation of organic matter) is a non-equilibrium process in which ^{12}C is strongly partitioned into organic matter, leading to two major crustal reservoirs of carbon; a reduced low-$\delta^{13}C$ reservoir and an oxidized

---

[1] Rare exceptions include carbon monoxide (relative to graphite), and nitric oxide and ammonium ion for the nitrogen system.

high-δ^{13}C reservoir. Assuming that most carbon in the surficial or crustal reservoir originated as volcanic CO_2 emissions, the massive biologically induced reduction of carbon over time led to a concomitant production of free oxygen (O_2 gas) critical to most life-forms on Earth today.

7.2 THE CARBON CYCLE

The carbon cycle is complex and has been studied at many scales. The mantle reservoir, which swamps all others in size, is unimportant in a consideration of changes in atmospheric $p(CO_2)$ related to short-term anthropogenic contributions. Transfer to and from the mantle reservoir is simply too slow. Likewise, the abundance of CO_2 in the atmosphere is minuscule compared with the major reservoirs and can be ignored when one is concerned with the long-term carbon budget. However, the transfer of CO_2 between reservoirs can be accomplished via the atmosphere, making it an important *flux* between reservoirs. In addition to size, therefore, fluxes into and out of the different reservoirs are of critical importance.

The major reservoirs for carbon in the crust are sedimentary inorganic carbon in the form of carbonates (e.g., limestones, dolomites), organic carbon, and carbon hosted in crystalline rocks. Reservoirs, their abundance, fluxes, and their δ^{13}C values are given in Table 1 and illustrated in Figs. 7.1 and 7.2.

As a first approximation, we can assume that the carbon cycle is a balanced, steady-state system. Fluxes into and out of each reservoir are more or less the same, as seen in Fig. 7.1, with the anthropogenic flux of fossil fuel CO_2 causing a minor, short-term imbalance. Although atmospheric CO_2 is small in amount, the fluxes involving it are immense, so that equilibrium between oceans and terrestrial organic carbon is maintained at a near steady state. There is a huge annual incorporation of CO_2 during plant respiration (122×10^{15} g C/year) which is nearly balanced by the decomposition of plants and organic carbon in soils. The other large flux is between the atmosphere and surface ocean, where CO_2 is transferred to dissolved bicarbonate and back again. Finally, the flux between the shallow ocean and deep oceans is large and nearly balanced. The fluxes between the other reservoirs—carbonates, the mantle, and organic carbon in sedimentary rocks—are orders of magnitude slower and have relevance only in studies of long-term variation.

TABLE 7.1: Mass and carbon isotope composition of major carbon reservoirs.

Reservoir	Mass/10^{15} gC	Average δ^{13}C (‰ vs. PDB)
Atmosphere (@ 290 PPM)	775	−6 to −7
Ocean (TDC)	35,000	0
(DOC)	1,000	−20
(POC)	3	−22
Land plants	1,600	−25 (−12 for C_4 plants)
Soil humus	1,000–3000	−25 (−12 for C_4 plants)
Sedimentary inorganic C (carbonates)	60,000,000	0 to 1
Organic carbon	15,000,000	−23
Continental silicic crust	7,000,000	−6
Mantle	342,000,000	−5 to −6

(After Anderson and Arthur, 1983, and Des Marais, 2001.)

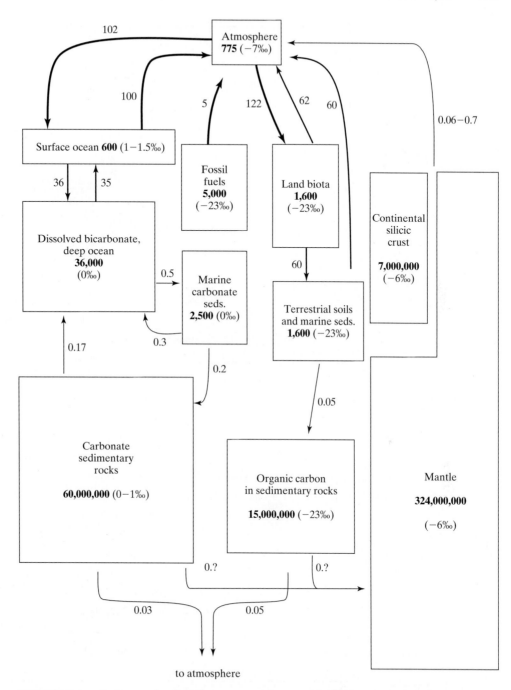

FIGURE 7.1: Carbon cycle, showing amounts, fluxes, and $\delta^{13}C$ values of different reservoirs. Abundance is shown in bold in 10^{15} g. Flux is shown in italics in 10^{15} g/yr. $\delta^{13}C$ values are in parentheses. Flux arrow thickness is proportional to relative rate.

152 Chapter 7 Carbon in the Low-Temperature Environment

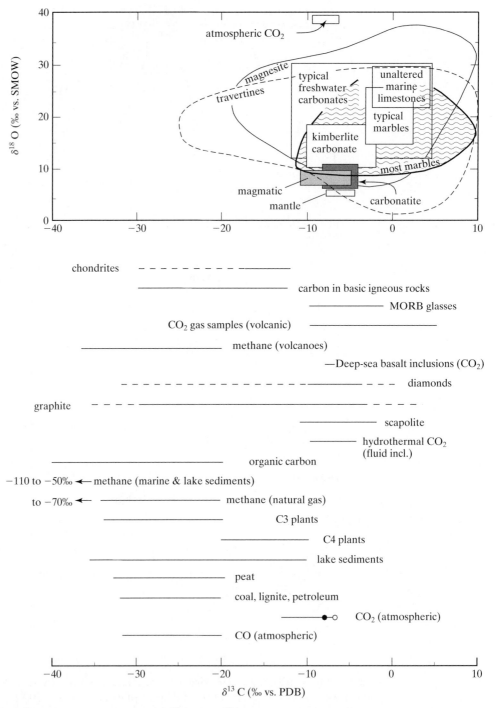

FIGURE 7.2: General range of $\delta^{13}C$ (and $\delta^{18}O$) values of selected carbon-bearing materials. The modern average $\delta^{13}C$ value of the atmosphere is shown by the filled circle; the pre-anthropogenic value is the open circle.

7.2.1 Carbon Isotope Budget of the Earth

Fractionation between oxidized inorganic carbon species is generally small. In contrast, fractionation between inorganic and biologic organic carbon is very large. In fact, isotopically light carbon is a characteristic signature of life. The $\delta^{13}C$ values of 3.7-Ga carbonates from Greenland have been used to postulate a gradual buildup of organic matter through time (e.g., Schidlowski et al., 1979; Mojzsis et al., 1996). If all terrestrial carbon has come from the mantle with minimal fractionation, then the average $\delta^{13}C$ values of the terrestrial carbon reservoirs must equal those of the mantle. The formation of light organic carbon leads to two distinct reservoirs: reduced organic carbon, with low $\delta^{13}C$ values; and inorganic oxidized carbon, with high $\delta^{13}C$ values. The two main terrestrial reservoirs—carbonate sedimentary rocks and organic carbon—should equal $-6‰$ (see 7.3.1), according to the equation

$$x(\delta^{13}C)_{\text{carbonate}} + (1-x)(\delta^{13}C)_{\text{organic carbon}} = -6‰, \tag{7.1}$$

where x is the fraction of carbonate relative to total carbon in both reservoirs. Under the assumption of a $\Delta^{13}C$ (carbonate–organic carbon) value of 23‰, equation (7.1) becomes

$$x(\delta^{13}C)_{\text{carbonate}} + (1-x)(\delta^{13}C_{\text{carbonate}} - 23) = -6‰, \tag{7.2}$$

resulting in a carbonate fraction of 0.74 (assuming that $\delta^{13}C_{\text{carbonate}} = 0$), in good agreement with the data in Table 1. (A similar calculation by Hoefs, 1982, gave a $\delta^{13}C$ value of $-5.5‰$ for "juvenile" carbon).

7.3 CARBON RESERVOIRS

General ranges for $\delta^{13}C$ values of typical reservoirs are shown in Figure 7.2 and described in further detail in the sections that follow.

7.3.1 Mantle

Mantle carbon enters the crustal reservoir mainly from midocean spreading ridges. The $\delta^{13}C$ value of CO_2 from the mantle is ~ -6 to $-5‰$. The $\delta^{13}C$ values of carbonatites and MORB are close to mantle value, with a minor light component (-22 to $-26‰$) for diamonds and MORB (Deines, 2002). The explanation for the low values may be related to the subduction of organic carbon or degassing processes in the mantle. (See Section 11.2.2 for more details.)

7.3.2 Plants

The formation of reduced organic carbon occurs by the reduction of CO_2 during photosynthesis.[2] Other forms of reduced carbon, from petroleum, to animals, to black shales, ultimately can trace their origin back to photosynthesis. Therefore, we begin by reviewing the mechanisms and isotopic fractionations that occur during photosynthesis. Terrestrial plants derive their carbon from the photosynthetic fixation of atmospheric CO_2, as does marine plankton. Other aquatic plants fix carbon from dissolved HCO_3^-.

---

[2] Minor CO_2 reduction also occurs by chemotrophs, which utilize chemical energy rather than light energy.

Photosynthetic organisms remove ~10% of the CO_2 in the atmosphere each year, a percentage that is balanced by the decomposition of plant material at a later stage. The simple glucose-producing reaction involves the reduction of CO_2 and the production of O_2, using water as a hydrogen donor:

$$6\,CO_2 + 12\,H_2O \longrightarrow C_6H_{12}O_6 + 6\,H_2O + 6\,O_2. \qquad (7.3)$$

The overall reaction is divided into two separate mechanisms: the "light reaction" and the "dark reaction." The light reaction involves the transfer of electrons from the donor molecule, water, to the acceptor molecule, nicotinamide adenine dinucleotide phosphate (NADP), an important coenzyme in the plant cell. The dark reaction uses NADPH and adenosine triphosphate (ATP) to fix reduced C and transfer it into carbohydrates in the Calvin–Benson cycle (C_3 cycle).[3] The important step in terms of stable isotope chemistry is carboxylation, which involves the addition of CO_2 to the acceptor molecule, ribulose bisphosphate (RuBP), catalyzed by the enzyme ribulose bisphosphate carboxylase–oxygenase (Rubisco). It is the partial removal of CO_2 during carboxylation that leads to the major carbon isotope fractionation in plants.

An additional process that occurs in plants is *photorespiration*, wherein O_2 is taken in and CO_2 is generated. In this process, Rubisco catalyzes the oxygenation of RuBP, ultimately releasing CO_2, an energy-consuming process. Two other photosynthetic pathways, called the C_4 dicarboxylic acid pathway (C_4 pathway) and the crassulacean acid metabolism (CAM), have evolved that limit the deleterious effects of photorespiration. In these processes, CO_2 is concentrated at the active site of Rubisco to enhance the efficiency of photosynthesis. Photorespiration is mostly eliminated in plants that use these photosynthetic pathways.

A model of the C_4 pathway was first published in 1965 and 1966 (Hatch and Slack, 1966), with $^{14}CO_2$ used as an isotopic tracer to identify the products of photosynthesis in sugarcane. The first labeled products are four-carbon acids (oxaloacetate and then malate), rather than the three-carbon acid 3-phosphoglyceric acid (3PGA) in the C_3 pathway. Malate enters the bundle sheath cells and is decarboxylated, releasing CO_2, which is used by the Calvin cycle as in C_3 plants. By concentrating CO_2 at the site of carboxylation (Rubisco), C_4 (and CAM) minimizes photorespiration, a benefit under conditions of low stomatal conductance.[4] Low stomatal conductance is beneficial in hot arid environments, where minimizing water loss is critical, and under conditions of low $p(CO_2)$. The disadvantage is that C_4 photosynthesis is less efficient than C_3, so there is a competition between plants that use one or the other pathway. The basic rules that are useful for paleoclimate reconstruction are the following (Ehleringer et al., 1997):

- C_4 plants do well under conditions of low $p(CO_2)$.
- C_4 plants have high water efficiency and so are tolerant to high temperatures and aridity. There is a nearly perfect correlation between % C_4 flora in a community and minimum growing-season temperature (Teeri and Stowe, 1976).

Only about 0.4% of angiosperms are C_4 plants, but they account for 18% of total global productivity. C_4 plants include important crops such as maize, sugarcane, sorghum, and tropical pasture grasses. The C_3 pathway is used by trees, most shrubs, herbs, cool-weather grasses, and aquatic plants. C_3 plants are favored in high latitudes, cooler climates, and, because

---

[3] Melvin Calvin won the Nobel Prize in Chemistry in 1961 for his work on the assimilation of CO_2 by plants.
[4] Stomata are the pore openings of the epidermal layer of plant tissue where transpiration occurs.

C_4 plants are so successful in regions with summer rainfall, regions with more arid summers. Forests, woodlands, and high-latitude grasses are generally C_3.

The different photosynthetic pathways were first recognized by stable isotope geochemists on the basis of two distinct populations of $\delta^{13}C$ values in plants (Bender, 1968; Smith and Epstein, 1970). The typical range of C_3 plants is -33 to $-23‰$, with an average of -27 to $-26‰$ (Fig. 7.3). The $\delta^{13}C$ values of C_4 plants are about $13‰$ higher than those of C_3 plants, ranging from -16 to -9 and averaging -13 to $-12‰$. There is no evidence of C_4 plants prior to the Cenozoic (Cerling and Quade, 1993; Cerling et al., 1993).

Aquatic plants derive their carbon from dissolved carbon in water. The range of $\delta^{13}C$ values varies between the different types of plant. The $\delta^{13}C$ values of algae range from -22 to $-10‰$, plankton from -31 to $-18‰$, and kelp from less than -20 to $-10‰$. Most warm-water plankton have a $\delta^{13}C$ value of -22 to $-17‰$. There is a temperature-dependent fractionation of the $\delta^{13}C$ value of plankton (Sackett et al., 1965; Degens et al., 1968). Because plankton get a significant portion of their carbon from dissolved CO_2, as opposed to bicarbonate, the temperature dependence is a function of availability. There is no need to postulate temperature-sensitive enzymatic reactions to explain observed fractionations with change in water temperature. When molecular CO_2 is sufficiently abundant, there is constant fractionation of about $19‰$ between CO_2 and cells over all reasonable temperatures. The average $\delta^{13}C$ value of marine plants is $\sim -20‰$.

The carbon isotope fractionation attending photosynthesis was first outlined by Park and Epstein (1960) who noted that two steps involved significant fractionation. These were the preferential uptake of ^{12}C from air CO_2 and the conversion of dissolved CO_2 by Rubisco

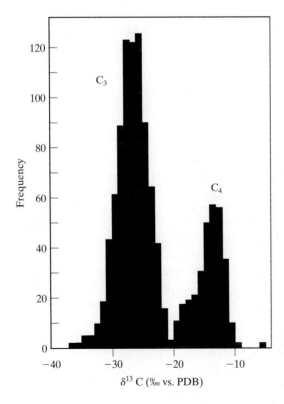

FIGURE 7.3: $\delta^{13}C$ values of C_3 and C_4 terrestrial plants. The difference of $13‰$ between the two groups makes for unambiguous identification. (After Deines, 1980.)

to reduced carbon—A simple expression relating the $\delta^{13}C$ value of a C_3 plant to that of CO_2 in air is given by Farquhar et al. (1989):

$$\Delta^{13}C_{(CO_2\ in\ air\ -\ plant)} = a\frac{p_a - p_i}{p_a} + b\frac{p_i}{p_a}. \tag{7.4}$$

Here, $\Delta^{13}C$ is the overall discrimination during C assimilation, a (4.4‰) is the discrimination associated with gas phase diffusion in air, b (29‰) is the discrimination associated with carboxylation (mainly by Rubisco), and p_a and p_i are the partial pressures of CO_2 in the bulk air and intercellular air spaces, respectively. If we take $p_i = 0.7 p_a$ as a typical ratio, then $\Delta = 21.6$. If the $\delta^{13}C$ value of air is $-7.9‰$,[5] plants will form with a $\delta^{13}C$ value of $-28.9‰$ in good agreement with typical values for modern C_3 plants. With lower internal pressure, the discrimination in the carboxylation step is reduced, and the overall fractionation associated with photosynthesis decreases.

Equation (7.4) can be used to evaluate the effects of different environmental conditions on the $\delta^{13}C$ values of plants. Under conditions of low productivity where fixation is limited by the intensity of Rubisco conversion (e.g., low light, low nutritional levels), p_i approaches p_a. From equation (7.4), fractionation approaches a constant value of 29‰ due to the discrimination during Rubisco assimilation. At the other extreme, when diffusion is limiting due to drought conditions (stomates close, and very little CO_2 can enter the cells), for example, all internal CO_2 is assimilated by Rubisco, p_i/p_a approaches low values, and fractionation is controlled by the different rates of diffusion of $^{12}CO_2$ and $^{13}CO_2$, leading to higher $\delta^{13}C$ values of plants.

C_4 plants differ by transferring CO_2 to the bundle sheath cells, where Rubisco carboxylation fixes CO_2. If all CO_2 were incorporated by Rubisco, there would be no fractionation associated with Rubisco carboxylation. However, there is a finite leakage of CO_2 out of the bundle sheath cells, leading to carbon isotope discrimination during Rubisco fixation. Overall, the discrimination is given by (Farquhar, 1983)

$$\Delta = a\frac{p_a - p_i}{p_a} + (b_4 + b_3\theta)\frac{p_i}{p_a} = a + (b_4 + b_3\theta - a)\frac{p_i}{p_a}. \tag{7.5}$$

In equation (7.5), the b term is expanded to include two fractionation factors: b_3 and b_4. The b_3 term is similar to the b term in equation (7.4) (discrimination by Rubisco), and b_4 ($-5.7‰$) is the discrimination associated with the fixation of gaseous CO_2 in the outer layer of photosynthetic cells by phosphoenolpyruvate (PEP) carboxylation (the C_4 acid step unique to C_4 plants). θ is the fraction of carbon fixed by PEP carboxylation that ultimately leaks out of the bundle sheath cells and contributes to the discrimination at Rubsico. If $\theta = 0$, all carbon fixed by PEP carboxylation is refixed by Rubisco, so that the Rubisco discrimination ($b_3\theta$) equals zero. Typical values of p_i in C_4 plants are in the range from 0.3 to 0.4, and $\theta = 0.3$, giving a Δ value of 4‰, or a $\delta^{13}C$ value of $-11.9‰$ for the plant, in good agreement with measured values. Note that when $b_4 + b_3\theta = a$, $\Delta = a$ and is independent of $p(CO_2)$, as has been observed in experimental studies (Fig. 7.4).

The sensitivity of C_3 plants to $p(CO_2)$ has led researchers to propose using $\delta^{13}C$ of plants as a proxy for $p(CO_2)$ variations in the recent past (e.g., Marino and McElroy, 1991;

---

[5]The Holocene, preindustrial value was $-6.5‰$.

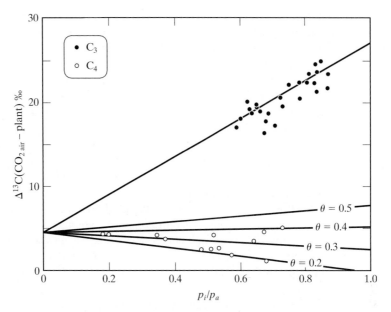

FIGURE 7.4: Variation in isotopic composition of C_3 and C_4 plants as a function of p_i/p_a. Curves are drawn on the basis of equations (7.4) and (7.5), with $a = 4.4, b = b_3 = 27$, $b_4 = -5.7$, and varying θ. The insensitivity of C_4 plants to changing $p(CO_2)$ indicates that $a \approx b_4 + b_3\theta$. (After Farquhar et al., 1988.)

Feng and Epstein, 1995). Similarly, $\delta^{13}C$ values of marine phytoplankton have been used to estimate paleo-$p(CO_2)$ levels (Popp et al., 1989; Freeman and Hayes, 1992).

Local carbon isotope variations occur in dense forest canopies, where plants preferentially *incorporate* ^{12}C, whereas the decay of leaf litter rereleases the ^{12}C-enriched CO_2. The net "canopy effect" causes biomass formed close to the forest floor to have $\delta^{13}C$ values that are approximately 5‰ less than those at the canopy top (Cerling et al., 2004).

7.3.3 Organic Carbon in Sediments

Soil and sedimentary organic carbon reflect the biogenic source. This becomes evident when one considers that coal and disseminated kerogen (i.e., the insoluble organic matter in sediments) have $\delta^{13}C$ values that overlap the range defined by preindustrial C_3 plants. Soils in sugarcane fields have $\delta^{13}C$ values that are consistent with a C_4 plant community. Most petroleum deposits have $\delta^{13}C$ values between -32 and -21‰ similar to values of marine plankton. The isotopic range of most petroleum deposits is -30 to -27‰ compared with that of coal (-26 to -23‰); the difference is consistent with petroleum having a marine sediment origin (Deines, 1980). Keep in mind that modern marine and terrestrial biomass have distinct isotopic ranges.

There are only minor isotopic shifts associated with the preferential oxidation of different components of organic matter. The $\delta^{13}C$ value of soil CO_2 generally matches that of the soil organic matter, indicating little isotopic fractionation during decay. Increasing diagenesis (and catagenesis at higher temperatures) leads to the loss of CO_2, H_2O, and CH_4. H/C ratios decrease with increasing thermal maturity,[6] until a value of zero is reached for graphite, the

---

[6]In a straight-chain *n*-alkane hydrocarbon molecule, the H/C ratio is $2 + 2/C$. In the extreme case of graphite, the H/C ratio is 0.

ultimate product of metamorphism. Inorganic thermal breakdown of the macromolecular carbon skeleton of kerogens favors the breaking of slightly weaker $^{12}C-^{12}C$ bonds, thereby liberating ^{13}C-depleted smaller molecules (e.g., methane and CO_2), such that ^{13}C is slightly concentrated in the residual kerogen.

C/N ratios can be used to distinguish algal and land-plant origins of organic matter. The lower C/N ratios in aqueous plants are related to the absence of cellulose and lignin, in contrast to land plants, in which lignocellulosic structural biopolymers are needed to fight the effects of gravity. In combination with $\delta^{13}C$ values, different sources for organic matter in lake sediments can be identified (Fig. 7.5).

The power of using C/N ratios and $\delta^{13}C$ values in lake sediments to determine origins is illustrated in a core from Lake Baikal in southern Siberia. Lake Baikal is extremely deep and has a record of over 9,000 years of continuous sedimentation. Sediment type, C/N ratios, and $\delta^{13}C$ values all change dramatically at ~6.8 thousand years before the present (Fig. 7.6). The low C/N ratios in the younger sediments and $\delta^{13}C$ values averaging $-28‰$ indicate a very high component of lacustrine algae with a minor vascular plant material component. Older sediments have a much larger land plant component, indicating lower productivity in the lake. The higher $\delta^{13}C$ values suggest a significant C_4 component (Qui et al., 1993).

The use of organic carbon in sedimentary sequences for chemostratigraphic reconstruction is complicated by the wide range of source material and the preferential breakdown of certain components. One way of separating the various components has been to analyze single compounds, using compound-specific gas chromatography techniques. Hayes et al. (1989) found

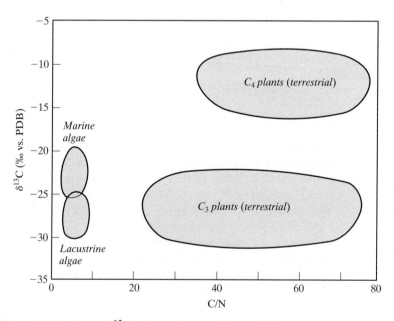

FIGURE 7.5: **C/N ratios vs. $\delta^{13}C$ values of organic matter.** The absence of cellulose in aquatic plants and algae leads to much lower C/N ratios than those for terrestrial plants. Minor diagenesis will not appreciably affect either C/N or isotope ratios, so that detrital material can be traced in sediments. (After Meyers, 1994.)

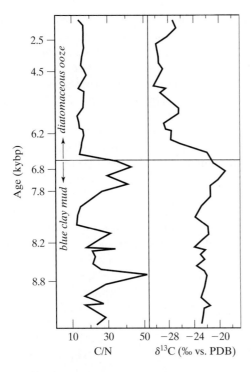

FIGURE 7.6: Variations in C/N ratio and $\delta^{13}C$ value in a sediment column from the northern basin of Lake Baikal in southern Siberia. The change in sediment type ~6.5 thousand years ago is due to a reduction in glacial flour contribution and more in situ algal production. (After Qui et al., 1993.)

that by analyzing only the geoporphyrin fraction, they were able to discriminate primary carbon of marine photosynthetic origin from that derived from nonphotosynthetic sources. The geoporphyrin fractions were consistently depleted in ^{13}C relative to total organic carbon by up to 7‰ and provided far more information than bulk organic carbon alone. Additional work has shown that the organic carbon (as well as biogenic carbonates) could be used to evaluate differences in productivity in the past (e.g., Popp et al., 1989).

The $\delta^{13}C$ value of particulate organic carbon (POC) transferred by rivers is a function of the source. Values for C_3-based catchments are in the range from −25 to −30‰ (Sackett and Thompson, 1963; Strain and Tan, 1979; Spiker, 1981). POC derived from a high percentage of C_4 plants obviously has a higher value than that derived from a mainly C_3 source.

The $\delta^{13}C$ values of total dissolved carbon (TDC), as opposed to particulate organic matter, may also have significant contributions from dissolved carbonate rocks, in addition to the oxidation of organic material. Rivers draining carbonate-poor catchments typically have $\delta^{13}C$ values around −20‰ while carbonate-rich drainages have $\delta^{13}C$ values closer to −10 to −11‰. Under cool conditions, less organic matter is oxidized, and the $\delta^{13}C$ value of river waters reflects a larger contribution of dissolved carbonate. Estimated average global TDC values range from −6 to −9‰ (Anderson and Arthur, 1983) to ~−5‰ (Kump, 1991), a number that is used in considering the global carbon cycle (Section 7.4).

7.3.4 Methane

The $\delta^{13}C$ values of methane are invariably far lower than those of their source material (Rosenfeld and Silverman, 1959). Depending on the mechanism of methane formation, $\delta^{13}C$

values of methane can be as low as $-100‰$ (Fig. 7.7) and form organic matter via methanotrophic organisms with very low $\delta^{13}C$ values (Kaplan and Nissenbaum, 1966). The oxidation of light methane will generate CO_2 and carbonates with anomalously low $\delta^{13}C$ values (Irwin et al., 1977). In freshwater systems, CH_4 is formed by acetate fermentation (e.g., $CH_3COOH \rightarrow CH_4 + CO_2$), wherein the carbon in methane is derived from the methyl group of acetate. In saline environments, methane is produced by the reduction of CO_2 (e.g., $CO_2 + 8(H) \rightarrow CH_4 + 2H_2O$). In the first case, carbon isotope fractionation between acetate and newly formed CH_4 is small. In contrast, a large fractionation between CO_2 and methane is evolved during CO_2 reduction, so that methane formed by this mechanism has much lower $\delta^{13}C$ values (Whiticar et al., 1986; Whiticar, 1999).

7.3.5 Atmospheric CO_2

The average global $\delta^{13}C$ value of CO_2 in the atmosphere has changed from $-6.7‰$ in 1956, to $-7.9‰$ in 1982 (Keeling et al., 1979), to less than $-8‰$ today as a result of the burning of fossil fuels. Variations at a longer timescale have been measured by using C_4 plants[7] and ice cores (Fig. 7.8). Ancient variations in atmospheric $\delta^{13}C$ are more difficult to determine, but can be estimated from marine carbonates. The global atmospheric $\delta^{13}C$ value is related to the

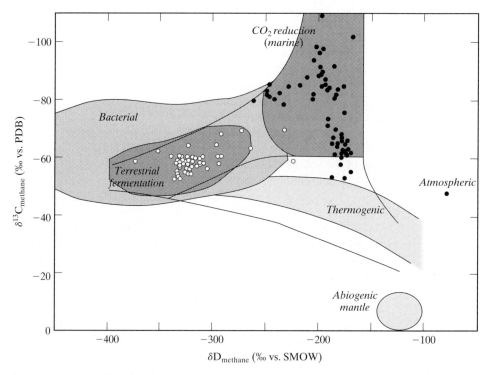

FIGURE 7.7: Combined carbon–hydrogen isotope plot of methane. With the use of combined carbon and hydrogen isotope ratios, formation from terrestrial fermentation, marine CO_2 reduction, and thermal cracking can easily be distinguished. (After Whiticar et al., 1986, and Whiticar, 1999.)

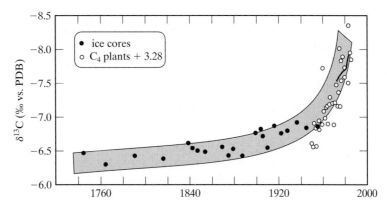

FIGURE 7.8: Variations in $\delta^{13}C$ value of atmospheric CO_2 as a function of age. Data are from direct measurements of atmospheric CO_2 (line starting in 1970), ice-core air inclusions (solid circles), and measured $\delta^{13}C$ values of C_4 maize (open circles). (After Marino and McElroy, 1991.)

overall global carbon cycle, where different sources and sinks and their respective fluxes lead to a quasi-steady-state system.

There are also variations associated with the burning of fossil fuels near large industrial, heavily populated areas. Diurnal, seasonal, and hemispheric variations (Northern Hemisphere vs. Southern Hemisphere) are also found to exist, due to different natural sources and sinks of CO_2 (e.g., Ciais et al., 1995; Clark-Thorne and Yapp, 2003).

7.4 $\delta^{13}C$ VALUES OF CARBONATES

7.4.1 Introduction

Most marine carbonates form in near equilibrium with dissolved inorganic carbon, primarily HCO_3^-. Nonequilibrium, related to *vital* effects (see Section 7.4.3), can be corrected for and, with proper characterization, does not limit the information that is available from carbon isotope data from carbonates. The fractionation during carbonate precipitation is small and relatively insensitive to temperature (unlike the oxygen isotope fractionation), so that $\delta^{13}C$ values of ancient marine carbonates reflect the $\delta^{13}C$ value of dissolved inorganic carbon in the past. Locally, and on relatively short timescales, the $\delta^{13}C$ value of dissolved inorganic carbon will be related to productivity, ocean circulation, weathering, and other factors, such as a rapid input of exotic carbon sources. Over the 10–100-million-year scale, the global $\delta^{13}C$ value of dissolved inorganic carbon varies in relation to the relative proportions of the two major carbon reservoirs—organic carbon and carbonate—and because the residence time of dissolved inorganic carbon in the ocean is long, major carbon isotope variations in marine sequences can be correlated over a wide geographic scale (Scholle and Arthur, 1980).

---

[7]The $\delta^{13}C$ value of C_4 plants is relatively insensitive to $p(CO_2)$, as seen in Fig. 7.4.

These ideas were formulated in part by Keith and Weber (1964) and Weber et al. (1965) 40 years ago. The use of carbon isotope geochemistry in the reconstruction of paleoclimate is, without doubt, one of the most important tools in the paleoclimatologist's modern arsenal.

7.4.2 General Characterization of Carbonates

A certain number of basic rules apply to carbonates:

1. The average $\delta^{18}O$ and $\delta^{13}C$ values of unaltered marine carbonates are close to 0‰. The $\delta^{13}C$ value of carbonates in equilibrium with surface waters is 2–4‰. Lower values are related to vital effects and diagenesis. Deepwater limestones have slightly positive $\delta^{18}O$ values because they form under cold conditions in which the fractionation between carbonates and water is larger.
2. The source of the carbon for marine limestones is dissolved bicarbonate, with a $\Delta^{13}C_{\text{carbonate–dissolved bicarbonate}}$ of 1 to 2‰. Average TDC (predominantly HCO_3^-) for the bulk ocean has a $\delta^{13}C$ value of near 0‰. $\delta^{13}C$ values of TDC in the shallow oceans are higher than those in the overall ocean by about 1 to 2‰ due to the actions of the biological pump (Box 7.1).
3. Terrestrial carbonates have low $\delta^{18}O$ values because they form in equilibrium with meteoric water. Tuffs and freshwater limestones incorporate a combination of dissolved inorganic carbon and atmospheric CO_2. If the dissolved inorganic carbon comes from the dissolution of marine carbonates, the $\delta^{13}C$ value will be near zero. In contrast, dissolved inorganic carbon sourced by the oxidation of organic matter will lead to strongly negative $\delta^{13}C$ values of precipitating carbonates. In general, freshwater limestones have negative $\delta^{13}C$ values, depending on the contribution from organic carbon.
4. The $\delta^{13}C$ value of soil carbonates is controlled by CO_2 respired from soil biota, as well as from decaying organic litter and soil organic carbon, so the $\delta^{13}C$ values of soil carbonates (paleosols) are an indicator of the type of plant (C_3 vs. C_4) living at (or shortly before) the time of carbonate formation. Soils formed in the presence of C_4 plants have $\delta^{13}C$ values 12 to 13‰ higher than those formed in predominantly C_3 plant communities. The $\delta^{18}O$ value is a function of local meteoric water, with secondary evaporation effects.
5. Diagenesis almost always lowers the $\delta^{18}O$ values of marine carbonates (O'Neil, 1987). $\delta^{13}C$ values also tend to become lower during diagenesis, but the effect is often far less pronounced. This is a logical outcome of the fact that most diagenetic aqueous fluids contain only trace amounts of dissolved carbon. (See Fig. 6.8.) Much larger fluid/rock ratios are needed to alter the $\delta^{13}C$ than the $\delta^{18}O$ value. In rare cases, $\delta^{13}C$ values of diagenetically altered marine carbonates will increase by reaction with *residual CO_2* following methanogensis (removal of light carbon).

7.4.3 The Vital Effect

Certain organisms consistently deposit their carbonate shells out of carbon or oxygen (or both) isotope equilibrium with ambient fluids. Known as the *vital effect*, this phenomenon has been explained either by metabolic processes or by a combination of *metabolic effects* and *kinetic effects* (McConnaughey, 1989a; 1989b). Stable isotope relations observed in biogenic carbonates have fascinated isotope geochemists and biologists alike since the inception of the discipline. With appropriate caveats, it is reasonable to state that biogenic carbonate is commonly deposited in or near oxygen isotope equilibrium with ambient waters. Even when disequilibrium is the rule for a given organism, the departure from equilibrium is often fairly constant, and the temperature sensitivity of the disequilibrium fractionation is nearly the same as that of equilibrium fractionations. Thus, relative temperatures can often be determined with a good degree of confidence in these cases, a fact that explains in part the success of the oxygen isotope paleotemperature method.

In stark contrast to the oxygen isotope case, biogenic carbonate is rarely deposited in carbon isotope equilibrium with dissolved carbon in environmental waters. In addition, the overall magnitude of carbon isotope disequilibrium is greater than that for oxygen, because the carbon reservoir (HCO_3^-) in these systems is vastly smaller than the oxygen reservoir (H_2O). In all but a few cases, the direction of disequilibrium is such that ^{12}C is preferentially incorporated into the shells so that $\delta^{13}C$ values are more negative than equilibrium values. Departures from equilibrium precipitation are shown for many extant organisms in Fig. 7.9. The isotopic systematics illustrated in this figure have not changed significantly since 1991, the year that was published.

Several basic rules apply. Slowly precipitating carbonates are most likely to be in equilibrium. There are many shells that do not precipitate carbonates in equilibrium with water, but others do that are lighter than equilibrium values by 4 and 14‰ for oxygen and carbon isotope ratios, respectively. Photosynthesis has no apparent effect on the $\delta^{18}O$ value of newly forming carbonate material, but can lead to *higher* $\delta^{13}C$ values because organic matter formed during photosynthesis preferentially incorporates ^{12}C, leaving the inorganic carbon reservoir with higher $\delta^{13}C$ values.

The *kinetic* aspect of vital effects concerns the relative rates of both diffusion and chemical reactions of different isotopologues of dissolved carbonate species in the body fluids. Light isotopologues (both carbon and oxygen isotopes) diffuse faster than heavy isotopologues and react faster with other substances. As a consequence, $\delta^{18}O$ and $\delta^{13}C$ values of biogenic carbonate secreted *rapidly* with a kinetic isotope effect (1) are more negative than equilibrium values and (2) covary. (See Fig. 7.10.) The word *rapidly* is emphasized because organisms whose isotopic ratios reflect a kinetic process must precipitate their carbonate almost immediately after some key reaction because the normally rapid oxygen isotope exchange between HCO_3^- and H_2O does not take place. Information like this is helpful to marine biologists and biochemists attempting to understand physical and chemical controls on skeletogenesis in a variety of species. The difference between $\delta^{18}O/\delta^{13}C$ relations governed by both kinetic and metabolic effects and those governed by kinetic effects alone are shown nicely for two species of coral in the figure.

Metabolic processes have little effect on oxygen isotope ratios, but they can have a strong effect on carbon isotope ratios of dissolved inorganic carbon (DIC) in the internal pool of body fluids that are used for calcification. The processes of interest cause the *removal* of light CO_2 from this pool during photosynthesis and the *addition* of light CO_2 to the pool from respired CO_2. Photosynthetic organisms like algae preferentially use isotopically light $(CO_2)_{aq}$ to synthesize organic matter in soft tissues, thereby leaving behind isotopically heavy carbon in the internal DIC pool. Thus, if photosynthesis is a viable control on $^{13}C/^{12}C$ ratios in these systems, $\delta^{13}C$ values of skeletal parts of photosynthetic organisms should be more positive than those of nonphotosynthetic organisms living in similar environments. In fact, $\delta^{13}C$ values of shells of hermatypic corals (those containing symbiont zooxanthellae) usually are higher than those of ahermatypic corals. In addition, the higher the rate of photosynthesis or growth (e.g., at higher illumination), the higher are the $\delta^{13}C$ values of certain corals. But just the reverse effect has been noted for large foraminifera in natural settings: The greater the growth rate, the lower is the $\delta^{13}C$ of shell carbonate. Clearly, the effects are complicated.

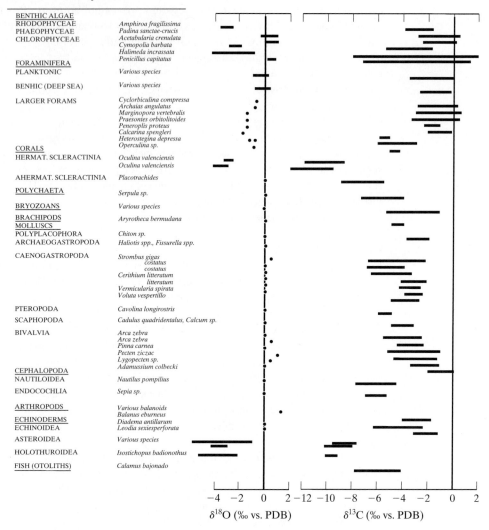

FIGURE 7.9: Summary of differences between equilibrium and measured $\delta^{18}O$ and $\delta^{13}C$ values for various carbonate-secreting organisms. (Reprinted from Wefer and Berger, 1991, with permission from Elsevier.)

Respired CO_2 has lower $\delta^{13}C$ values than that of ambient DIC and can become a constituent of the internal carbon pool in some organisms. For example, Lee and Carpenter (2001) provided good evidence that carbon from respired CO_2 is incorporated into biogenic carbonate by certain taxa of marine calcareous algae. When metabolic factors dominate the vital effect of a population of organisms, $\delta^{13}C$ values can vary widely while $\delta^{18}O$ values remain relatively constant.[8]

---

[8] CO_2 and H_2O are usually in approximate oxygen isotope equilibrium in body fluids due to the presence of the carbonic anhydrase enzyme, which catalyzes the exchange reaction.

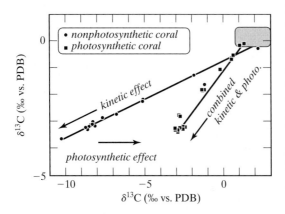

FIGURE 7.10: Stable isotope data for the photosynthetic coral Pavona clavus (squares) and the nonphotosynthetic coral Tubastrea sp. (circles). The covariation of $\delta^{18}O$ and $\delta^{13}C$ values for the ahermatypic coral is the normal *kinetic* relation. The departure to higher $\delta^{13}C$ values for the photosynthesizing coral is explained by the metabolic relation arising from the preferential incorporation of ^{12}C during photosynthesis, causing an elevation of the $\delta^{13}C$ values of the remaining inorganic carbon. (Reprinted from McConnaughey, 1989b, with permission from Elsevier.)

7.4.4 Carbonate Speciation Effects

In addition to vital effects related to kinetic and metabolic effects, isotope fractionations can be controlled by *speciation* effects. Through laboratory culturing experiments, Spero et al. (1997) showed that pH-dependent variations in carbonate speciation in seawater can bear significantly on carbon and oxygen isotope ratios of shells secreted by certain foraminifera and, presumably, other organisms as well. With increasing pH and a concomitant increase in [CO_3^{2-}], both $\delta^{18}O$ and $\delta^{13}C$ values steadily decrease below values predicted by equilibrium relations.[9] The magnitude of the effect is sufficient to lower $\delta^{18}O$ and $\delta^{13}C$ values of *O. universa* shells by 0.2 and 0.6‰, respectively, with an increase of only 0.2 pH unit.

Zeebe (1999) explained these effects by considering equilibrium oxygen isotope fractionation factors between the various species involved in carbonate precipitation. The oxygen isotope fractionation between HCO_3^- and CO_3^{2-} is an astonishingly high 16‰.[10] With an increase in pH, the CO_3^{2-}/HCO_3^- ratio increases such that if the carbonate species are taken up by the organism in proportion to their relative abundance, the slope of the $\delta^{18}O$–[CO_3^{2-}] theoretical relation is −0.0024, a value that is indistinguishable from the value of −0.0022 observed in the laboratory for *O. universa*.

The speciation effect becomes particularly important when oxygen isotope analyses of foraminifera are used to reconstruct seawater temperatures during glacial periods. Several lines of evidence indicate that glacial oceans were more alkaline than modern oceans. If $p(CO_2)$ of the atmosphere was significantly lower in peak glacial times than during interglacial periods, as chemical analyses of inclusions in ice cores would suggest, the concentration of CO_3^{2-} might have increased enough to induce a significant lowering of the $\delta^{18}O$ of shell carbonate that could be incorrectly attributed to an increase in ocean temperature.

7.4.5 Controls on the $\delta^{13}C$ Value of Marine Carbonates over Long Timescales

In spite of the complicating vital effects just discussed, to a first approximation, the $\delta^{13}C$ values of marine carbonates are directly related to the value of total dissolved inorganic carbon (DIC, mainly HCO_3^-). The average whole-ocean TDC has a value of 0‰. The calcite–bicarbonate carbon isotope fractionation is small and relatively insensitive to temperature (Fig. 7.11), so that

---

[9][CO_3^{2-}] increases almost linearly with pH from a normal seawater pH value of 8.0 to a value of 8.6.
[10]There is also a large oxygen isotope fractionation of about 8 per mil between the analogous $H_2PO_4^-$ and PO_4^{3-}. Large fractionations between such similar species are not well understood.

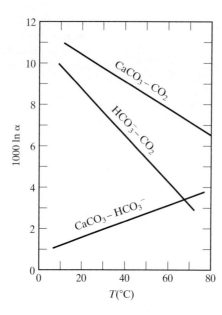

FIGURE 7.11: Carbon isotope fractionation curves for calcite, bicarbonate, and CO_2 gas. Data from Emrich et al. (1970).

measured $\delta^{13}C$ values of marine carbonates directly reflect the $\delta^{13}C$ value of TDC. The near-surface TDC has a $\delta^{13}C$ value that is 1 to 2‰ higher than that of the overall ocean, due to the effects of the biological pump (Box 7.1). These positive $\delta^{13}C$ values are about 9‰ higher than that of atmospheric CO_2, in excellent agreement with estimates for equilibrium fractionation between the two reservoirs (Mook et al., 1974). Given the reservoir sizes of the ocean and atmosphere, it is clear that the oceans control the $\delta^{13}C$ value of the atmosphere in the short term, and not the reverse[11] (excepting the geologically unusual condition associated with the anthropogenic burning of fossil fuels). The deep ocean is immense in relation to the shallow ocean. Just as the $\delta^{13}C$ value

---

BOX 7.1 The biological pump

There is a steady-state, nonequilibrium carbon isotope fractionation between TDC in the shallow and deep ocean due to high biological activity in the near-surface photic zone of the ocean. Photosynthesizing organisms incorporate carbon and nutrients during their growth in the shallow ocean. In doing so, they preferentially incorporate ^{12}C in their tissues, causing a ^{13}C enrichment in the remaining TDC of the shallow ocean. Upon their death, the remains of the organisms sink into the deep ocean, where they are oxidized, releasing light carbon back to the ocean. The continuous removal of light carbon from the shallow ocean in essence acts as a CO_2 pump, transferring light carbon from shallow to deep levels in the ocean and causing a 1 to 2‰ fractionation between the two ocean reservoirs. In times of reduced productivity, the intensity of the biological pump decreases, and the shallow oceans reequilibrate with the larger deep-ocean TDC reservoir. The transfer of carbonates from the shallow to deep ocean has a negligible effect because the $\delta^{13}C$ value of carbonates differs only slightly from that of TDC.

---

[11] The huge CO_2 fluxes between the atmosphere and plants make such fluxes a significant player in the $\delta^{13}C$ value of the ocean as well. The feedback mechanisms are complex and not totally understood.

of the atmosphere is controlled by the shallow ocean (and plants) on a yearly to decadal scale, the $\delta^{13}C$ value of the shallow ocean is controlled by the deep ocean on the thousand-year timescale.

On a timescale of 10^6–10^7 year, the $\delta^{13}C$ value of TDC in the ocean is a function of the proportions of organic and inorganic carbon in the crustal reservoir. TDC will be related to the proportions of inorganic and organic carbon by equation (7.2). Figure 7.12 shows the Phanerozoic secular carbon isotope curve for marine carbonates. Variations are related to tectonic activity (erosion or deposition of organic-rich material) and changing degrees of productivity. Note that the formation or dissolution of carbonates will not affect the $\delta^{13}C$ value of TDC because the isotopic compositions of the two reservoirs are almost the same. During the Carboniferous period, when extremely high amounts of organic carbon were buried, the $\delta^{13}C$ value of inorganic carbon and the TDC of the oceans reached their highest value in the Phanerozoic, and this is reflected in the carbonate record, which shows $\delta^{13}C$ values of carbonates going beyond $+4$‰

Because the reduction of inorganic carbon releases oxygen, the $p(O_2)$ of the atmosphere is closely tied to the amount of carbon in the reduced form.[12] The photosynthetic reduction of CO_2 releases O_2, thereby raising the $p(O_2)$ of the atmosphere (Berner, 1987). Accordingly, we see a close relationship between the long-term $\delta^{13}C$ value of marine carbonates and atmospheric $p(O_2)$ (Fig. 7.13).

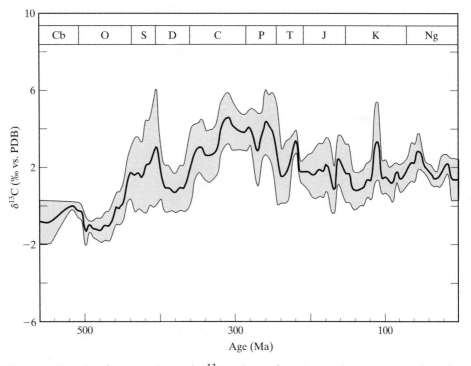

FIGURE 7.12: Secular variations of $\delta^{13}C$ values of marine carbonates. Data from low-magnesium carbonate shells. The shaded area is 1σ uncertainty for a Gaussian distribution. (After Veizer et al., 1999.)

---

[12]Oxygen stored in sulfate can be rereleased as O_2 during sulfate reduction. The removal of organic carbon by subduction also acts to shift the balance between the $\delta^{13}C$ value and $p(O_2)$.

168 Chapter 7 Carbon in the Low-Temperature Environment

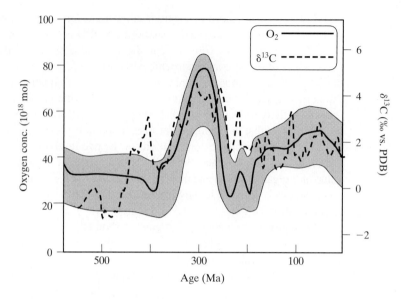

FIGURE 7.13: Correlation between the oxygen concentration of the atmosphere and the $\delta^{13}C$ value of marine carbonates. Curves from Berner and Canfield (1989) and Veizer et al. (1999).

The most extreme long-term variations are found in Proterozoic carbonates (Fig. 7.14). Carbon isotope values in marine carbonates range from slightly less than zero to over 10‰. If these excursions truly represent global events, they require a 20%- to greater-than-50% transfer of carbon between the inorganic and organic reservoirs (Des Marais, 2001), as well as concomitant fluctuations in atmospheric $p(O_2)$. It is almost certain that the large long-term deviations from isotopic steady-state conditions were caused by radical global changes in (i)

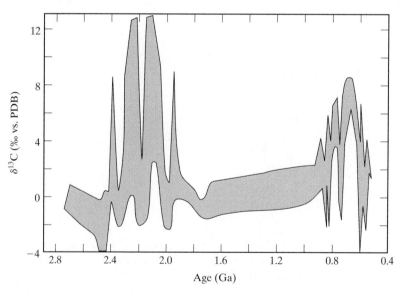

FIGURE 7.14: Carbon isotope excursions in Proterozoic carbonates. (After Des Marais, 2001.)

the ability of biota to reduce vast amounts of carbon and (ii) Earth's dynamic processes, which can temporarily shield reduced organic carbon from re-oxidation.

7.4.6 Variations in the δ^{13}C Values of Marine Carbonates at Short Timescales

Short-term "spikes" in δ^{13}C values of carbonates are related to catastrophic changes in ocean productivity or inputs of light carbon. Over 60 such worldwide events have been recognized (Holser et al., 1996). We must be sure to distinguish clearly between a local and a worldwide event. Locally, isolated shallow marine basins can exhibit sharp carbon isotope excursions due to local geographic effects, such as a changing circulation, overturn, changes in access to the open ocean, and the like. These do not have global significance. Because the residence time of TDC is long (1.8 ka and 55 ka for the shallow and deep oceans, respectively), complete homogenization on the worldwide scale should occur if an event affects the entire planet. To be sure that a carbon isotope excursion is truly of global significance, the event must be identified in geographically distinct carbonate sections throughout the world.

There are several mechanisms that can cause rapid global change in the δ^{13}C value of TDC. The first of these is a sudden loss of productivity due to a catastrophic event, such as the large impact at the Cretaceous/Tertiary (K/T) boundary, intense volcanic activity, or a rapid change in oceanic circulation. The sudden loss of productivity has been termed the "Strangelove Ocean" (Hsü and McKenzie, 1985), after the Stanley Kubrick classic film *Dr. Strangelove*. Loss of productivity shuts down the biological pump, and the shallow ocean quickly equilibrates with the deep ocean on a 100-year timescale (Fig. 7.15). The shallow ocean thus acquires the δ^{13}C values of the much larger deep-ocean reservoir. The recovery and start-up of the biological pump occurs on much longer timescales, ranging from 0.5 to 1 Ma, depending on the rapidity with which biological productivity resumes. The K/T boundary is the classic example of perturbation of the biological pump, wherein all of the features expected by catastrophic productivity loss are observed (Zachos and Arthur, 1986).

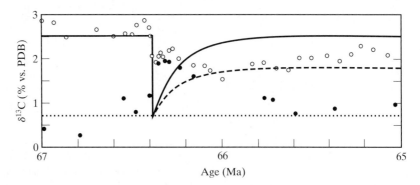

FIGURE 7.15: Variations in the δ^{13}C value of shallow-water unfilled foraminifera (P. petaloidea) and deep-water filled foraminifera (R. rotundata) across the K/T boundary. Convergence of the shallow- and deep-water foraminifera data indicate a loss of productivity and a shutdown of the biological pump. Lines show theoretical changes in carbonate δ^{13}C values across a catastrophic extinction boundary. The solid line represents the shallow-level carbonate trend with return to full recovery; the dashed line is the same, with an organic burial rate equal to 0.9 of the value before the event; the dotted line denotes the predicted deep-water carbonates. (See Zachos et al., 1989; Kump, 1991.)

Certainly, one of the best-refined examples of a global carbon isotope shift due to rapid climate change is that which occurs at the Paleocene/Eocene boundary (Fig. 7.16). Kennett and Stott (1991) recognized that there was a marked decrease in $\delta^{13}C$ values for *both* shallow- and deep-water foraminifera that occurred in just a few thousand years. This is a marked difference from the observations of K/T boundary sections, in which the $\delta^{13}C$ values of shallow and heavy carbon converge to a common value. The shifts can be explained in terms of a reduction in productivity. In the case of the Paleocene/Eocene boundary sections, there is a lowering of *both* planktic and benthic foraminifera $\delta^{13}C$ values. These data cannot be explained by the simple mixing of shallow and deep-ocean waters. To explain them requires the massive input of a ^{13}C-poor carbon source. Initial explanations for the isotopic anomaly at the Paleocene/Eocene were only partially satisfactory. Then, in 1995, Dickens et al. proposed that the dissociation of oceanic methane hydrate could explain not only the low $\delta^{13}C$ values seen in Figure 7.16, but also changes in the oxygen isotope ratios. The release of unstable methane hydrates (with $\delta^{13}C$ values averaging $-65‰$) related to changing ocean circulation patterns would cause a rapid lowering of the $\delta^{13}C$ values of the surface ocean and a more gradual change in the deep ocean as light CO_2 generated by the oxidation of methane was transferred through the thermocline. Similarly, the sharp decrease in $\delta^{18}O$ values in the planktic

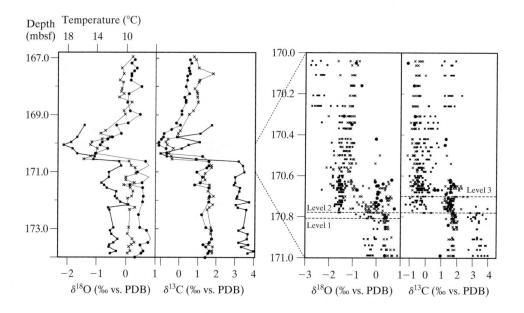

- ■ Surface-dwelling planktonic organisms
- × Thermocline-dwelling planktonic organisms
- • Benthic organisms

FIGURE 7.16: $\delta^{13}C$ and $\delta^{18}O$ values of benthic and planktic foraminifera across the Paleocene–Eocene boundary. Changes in $\delta^{18}O$ values indicate a warming of 5–8°C over a few thousand years. The abrupt lowering of $\delta^{13}C$ values signifies a switching off of the biological pump and the incorporation of light carbon, presumably from methane gas hydrates. (Reprinted from Jenkyns, 2003, with permission from the Royal Society of London.)

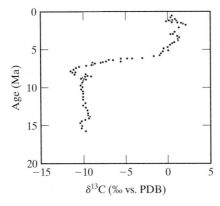

FIGURE 7.17: Carbon isotope shift related to a global shift from a C_3- to C_4-dominated plant community. The increased success of C_4 plants beginning ~7 Ma is related to a decrease in atmospheric $p(CO_2)$. Carbon isotope data are from paleosol carbonates from Pakistan. Similar trends are found in North and South America. (After Cerling et al., 1993.)

foraminifera data indicate a rapid warming of the surface ocean due to the increase in greenhouse gases, whereas the deep ocean took longer to respond.

7.5 $\delta^{13}C$ STUDIES OF TERRESTRIAL CARBONATES

Just about every type of carbonate precipitated on the land surface has been analyzed in order to extract (paleo)climatological information (e.g., Swart et al., 1993). Studies aimed at elucidating information about modern conditions include those on the source of CO_2 for deposition of travertines (Turi, 1986) and those on rainfall and evaporation from oxygen isotope ratios of calcretes (Rossinsky and Swart, 1993). Chronological information can be obtained from speleothems (Schwarcz et al., 1976; Baskaran and Krishnamurthy, 1993), stromatolites (Casanova and Hillaire, 1993), ostracods, soil carbonates (Cerling and Quade, 1993), and varved lake deposits (McKenzie, 1985), to name just a few.

The $\delta^{13}C$ value of pedogenic carbonate is dominated by the vegetation isotopic signal, with significant differences due to varying proportions of C_3 and C_4 plants (Cerling and Quade, 1993). Soil carbonates, fossil tooth enamel, and sediments in the Bengal Fan all show a remarkable worldwide increase of 10 to 12‰ between 7 and 5 Million years ago (Quade et al., 1989; Cerling et al., 1993; France-Lanord and Derry, 1994). This carbon isotope shift (Fig. 7.17) reflects the global expansion of C_4 biomass over a previously dominant C_3 plant community, due to a drop in atmospheric CO_2 levels. The cause of the global effect may be related to the initiation of the Asian monsoon system.

PROBLEM SET

1. During glacial times, the $p(CO_2)$ of the atmosphere dropped sufficiently to cause the (CO_3^{2-}) concentration in ocean surface waters to increase as much as 40 mmol/kg (Spero et al., 1997). Assuming that the effect of (CO_3^{2-}) concentration on $\delta^{13}C$ values of foraminifera carbonate from *Orbulina universa* is given by $\Delta(\delta^{13}C)/(CO_3^{2-} \text{ mmol/kg}) = -0.006$, how much would the $\delta^{13}C$ value of this species change between a glacial–interglacial cycle?

2. Assuming that there are $5,000 \times 10^{15}$ g of fossil fuels yet to be burned and that present consumption is 5.4×10^{15} g/yr (about 44 trillion pounds of CO_2/year),
   a. how much does the $\delta^{13}C$ value of the atmosphere change annually if all of the products of burning go into the atmosphere? What is wrong with this calculation?
   b. What will the $\delta^{13}C$ value of the atmosphere be when all of the available fossil fuels are burned? Assume that half will go into the shallow ocean, with the appropriate fractionation between $CO_2(g)$ and dissolved HCO_3^-.
3. Explain how the $\delta^{13}C$ value of plants changes under conditions of intense water stress.
4. There are three major crustal reservoirs of carbon: carbonates, organic carbon in sediments, and crystalline rocks. If the $\delta^{13}C$ values of carbonates and organic carbon were -1 and $-28‰$ respectively (instead of the estimates given in Table 1), what would be the $\delta^{13}C$ value of average crystalline rocks? Assume the abundances in Table 1 and a mantle value of $-6‰$.
5. Biogenic methane can have a $\Delta^{13}C$ value greater than $50‰$ relative to organic matter. Carbonates with the lowest and highest $\delta^{13}C$ value samples are related to methane. Explain how carbonates with superhigh and superlow $\delta^{13}C$ values can form as a result of methanogenesis.
6. Derive a simple two box model for the shallow-deep ocean system, including the two reservoirs and input and output fluxes. The input flux to the shallow ocean is riverine. Transfer between the two reservoirs is accomplished by a slow flux associated with diffusion. A more rapid transfer from the shallow to the deep ocean is accomplished by the biological pump. Outputs from both shallow and deep ocean are sedimentation of carbonates and organic matter (separate fluxes). (See Kump, 1991, for more details.) Use the following parameters:

   $\delta^{13}C_{\text{riverine input}} = -5‰$, $\delta^{13}C_{\text{carbonates}} = 0$, $\delta^{13}C_{\text{organic matter}} = -25‰$, $\Delta^{13}C_{\text{biol.pump}} = 2‰$.

7. Figure 7.15 shows how the $\delta^{13}C$ values of shallow and deep oceans change with time if the biological pump is suddenly "shut off." (Productivity stops.) Sketch $\delta^{13}C$-vs-time curves for relative deep-ocean and shallow-ocean carbonates for the following conditions (some of these may not be geologically reasonable, but never mind!):
   a. Decreased burial, constant productivity
   b. Decreased burial and decreased productivity
   c. Decreased productivity, constant burial

   (*Hint:* Keep the box model for question 6 in mind; see Holser, 1997, for further discussion.)
8. Explain, in one or two sentences for each entry, the $\delta^{13}C$–$\delta^{18}O$ values of the following carbonates and why they have those values:
   a. deep marine limestones
   b. shallow marine limestones
   c. freshwater limestones
   d. soil carbonates
   e. diagenetically altered marine carbonates
9. The atomic weight of carbon reported in a typical periodic table is 12.0107 amu. Given that the $^{13}C/^{12}C$ ratio of PDB is 0.0112372, what is the $\delta^{13}C$ value of "typical" carbon as defined by the periodic table ($^{12}C = 12$ amu; $^{13}C = 13.003354838$ amu).

ANSWERS

2. The atmosphere has 775×10^{15} gC with a $\delta^{13}C$ value of $-8‰$. If we add 5.4×10^{15} g to the atmosphere, the new $\delta^{13}C$ value will be $\dfrac{775 \times (-8) + 5.4 \times (-25)}{775 + 5.4} = -8.11‰$. It will change by $\sim 0.11‰$/year. Over 20 years, it should change by $\sim 2‰$. There will be some exchange with the oceans, so that the change should be slightly less than this.

3. The stomates will close down as much as possible to prevent water loss. As a result, only a small amount of CO_2 will be admitted into the cell. The internal partial pressure of CO_2 will be far less than the external pressure. From equation 7.4, it is clear that for low internal CO_2 pressures, the fractionation approaches the value for constant a.

4. The mass-balance equation for this problem is
$$\frac{(-28)(15) + (-1)(60) + x(7)}{15 + 60 + 7} = -6.$$
Thus, $x = -1.7‰$

5. Methane will have an extremely low carbon isotope ratio. If the methane is later oxidized to CO_2 or bicarbonate, it will form a carbonate with a very low $\delta^{13}C$ value. Likewise, the removal of methane from a system will leave behind CO_2 or bicarbonate with a very high $\delta^{13}C$ value. If this is converted to carbonate, it will have a high $\delta^{13}C$ value as well.

6. There are two boxes: shallow and deep ocean. Input flux to the shallow ocean is riverine. There is a diffusional exchange between shallow and deep ocean, but the flux of the biological pump (shallow to deep ocean) is larger than the diffusion exchange flux. Both shallow and deep ocean can lose carbon as either carbonates or organic matter.

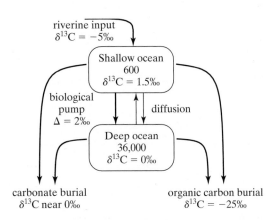

Answer to question 6.

7. **a.** The burial of carbonates has very little effect on the $\delta^{13}C$ values of the ocean, but the burial of organic matter removes light carbon from the ocean, leaving the ocean *heavier*. If burial shuts down, the $\delta^{13}C$ of the ocean will decrease due to light influx of riverine input until it has the same value as that of the river. As long as productivity is active, the biological pump is going to cause a difference between shallow and deep ocean.
   b. Similar to the answer to question a, but without productivity, the difference between shallow and deep ocean will disappear, with the shallow ocean quickly assuming the same $\delta^{13}C$ value as the deep ocean.
   c. In this case, the removal of light carbon continues, so that the overall $\delta^{13}C$ value of the ocean will not change. But without productivity, the difference between shallow and deep ocean disappears.

9. The atomic weight of carbon is given by $12x + 13.00335(1 - x) = 12.0107$. $x = 0.989336$. So the $^{13}C/^{12}C$ ratio of typical 'periodic table' carbon is $(1 - 0.989336)/0.989336 = 0.0107789$.

Using the definition of $\delta^{13}C$, we have $\frac{0.0107789 - 0.0112372}{0.0112372} \times 1000 = -40.78‰$, far lower than the Earth's average value.

174 Chapter 7 Carbon in the Low-Temperature Environment

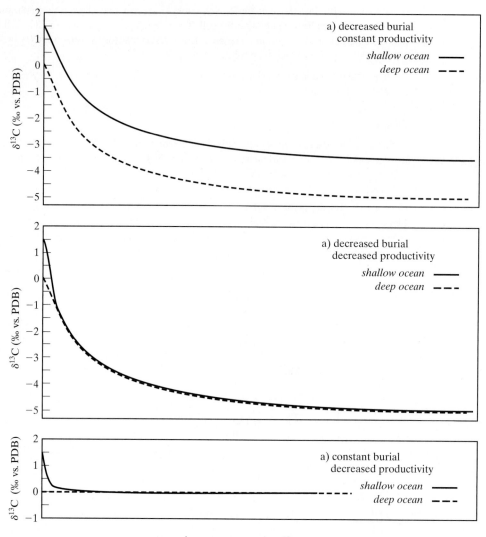

Answer to question 7.

REFERENCES

Anderson, T. F., and Arthur, M. A. (1983). Stable isotopes of oxygen and carbon and their application to sedimentologic and paleoenvironmental problems. In *Stable Isotopes in Sedimentary Geology*, Vol. 10 (ed. M. A. Arthur, T. F. Anderson, I. R. Kaplan, J. Veizer, and L. S. Land), pp. 1–151. Columbia, SC: SEPM Short Course.

Baskaran, M., and Krishnamurthy, R. V. (1993). Speleothems as proxy for the carbon isotope composition of atmospheric CO_2. *Geophysical Research Letters* **20**, 2905–2908.

Bender, M. M. (1968). Mass spectrometric studies of carbon-13 variations in corn and other grasses. *Radiocarbon* **10**, 468–472.

Berner, R. A. (1987). Models for carbon and sulfur cycles and atmospheric oxygen; application to Paleozoic geologic history. *American Journal of Science* **287**, 177–196.

Berner, R. A., and Canfield, D. E. (1989). A new model for atmospheric oxygen over Phanerozoic time. *American Journal of Science* **289**, 333–361.

Casanova, J., and Hillaire, M.-C. (1993). Carbon and oxygen isotopes in African lacustrine stromatolites; palaeohydrological interpretation. In *Climate Change in Continental Isotope Records*, Vol. 78 (ed. P. K. Swart, K. C. Lohmann, J. A. McKenzie, and S. Savin), pp. 123–133. Washington, DC: American Geophysical Union.

Cerling, T. E., Hart, J. A., and Hart, T. B. (2004). Stable isotope ecology in the Ituri forest. *Oecologia* **138**, 5–12.

Cerling, T. E., and Quade, J. (1993). Stable carbon and oxygen isotopes in soil carbonates. In *Climate Change in Continental Isotopic Records*, Vol. 78 (ed. P. K. Swart, K. C. Lohmann, J. McKenzie, and S. Savin), pp. 217–231. Washington, DC: American Geophysical Union.

Cerling, T. E., Wang, Y., and Quade, J. (1993). Expansion of C_4 ecosystems as an indicator of global ecological change in the Late Miocene. *Nature* **361**, 344–345.

Ciais, P., Tans, P. P., Trolier, M., White, J. W. C., and Francey, R. J. (1995). A large northern hemisphere terrestrial CO_2 sink indicated by the $^{13}C/^{12}C$ ratio of atmospheric CO_2. *Science* **269**, 1098–1101.

Clark-Thorne, S. T., and Yapp, C. J. (2003). Stable carbon isotope constraints on mixing and mass balance of CO_2 in an urban atmosphere: Dallas metropolitan area, Texas, USA. *Applied Geochemistry* **18**, 75–95.

Degens, E. T., Guillard, R. R. L., Sackett, W. M., and Hellebust, J. A. (1968). Metabolic fractionation of carbon isotopes in marine plankton—1. Temperature and respiration experiments. *Deep Sea Research* **15**, 1–9.

Deines, P. (1980). The isotopic composition of reduced organic carbon. In *Handbook of Environmental Isotope Geochemistry; Volume 1, The Terrestrial Environment, A*, Vol. 1 (ed. P. Fritz and J. C. Fontes), pp. 329–406. Amsterdam: Elsevier.

Deines, P. (2002). The carbon isotope geochemistry of mantle xenoliths. *Earth-Science Reviews* **58**, 247–278.

Des Marais, D. J. (2001). Isotopic evolution of the biogeochemical carbon cycle during the Precambrian. In *Stable Isotope Geochemistry*, Vol. 43 (ed. J. W. Valley and D. R. Cole), pp. 555–578. Washington, DC: Mineralogical Society of America.

Dickens, G. R., O'Neil, J. R., Rea, D. K., and Owen, R. M. (1995). Dissociation of oceanic methane hydrate as a cause of the carbon isotope excursion at the end of the Paleocene. *Paleoceanography* **10**, 965–971.

Ehleringer, J. R., Cerling, T. E., and Helliker, B. R. (1997). C_4 photosynthesis, atmospheric CO_2, and climate. *Oecologia* **112**, 285–299.

Emrich K., Ehalt, D. H., and Vogel, J. C. (1970). Carbon isotope fractionation during the precipitation of calcium carbonate. *Earth and Planetary Science Letters* **8**, 363–371.

Farquhar, G. D. (1983). On the nature of carbon isotope discrimination in C_4 plants. *Australian Journal of Plant Physiology* **10**, 205–226.

Farquhar, G. D., Ehleringer, J. R., and Hubick, K. T. (1989). Carbon isotope discrimination and photosynthesis. *Annual Review of Plant Physiology and Plant Molecular Biology* **40**, 503–537.

Farquhar, G. D., Hubick, K. T., Condon, A. G., and Richards, R. A. (1988). Carbon isotope fractionation and plant water-use efficiency. In *Stable Isotopes in Ecological Research*, Vol. 68 (ed. P. W. Rundel, J. R. Ehleringer, and K. A. Nagy), pp. 21–40. New York: Springer-Verlag.

Feng, X., and Epstein, S. (1995). Carbon isotopes of trees from arid environments and implications for reconstructing atmospheric CO_2 concentration. *Geochimica et Cosmochimica Acta* **59**, 2599–2608.

France-Lanord, C., and Derry, L. A. (1994). $\delta^{13}C$ of organic carbon in the Bengal Fan: Source evolution and transport of C_3 and C_4 plant carbon to marine sediments. *Geochimica et Cosmochimica Acta* **58**, 4809–4814.

Freeman, K. H., and Hayes, J. M. (1992). Fractionation of carbon isotopes by phytoplankton and estimates of ancient CO_2 levels. *Global Biogeochemical Cycles* **6**, 185–198.

Hatch, M. D., and Slack, C. R. (1966). Photosynthesis in sugarcane leaves: A new carboxylation reaction and the pathway of sugar formation. *Biochemical Journal* **101**, 103–111.

Hayes, J. M., Freeman, K. H., Hoham, C. H., and Popp, B. N. (1989). Compound-specific isotopic analyses: A novel tool for reconstruction of ancient biogeochemical processes. *Organic Geochemistry* **16**, 1115–1128.

Hoefs, J. (1982). Isotope geochemistry of carbon. In *Stable Isotopes* (ed. H.-L. Schmidt, H. Förstel, and K. Heinzinger), pp. 103–113. Amsterdam: Elsevier.

Holser, W. T. (1997). Geochemical events documented in inorganic carbon isotopes. *Palaeogeography, Palaeoclimatology, Palaeoecology* **132**, 173–182.

Holser, W. T., Magaritz, M., and Ripperdan, R. L. (1996). Global Isotopic Events. In *Global Events and Event Stratigraphy in the Phanerozoic* (ed. O. H. Walliser), pp. 63–88. Berlin and New York: Springer.

Hsü, K. J., and McKenzie, J. A. (1985). A "strangelove" ocean in the earliest Tertiary. *Chapman Conference on Natural Variations in Carbon dioxide and the Carbon Cycle*, 487–492. Washington, DC: American Geophysical Union.

Irwin, H., Curtis, C., and Coleman, M. (1977). Isotopic evidence for source of diagenetic carbonates formed during burial of organic-rich sediments. *Nature* **269**, 209–213.

Jenkyns, H. C. (2003). Evidence for rapid climate change in the Mesozoic–Palaeogene greenhouse world. *Philosophical Transactions of the Royal Society of London, A.* **361**, 1885–1916.

Kaplan, I. R., and Nissenbaum, A. (1966). Anomalous carbon-isotope ratios in nonvolatile organic matter. *Science* **153**, 744–745.

Keeling, C. D., Mook, W. G., and Tans, P. P. (1979). Recent trends in the $^{13}C/^{12}C$ ratio of atmospheric carbon dioxide. *Nature* **277**, 121–123.

Keith, M. L., and Weber, J. N. (1964). Carbon and oxygen isotopic composition of selected limestones and fossils. *Geochimica et Cosmochimica Acta* **28**, 1787–1816.

Kennett, J. P., and Stott, L. D. (1991). Abrupt deep-sea warming, palaeoceanographic changes and benthic extinctions at the end of the Palaeocene. *Nature* **353**, 225–229.

Kump, L. R. (1991). Interpreting carbon-isotope excursions; strangelove oceans. *Geology* **19**, 299–302.

Lee, D., and Carpenter, S. J. (2001). Isotopic disequilibrium in marine calcareous algae. *Chemical Geology* **172**, 307–329.

Marino, B. D., and McElroy, M. B. (1991). Isotopic composition of atmospheric CO_2 inferred from carbon in C4 plant cellulose. *Nature* **349**, 127–131.

McConnaughey, T. (1989a). ^{13}C and ^{18}O disequilibrium in biological carbonates: I. Patterns. *Geochimica et Cosmochimica Acta* **53**, 151–162.

McConnaughey, T. (1989b). ^{13}C and ^{18}O disequilibrium in biological carbonates: II. In vitro simulation of kinetic isotope effects. *Geochimica et Cosmochimica Acta* **53**, 163–171.

McKenzie, J. A. (1985). Carbon isotopes and productivity in the lacustrine and marine environment. In *Chemical Processes in Lakes* (ed. W. Stumm), pp. 99–118. New York: Wiley Interscience.

Meyers, P. A. (1994). Preservation of elemental and isotopic source identification in sedimentary organic matter. *Chemical Geology* **114**, 289–302.

Mojzsis, S. J., Arrhenius, G., McKeegan, K. D., Harrison, T. M., Nutman, A. P., and Friend, C. R. L. (1996). Evidence for life on Earth before 3,800 million years ago. *Nature* **384**, 55–59.

Mook, W. G., Bommerson, J. C., and Staverman W. H. (1974). Carbon isotope fractionation between dissolved bicarbonate and gaseous carbon dioxide. *Earth and Planetary Science Letters* **22**, 169–176.

O'Neil, J. R. (1987). Preservation of H, C, and O isotopic ratios in the low temperature environment. In *Stable Isotope Geochemistry of Low Temperature Fluids*, Vol. 13 (ed. T. K. Kyser), pp. 85–128. Saskatoon, SK, Canada: Mineralogical Association of Canada.

Park, R., and Epstein, S. (1960). Carbon isotope fractionation during photosynthesis. *Geochimica et Cosmochimica Acta* **21**, 110–126.

Popp, B. N., Takigiku, R., Hayes, J. M., Louda, J. W., and Baker, E. W. (1989). The post-Paleozoic chronology and mechanism of ^{13}C depletion in primary marine organic matter. *American Journal of Science* **289**, 436–454.

Quade, J., Cerling, T. E., and Bowman, J. R. (1989). Development of Asian monsoon revealed by marked ecological shift during the latest Miocene in northern Pakistan. *Nature* **342**, 163–166.

Qui, L., Williams, D. F., Gvorzdkov, A., and Karabanov, E. (1993). Biogenic silica accumulations and paleoproductivity in the northern basin of Lake Baikal during the Holocene. *Geology* **21**, 25–28.

Revelle, R. (1985). The scientific history of carbon dioxide. In *The Carbon Cycle and Atmospheric CO_2: Natural Variations, Archean to Present*, Vol. 32 (ed. E. T. Sundquist and W. S. Broecker), pp. 1–4. Washington, DC: American Geophysical Union.

Rosenfeld, W. D., and Silverman, S. R. (1959). Carbon isotope fractionation in bacterial production of methane. *Science* **130**, 1658–1659.

Rossinsky, V., Jr., and Swart, P. K. (1993). Influence of climate on the formation and isotopic composition of calcretes. In *Climate Change in Continental Isotopic Records*, Vol. 78 (ed. P. K. Swart, K. C. Lohmann, J. A. McKenzie, and S. Savin), pp. 67–75. Washington, DC: American Geophysical Union.

Sackett, W. M., Eckelmann, W. R., Bender, M. L., and Be, A. W. M. (1965). Temperature dependence of carbon isotope composition in marine plankton and sediments. *Science* **177**, 52–56.

Sackett, W. M., and Thompson, R. R. (1963). Isotopic organic carbon composition of recent continental derived clastic sediments of eastern Gulf Coast, Gulf of Mexico. *Bulletin of the American Association of Petroleum Geologists* **47**, 525–528.

Schidlowski, M., Appel, P. W. U., Eichmann, R., and Junge, C. E. (1979). Carbon isotope geochemistry of the 3.7×10^9 yr-old Isua sediments, West Greenland; implications for the Archaean carbon and oxygen cycles. *Geochimica et Cosmochimica Acta.* **43**, 189–200.

Scholle, P. A., and Arthur, M. A. (1980). Carbon isotope fluctuations in Cretaceous pelagic limestones: Potential stratigraphic and petroleum exploration tool. *The American Association of Petroleum Geologists Bulletin* **64**, 67–87.

Schwarcz, H. P., Harmon, R. S., Thompson, P., and Ford, D. C. (1976). Stable isotope studies of fluid inclusions in speleothems and their paleoclimatic significance. *Geochimica et Cosmochimica Acta* **40**, 657–665.

Smith, B. N., and Epstein, S. (1970). Two categories of $^{13}C/^{12}C$ ratios for higher plants. *Plant Physiology* **47**, 380–383.

Spero, H. J., Bijma, J., Lea, D. W., and Bemis, B. E. (1997). Effect of seawater carbonate concentration on foraminiferal carbon and oxygen isotopes. *Nature* **390**, 497–500.

Spiker, E. C. (1981). Carbon isotopes as indicators of the source and fate of carbon in rivers and estuaries. *Carbon Dioxide Effects Research and Assessment Program: Flux of Organic Carbon by Rivers to the Oceans; Report of a Workshop*. U.S. Dept. of Energy, Office of Energy Research, 75–108.

Strain, P. M., and Tan, F. C. (1979). Carbon and oxygen isotope ratios in the Saguenay Fjord and the St. Lawrence Estuary and their implications for paleoenvironmental studies. *Estuarine Coastal Marine Science* **8**, 119–126.

Swart, P. K., Lohmann, K. C., McKenzie, J. A., and Savin, S. (1993). *Climate Change in Continental Isotopic Records*. In *Geophysical Monograph*, Vol. 78, p. 374.

Teeri, J. A., and Stowe, L. G. (1976). Climatic patterns and the distribution of C_4 grasses in North America. *Oecologia* **23**, 1–12.

Turi, B. (1986). Stable isotope geochemistry of travertines. In *The Terrestrial Environment, B*, Vol. 2 (ed. P. Fritz and J. C. Fontes), pp. 207–238. Amsterdam: Elsevier.

Veizer, J., Ala, D., Azmy, K., Bruckschen, P., Buhl, D., Bruhn, F., Carden, G. A. F., Diener, A., Ebneth, S., Godderis, Y., Jasper, T., Korte, C., Pawellek, F., Podlaha, O. G., and Strauss, H. (1999). $^{87}Sr/^{86}Sr$, $\delta^{13}C$ and $\delta^{18}O$ evolution of Phanerozoic seawater. *Chemical Geology* **161**, 59–88.

Weber, J. N., Bergenback, R. E., Williams, E. G., and Keith, M. L. (1965). Reconstruction of depositional environments in the Pennsylvanian Vanport basin by carbon isotope ratios. *Journal of Sedimentary Petrology* **35**, 36–48.

Wefer, G., and Berger, W. H. (1991). Isotope paleontology: Growth and composition of extant calcareous species. *Marine Geology* **100**, 207–248.

Whiticar, M. J. (1999). Carbon and hydrogen isotope systematics of bacterial formation and oxidation of methane. *Chemical Geology* **161**, 291–314.

Whiticar, M. J., Faber, E., and Schoell, M. (1986). Biogenic methane formation in marine and freshwater environments; CO_2 reduction vs. acetate fermentation; isotope evidence. *Geochimica et Cosmochimica Acta* **50**, 693–709.

Zachos, J. C., and Arthur, M. A. (1986). Paleoceanography of the Cretaceous/Tertiary boundary event; inferences from stable isotopic and other data. *Paleoceanography* **1**, 5–26.

Zachos, J. C., Arthur, M. A., and Dean, W. E. (1989). Geochemical evidence for suppression of pelagic marine productivity at the Cretaceous/Tertiary boundary. *Nature* **337**, 61–64.

Zeebe, R. E. (1999). An explanation of the effect of seawater carbonate concentration on foraminiferal oxygen isotopes. *Geochimica et Cosmochimica Acta* **63**, 2001–2007.

CHAPTER 8

LOW-TEMPERATURE MINERALS, EXCLUSIVE OF CARBONATES

8.1 INTRODUCTION

Although carbonates are the most analyzed of all sedimentary phases, they are not the only option open to the stable isotope geochemist interested in reconstructing conditions or processes occurring in the past. A large number of other materials are available, notably phosphates, cherts, iron and manganese oxides, and clay minerals. The benefits and disadvantages, as well as the information that can be retrieved from the non-carbonate low-temperature minerals, varies widely. Phosphates and cherts are far more resistant to diagenesis than carbonates are. Clay minerals contain both oxygen and hydrogen and are of both marine and terrestrial origin. Bone or tooth apatite is now widely used to reconstruct temperatures in continental settings *and* to improve our understanding of extinct (and extant) animal physiology and behavior.

8.2 PHOSPHATES

Phosphates form over a wide range of pressures and temperatures and are somewhat of a "garbage-can" mineral, incorporating a wide range of elements. Apatite is the most common phosphate mineral, consisting of the end members (1) fluor-apatite, $Ca_5(PO_4)_3F$; (2) chlorapatite, $Ca_5(PO_4)_3Cl$; (3) hydroxyapatite, $Ca_5(PO_4)_3(OH)$; and (4) carbonate apatite, $Ca_5(PO_4, CO_3, OH)_3(F, OH)$.[1] Apatite is found in almost all types of igneous and metamorphic rock, in sedimentary rocks, and as the hard parts of animals, both shell material and bones. In the next section, we are concerned only with low-temperature carbonate apatite, or

---

[1]Driessens (1980) gives the general formula $Ca_{10}(PO_4)_6(CO_3)_{0.15}(Cl)_{0.1}(OH)_{1.6}$ for tooth enamel and discusses chemical variations in detail.

"biological" apatite. One important distinction between phosphates and carbonates from a stable isotope point of view is that carbonate can form directly from water, whereas low-temperature inorganic phosphate formation is extremely slow unless mediated by biochemical enzyme-catalyzed reactions.

8.2.1 Analytical Techniques

Tudge (1960) was the first person to tackle phosphates for oxygen isotope analysis. He recognized that a major limitation of the carbonate paleothermometer was that one could not measure the $\delta^{18}O$ value of the water from which the carbonate precipitated. Therefore, one had to "guess" this value. Tudge knew that two phases which precipitated at the same time and in equilibrium could be used for paleotemperature estimates that would be independent of the $\delta^{18}O$ value of water. Urey had calculated that the $^{18}O/^{16}O$ fractionation between the CO_3^{2-} and PO_4^{3-} ion was large, so that syngenetic carbonates and phosphates should act as an excellent paleothermometer (Fig. 8.1). Part of the problem is that analysis by fluorination of phosphates is difficult. Earlier attempts by Craig and Steinberg[2] to fluorinate $Ca_3(PO_4)_2$ were unsuccessful. Tudge developed a procedure whereby biogenic phosphate was dissolved in acids and ultimately reprecipitated as $BiPO_4$, which was then fluorinated to yield O_2 gas. His technique was used successfully for some 30 years, until a

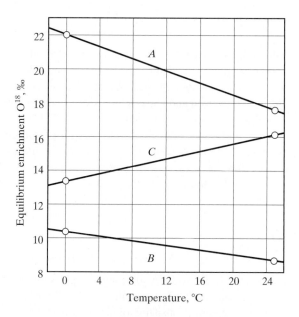

FIGURE 8.1: Reproduction of Urey's calculations of ^{18}O enrichment in phosphate relative to carbonate from Tudge (1960). Curve A is for $CO_3^{2-}{}_{(g)}-H_2O_{(l)}$, curve B is for $PO_4^{3-}{}_{(g)}-H_2O_{(l)}$, and curve C is for $PO_4^{3-}{}_{(g)} - CO_3^{2-}{}_{(g)} + 25‰$. Unfortunately, the large fractation predicted by Urey is not seen in biological phosphate–carbonate pairs. (Reprinted with permission from Elsevier.)

---

[2]The work of Craig and Steinberg is discussed in Tudge, but not published.

more straightforward method of producing Ag_3PO_4 was published (Crowson et al., 1991). Originally converted to O_2 by fluorination, Ag_3PO_4 is now routinely reduced with graphite for analysis as CO_2 or CO (O'Neil et al., 1994; Vennemann et al., 2002). In a majority of oxygen and carbon isotope studies of modern or relatively young biogenic apatite, the CO_3^{2-}-bound oxygen and carbon can be measured by means of the analytically painless phosphoric acid technique normally reserved for carbonates.[3] Oxygen in the carbonate ion in apatite is far easier to analyze than oxygen in the phosphate ion, but is also more easily reset during diagenesis.

8.2.2 Applications to Marine Paleothermometry

Tudge never pursued the phosphate thermometer, and the idea remained dormant until pioneering work by Longinelli, starting with a publication in *Nature* (1965). Longinelli measured phosphate from shells of living marine organisms, assigning an average growth temperature to each sample (Longinelli and Nuti, 1973) and thereby deriving the fractionation equation.[4]

$$t(°C) = 111.4 - 4.3(\delta_p - \delta_w + 0.5) \tag{8.1}$$

Like Epstein's carbonate–water equation of 1953, Longinelli's phosphate–water equation has stood the test of time. In this equation, δ_p and δ_w are the $\delta^{18}O$ values of the phosphate and water on the same scale (e.g., SMOW). Typical values for δ_p of marine phosphates are 20‰ vs. SMOW, corresponding to a temperature of 23°C.

Unfortunately, the temperature dependence of phosphate–water fractionation is almost identical to that of carbonate–water fractionation. There is a ~9‰ fractionation between carbonates and phosphates, but it remains constant over the entire temperature range of biological activity. The idea of using combined phosphate–carbonate for paleothermometry was dead. However, phosphates could still be used as a paleothermometer if one made the same assumptions about water composition as had been done for carbonates. This was certainly not ideal, but the resilience of phosphates to diagenetic alteration makes them an attractive mineral for reconstructing paleoclimates.

In a series of three papers, Kolodny, Luz, and Shemesh revisited biogenic phosphates in the early to mid-eighties. In Kolodny et al. (1983), they posed a nagging question: If the exchange rate between inorganic aqueous solutions and PO_4^{3-} ions is so slow, then do organisms incorporate phosphate in oxygen isotope equilibrium with ambient water, or do they simply inherit the $\delta^{18}O$ value of the phosphate ion from food? The authors used several approaches to address this question. First, they measured the $\delta^{18}O$ values of the bones of fish living at different depths in Lake Baikal. Presumably, the $\delta^{18}O$ value of water and dissolved phosphate would be the same at all depths, but the lake is thermally stratified. If the phosphate oxygen incorporated by an organism was in equilibrium with ambient water due to enzyme-catalyzed exchange reactions, then there should be a regular increase in the $\delta^{18}O$ value of biological phosphate with depth (lower temperature, larger fractionation). If phosphate was incorporated without reequilibration, the $\delta^{18}O$ value of the fish bones should be constant, regardless of temperature. The authors found that

---

[3]The equilibrium $\Delta^{18}O(CO_3^{2-}-PO_4^{3-})$ value for biogenic phosphate formed at 37°C is ~8.7‰ (Bryant, Koch, Froelich, Showers, and Genna, 1996).

[4]The equation was incorrectly stated in the original paper. It was modified by Friedman and O'Neil (1977) to yield a new $\alpha_{CO2-H2O}$ value. Kolodny et al. (1983) derived the empirical calibration equation $113.3 - 4.38(\delta_p - \delta_w)$, which is nearly identical to the original corrected equation.

the $\delta^{18}O$ values of fish bones did indeed increase, from 6.3‰ near the surface to 9.2‰ in the deepest levels, consistent with measured temperature variations in the lake.

A second test was to raise fish living in otherwise identical conditions, with food of different $\delta^{18}O_{phosphate}$ values. Fish meal varied by 5‰ between fisheries, but the $\Delta^{18}O_{bone\ phosphate-water}$) values were the same. From these results, it is clear that, during apatite formation, phosphate oxygen bonds are broken through enzymatic reactions and complete isotopic exchange with ambient (body) water occurs. With their fears abated, Kolodny, Luz, and Shemesh were ready to tackle problems of paleoclimate.

Work on phosphorites, fish teeth and bones, and conodonts shows clear trends toward lower $\delta^{18}O$ values in older samples. The trend could be due to changes in the $^{18}O/^{16}O$ ratio of the ocean with time, changing ocean temperatures, or diagenesis. Shemesh et al. (1983) made a strong case against diagenesis and demonstrated the resistance of phosphorites to diagenesis by comparing cherts, carbonates, and phosphates from the Campanian Mishash formation in Israel. The $\delta^{18}O$ values of cherts and carbonates cover a large range, with only the very highest values consistent with unaltered material. All other samples appear to have been reset. In contrast, the phosphates preserve the expected $\delta^{18}O$ value of formation, with very little variation (Fig. 8.2).

The trend for marine phosphates plotted against age is remarkably similar to the marine carbonate curve, reinforcing the validity of the carbonate curve (Fig. 8.3). The major difference is that the carbonate curve excludes many samples that have been altered. The data used to generate the carbonate curve are the "cream of the crop"—those few samples which

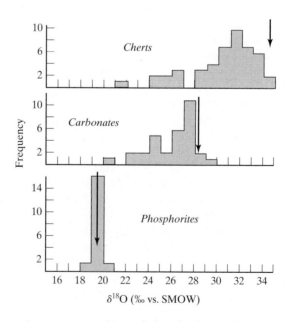

FIGURE 8.2: Oxygen isotope composition of phosphorites, carbonates, and cherts from the Campanian Mishash formation, Israel. The vertical arrows represent expected $\delta^{18}O$ values for each mineral in equilibrium with marine waters. Both carbonates and cherts are skewed to lower $\delta^{18}O$ values, consistent with diagenesis. The phosphates appear to preserve the original equilibrium $\delta^{18}O$ values, free from diagenesis. (After Shemesh et al., 1983.)

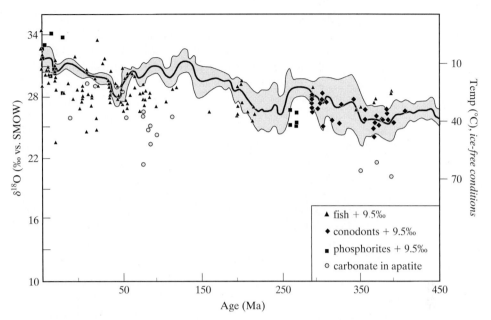

FIGURE 8.3: Compilation of marine phosphate analyses as a function of time. 9.5‰ was added to all points in order to generate a curve directly comparable to that for carbonates (the continuous curve). The data on carbonates in apatite are the $\delta^{18}O$ values of the carbonate component in apatite. Unlike the phosphate, which appears to preserve primary values, the carbonate component in old samples is clearly reset. (Data from Shemesh et al., 1983; Luz, Kolodny, and Kovach, 1984; Kolodny and Luz, 1991; Lécuyer et al., 1993.)

most likely preserve primary $\delta^{18}O$ values. In contrast, the phosphate data include all analyses, clearly illustrating the resistance of phosphates to diagenesis.

8.2.3 Application to Mammals: Theory

All of the aforementioned examples are for marine samples, with the ultimate goal of defining changes in temperature or the $\delta^{18}O$ value of the ocean through time. Longinelli (1973) first proposed analyzing bones or teeth from mammals, with a different goal in mind. Rather than using the data for marine paleothermometry, as is the case with marine ectotherms,[5] he envisioned using the oxygen isotope data to determine meteoric water values in ancient continental environments. The $\delta^{18}O$ value of meteoric water is a function of temperature, latitude, altitude, etc. (see Chapter 4), so the data from mammals would ultimately provide information about the terrestrial paleoclimate. The basic idea is simple: The body temperature of a mammal is constant; consequently, the $\delta^{18}O$ values of their apatite, in either bones or teeth, is a function only of the $\delta^{18}O$ value of their body water, which in turn is controlled by drinking (meteoric) water. Thus, phosphate is an indirect proxy for ancient meteoric water values. As

---

[5]Ectotherms are what is commonly referred to as cold-blooded organisms, those which are unable to internally regulate their body temperature. Endotherms, comprising mammals and birds, are warm-blooded creatures. Homeotherms are all animals that regulate their body temperature, regardless of metabolism.

intriguing as this idea sounds, it was another decade before anyone paid it serious attention, perhaps because the problems were viewed more as biological ones, whereas most stable isotope laboratories were housed in geological institutes. Today, there are at least a dozen papers per year on the subject! Questions addressed include paleoclimatology, paleophysiology, paleoecology, and more. Given page constraints, only a few major considerations and examples are mentioned here.

Longinelli (1984) laid out the basic ideas and premises having to do with mammalian bone phosphate: (1) the body temperature of mammals is constant, (2) the mean isotopic composition of environmental water (i.e., meteoric water) is the main variable controlling the oxygen isotope composition of body water, and (3) bone or tooth phosphate occurs in equilibrium with body water. If these conditions are met, then mammalian phosphate is an indirect proxy for the $\delta^{18}O$ value of local meteoric water at the time the animal lived. Longinelli tested his idea by measuring blood samples from pigs, wild boar, deer, and humans from various regions. The $\delta^{18}O$ values of any given population were constant and varied linearly with respect to meteoric water. The slopes of the best-fit lines were all less than unity, varying from species to species. For pigs and humans, the relationships are, respectively,

$$\delta^{18}O_{blood} = 0.88 \delta^{18}O_{meteoric\ water} + 2.1 \quad \text{(pigs)} \tag{8.2}$$

and

$$\delta^{18}O_{blood} = 0.60 \delta^{18}O_{meteoric\ water} + 0.68 \quad \text{(humans)}. \tag{8.3}$$

For all species, the fractionation between blood and bone is constant, as expected. Following Longinelli's work, Luz, Kolodny, and Horowitz (1984) provided a quantitative explanation for the relationship between the $\delta^{18}O$ value of body water, and drinking water, using simple box models of inputs and outputs.[6] Figure 8.4 illustrates the system. An organism has oxygen inputs from three sources: drinking water, atmospheric oxygen, and food oxygen. The $\delta^{18}O$ values of drinking water and food oxygen are closely related to that of local meteoric water, while atmospheric oxygen is constant the world over. The major outputs are fluid loss (urine,

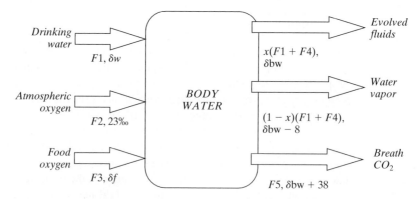

FIGURE 8.4: Box model relating $\delta^{18}O$ value of body water to those of drinking water and food. (After Luz et al., 1984a.)

---

[6]More detailed calculations have since been made (e.g., Ayliffe and Chivas, 1990; Bryant and Froelich, 1995; Kohn et al., 1996), but results of these later models vary little from those obtained in the original work.

sweat, etc.), water vapor in breath, and respired CO_2. Oxygen is also exhaled, but it is unused oxygen, with the same $\delta^{18}O$ value coming in as going out, so it does not enter into the overall flux calculations. Evolved fluids are assumed to have the $\delta^{18}O$ value of body water, whereas the $\delta^{18}O$ values of water vapor and CO_2 are related to that of body water by known and constant fractionation factors at 37°C. Fluxes normalized to O_2 are given for each input and output. Note that the amount of evolved fluid is equal to the original fluid plus a contribution from metabolic water ($F4$).

If the animal remains a constant size, then the inputs and outputs must equal each other. The overall isotopic composition of what is going in must equal that going out; otherwise the $\delta^{18}O$ value of the organism would change. A mass balance is given by

$$F1\delta w + 23F2 + F3\delta f = x(F1 + F4)\delta bw$$
$$+ (1 - x)(F1 + F4)(\delta bw - 8) + F5(\delta bw + 35). \quad (8.4)$$

Equation (8.4) can be rearranged to solve for the $\delta^{18}O$ value of body water:

$$\delta bw = \frac{F1}{F1 + F4 + F5}\delta w + \frac{8(1 - x)(F1 + F4) + 23F2 - 38F5 + F3\delta f}{F1 + F4 + F5}. \quad (8.5)$$

Equation (8.5) is of the same form as equations (8.2) and (8.3) for constant δf. The slope is defined by the proportion of drinking water related to oxygen generated by metabolic processes. The more intense the metabolism (or, conversely, the less an animal drinks), the shallower the slope relating $\delta^{18}O_{\text{body water}}$ to $\delta^{18}O_{\text{meteoric water}}$. In natural systems, the $\delta^{18}O$ value of food will depend on the $\delta^{18}O$ value of meteoric water, especially for herbivores. If we let $\delta f = \delta w + K$, then equation (8.5) becomes

$$\delta bw = \frac{F1 + F3}{F1 + F4 + F5}\delta w + \frac{8(1 - x)(F1 + F4) + 23F2 - 38F5 + F3(K)}{F1 + F4 + F5} \quad (8.6)$$

and we also have

$$\delta bw = \frac{(F_{\text{total input}}) - F2}{(F_{\text{total input}})}\delta w + \frac{8(1 - x)(F1 + F4) + 23F2 - 38F5 + F3(K)}{F1 + F4 + F5}. \quad (8.7)$$

In equation (8.7), we see that the relationship between body water and drinking water is controlled by the proportion of atmospheric O_2 to the total input flux. Because the $\delta^{18}O$ value of atmospheric oxygen is constant, the higher the proportion of oxygen in the total input flux, the shallower will be the slope relating δbw to δw.

The distinction between equations (8.5) and (8.6) becomes clear when one considers animals that do not drink or that drink very little. There is a strong correlation between $\delta^{18}O$ and δD values of rabbits, kangaroos, and deer (in their collagen) with relative humidity (Ayliffe and Chivas, 1990; Cormie et al., 1994; Huertas et al., 1995). These herbivores all derive a significant quantity of their water from plants, which, in turn, undergo varying amounts of evaporative water loss as a function of relative humidity. The result is that this "humidity signal" is recorded in the animal's bones. The extreme case is seen in the kangaroo rat, an animal that does not drink water, deriving all of its water from the metabolism of food!

8.2.4 Sample Applications

Stable isotope analysis of mammalian bone and tooth enamel has been used to address a wide range of applications. One of the first was by Koch et al. (1989), who succeeded in demonstrating the power of using the oxygen isotope geochemistry of mammalian apatite to address paleoclimate-related problems and to show that CO_3^{2-}-bound oxygen in apatite retains its initial $\delta^{18}O$ value, at least for modern and fairly recent samples.[7] (The obvious advantage is that the carbonate component of apatite is analyzed with the relatively simple phosphoric acid technique.) Koch and his colleagues found that seasonal variations in mastodont and mammoth tusks were preserved, with low $\delta^{18}O$ values correlating with cold-season growth.

Since this work, a number of other studies have examined seasonal variations by making high spatial resolution measurements on tooth enamel. Tooth enamel is generally the phase of choice, because it does not recrystallize during the life of the animal and it is far more resistant to diagenesis than bone, possessing a more coarsely crystallized structure and lower organic content. The crystallinity of tooth enamel is unchanged over periods in excess of 1 million years (Ayliffe et al., 1994), whereas bone recrystallization occur in a matter of years (Tuross et al., 1989). The high organic content of bone also presents the opportunity for diagenetic alteration during early bacterial activity (e.g., Blake et al., 1997). In Triassic samples, only tooth enamel appears to preserve original $\delta^{18}O$ values, while bone is clearly reset (Sharp et al., 2000). High-spatial-resolution studies have now been made on horse, beaver, sheep, and dinosaur teeth, with the researchers being able to constrain seasonal climate variations in the past and place limits on migration routes and distances. Even pinpointing the timing of major evolutionary changes has been done with apatite analyses, as demonstrated by the next example.

One question that has been of great interest to evolutionary paleontologists is when cetaceans (whales, dolphins, and porpoises) changed their metabolism from one in which only freshwater was ingested to the modern mechanism whereby their osmoregulatory system can cope with the excess salt load ingested by drinking pure seawater. From fossil localities and their associated sedimentary formations, the relative timing of the evolution of these animals from land to ocean was well known, but when the cetacean osmoregulatory system evolved was not. With this in mind, Thewissen et al. (1996) measured $\delta^{18}O$ values of cetacean teeth from Eocene Tethyan cetaceans. The $\delta^{18}O$ values of modern freshwater and marine cetaceans are distinct (Fig. 8.5). The authors found a jump between *Ambulocetus natans* and *Indocetus* sp., the former having $\delta^{18}O$ values consistent with freshwater and the latter with values clearly related to marine water. This result was unexpected, because *Ambulocetus* is found in sedimentary beds of unambiguous marine origin. The authors concluded that *Ambulocetus* ingested only freshwater or that it lived in freshwater while its teeth were mineralizing, only later migrating to the ocean. Metabolic studies have also been made that were aimed at gaining a better understanding of whether dinosaurs were homeotherms (Barrick and Showers, 1994; Fricke and Rogers, 2000). Interpretations are complex, but suggest that the dinosaurs may have been homeotherms.

---

[7]What defines a "recent" sample in terms of whether diagenesis has or has not occurred is quite variable, depending on conditions and material. Chemical changes in bone can occur in only a few years, with major crystallographic changes occurring over thousands to millions of years. How and when this might affect the $\delta^{18}O$ value of PO_4^{3-} or CO_3^{2-} oxygen is less clear. Pliocene, and perhaps Miocene samples often preserve initial $\delta^{18}O$ values in the CO_3^{2-} site of phosphate.

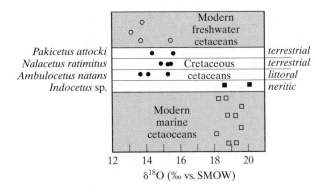

FIGURE 8.5: $\delta^{18}O$ values of modern freshwater and marine cetaceans and samples from the Eocene Tethys showing the transition from freshwater to marine conditions. (After Thewissen et al., 1996.)

The primary use of oxygen isotope analyses of mammalian apatite has been to reconstruct paleometeoric water values, which themselves tell us a great deal about paleoclimatological conditions. Examples of topics that have been addressed include glacial–interglacial transitions (Ayliffe et al., 1992), paleoaltitude estimates (Fricke, 2003), migration patterns (Koch et al., 1995), and seasons of growth (Bryant, Froelich, Showers, and Genna, 1996; Fricke and O'Neil, 1996; Sharp and Cerling, 1997; Balasse et al., 2003).

Carbon isotope variations in the carbonate site in apatite have been used to reconstruct paleoclimatic conditions and paleodiets. First-order variations in $\delta^{13}C$ values of mammalian carbonate are related to the proportion of C_3 and C_4 plants in the diet of the animal. C_3 plants have an average $\delta^{13}C$ value of $-26.7‰$, while C_4 plants have an average $\delta^{13}C$ value of $-12.5‰$. (See Chapter 7.) The $\delta^{13}C$ value of tooth enamel is approximately 14‰ higher than that of the food source (Cerling et al., 1998). Relative to C_4 plants, C_3 plants are favored under conditions of high atmospheric $p(CO_2)$ and low daytime growing-season temperatures. Cerling et al. (1993) found a global change in the $\delta^{13}C$ values of fossil horse teeth around 8 Ma. Prior to that time, the $\delta^{13}C$ value of horse teeth (and the teeth of other grazing animals) was restricted to a range from -15 to $-8‰$, with the mode at $-10‰$, typical of a pure C_3 diet (Fig. 8.6). In more recent samples, there is a bimodal distribution. The C_3 diet is still seen, but a second mode centered around 0 to 2‰ emerges, distinctive of a pure C_4 diet. Over this same time interval, major evolutionary changes in flora and fauna occurred. Changes in vegetation due to lower CO_2 levels in the atmosphere (CO_2 starvation), and extinctions of flora and fauna appear to be closely related.

Dietary reconstruction using stable isotopes is important for archaeological studies. The early hominid diet is constrained by the carbon isotope composition of tooth enamel. There had been a general consensus that the 3-Ma hominid *Australopithecus africanus* lived on a diet of fruits and leaves, in agreement with the belief that they occupied heavily wooded habitats. If this were indeed the case, the $\delta^{13}C$ values of tooth enamel would be consistent with a nearly pure C_3 diet, as is seen in modern primates. Instead, the $\delta^{13}C$ values from *A. africanus* suggest a mixed C_3 and C_4 diet, typical of carnivores and mixed C_3-C_4 feeders (Fig. 8.7).

One final example addresses a more recent archaeological problem. Domestic dog vertebrae from the Mesolithic site of Star Carr, Yorkshire, England, have $\delta^{13}C$ values indicating a marine diet. Apparently, these dogs fed predominantly on scraps from coastal people who made periodic trips inland (Clutton-Brock and Noe-Nygaard, 1990). Their dependency upon the people is clearly established from the carbon isotope data.

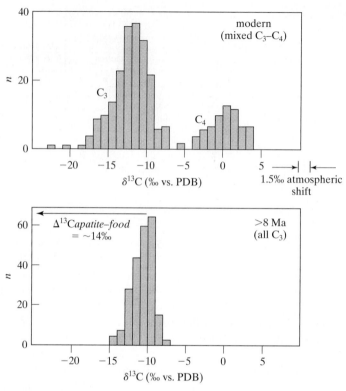

FIGURE 8.6: Carbon isotope composition of modern mammals, showing a mostly bimodal C_3–C_4 dietary preference, and of fossil mammals older than 8 Ma, which show evidence of a C_3 diet only. The $\delta^{13}C$ value of the diet is ~14‰ lower than that of carbonate. (After Cerling et al., 1998.)

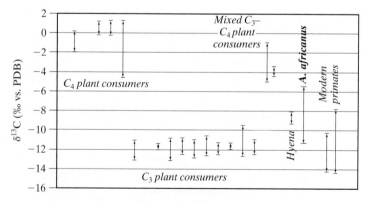

FIGURE 8.7: Carbon isotope composition of different species from the 3-Ma Makapansgat Member 3, South Africa. C_3 and C_4 plant eaters are distinct from one another. Mixed C_3–C_4 consumers and carnivores are intermediate, as is the hominid *A. africanus*. Modern primates (C_3 consumers) are also shown. (After Sponheimer and Lee-Thorp, 1999.)

8.3 CHERTS

Cherts are most commonly composed of microcrystalline quartz replacing authigenic silica, either opal or quartz. Cherts do not form by direct precipitation from ocean water—at least, not since the Phanerozoic. Dissolved silica concentrations are and were simply too low. Instead, silica-secreting organisms concentrate silica, which later reprecipitates or recrystallizes to form chert beds or concretions. Pre-Tertiary deposits formed from silica-secreting radiolarians and sponges. Extensive deep-ocean layers of biogenic siliceous oozes of mid-Tertiary and younger ages correspond with an explosion of diatom populations. Very old samples of thick chert beds, such as the Precambrian iron formations, may have been deposited directly from seawater, but no one knows.

8.3.1 Application to Precambrian Chert Deposits

Perry (1967) made the first systematic stable isotope study of Precambrian cherts. He analyzed massive cherts in which delicate microfossil remains were still evident, a good indication, he believed, that recrystallization and diagenesis had been minimal. (See, however, the discussion of neomorphism in Section 6.4.3.) What Perry found was a systematic and dramatic decrease in the $\delta^{18}O$ values with increasing age (Fig. 8.8), which he attributed to changing $\delta^{18}O$ values of ancient oceans. Later, more detailed work on Precambrian cherts (Knauth and Epstein, 1976; Knauth and Lowe, 1978) indicated that the very low $\delta^{18}O$ values were more likely artifacts of diagenesis. Lithographic facies and stratigraphic position strongly

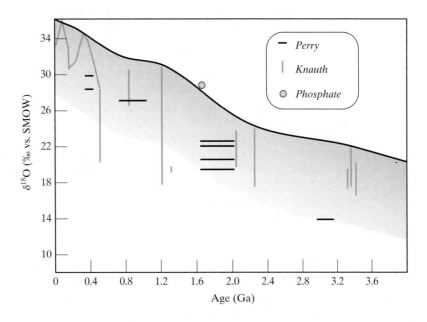

FIGURE 8.8: **Variation in $\delta^{18}O$ values of ancient cherts.** The shaded area represents the secular trend, with the highest $\delta^{18}O$ values being the least altered by diagenesis. Perry: Perry, 1967; Perry et al., 1978; Knauth: Knauth and Epstein, 1976; Knauth and Lowe, 1978. Phosphate datum from Shemesh et al. (1983), corrected for $\Delta^{18}O$ (silica–phosphate).

correlate with $\delta^{18}O$ values, consistent with at least partial diagenesis or precipitation from meteoric water. Nevertheless, the $\delta^{18}O$ values of the least-altered Precambrian samples were 21–22‰ at 3.4 Ga, significantly lower than Phanerozoic equivalents of 30‰ or more. The interpretation from the Knauth group was that the $\delta^{18}O$ value of the oceans was unchanged, but temperatures in the Archean were at least 70°C.

Perry had suggested lower $\delta^{18}O$ values of the ocean, Knauth higher temperatures. Each conclusion is profound in its significance. Perry pointed out that tillites (products of glaciers) at 2 Ga were incompatible with overall high temperatures. Knauth, using the argument that the oceans are buffered by hydrothermal processes at midocean ridges, felt that a better explanation would be high Earth-surface temperatures. The tillites, it was argued, could have been a transient feature. How hot is 70°C? Quite simply, it is scalding. For temperatures to change from steaming hot to glacial and back again is impossible today with effective feedback mechanisms in place from plants, but could it have happened in the past? The lack of a biological component to buffer $p(CO_2)$ could allow rapid changes in temperature. This explanation is supported by wild shifts in $\delta^{13}C$ values of ancient carbonates (Fig. 7.15). Clearly, some major climate shifts were happening in the ancient past. Regardless of which explanation is correct, the chert data are some of the most compelling evidence for major temperature or isotopic differences in the ancient past.

8.3.2 Application to Phanerozoic Cherts

Phanerozoic cherts have been analyzed by combining stable oxygen *and* hydrogen isotope data (Knauth and Epstein, 1976). For groups of data organized by age, the combined $\delta^{18}O$–δD values define an array with a $\delta D/\delta^{18}O$ slope of 8, equal to that of present meteoric water (Fig. 8.9). Although parallel to the meteoric water line, the data cannot be explained simply in terms of postformational exchange. If that were the case, then samples should all plot more or less on the same line parallel to the meteoric water line.

Instead, Knauth and Epstein suggested that these were primary features, indicative of their temperature of formation. Samples forming in equilibrium with ocean water would have unique δD–$\delta^{18}O$ values, depending on the temperature of formation. Line A in Fig. 8.9 defines the locus of samples in equilibrium with ocean water; higher temperatures plot at lower $\delta^{18}O$ and higher δD values.[8] For each age group, distinct temperature ranges are given, highest in the lower Paleozoic and Triassic, lowest in the post-Jurassic. The field for each age group, trending down from Line A with a slope parallel to the meteoric water line, is interpreted as samples having formed in equilibrium with meteoric water (or a mixture of meteoric and ocean water) *at the average global temperature for that period of time*. The data presented in Fig. 8.9 indicate very high temperatures, approaching 40°C in the Lower Paleozoic and Triassic. These values are at the limit of what might be expected for the temperature tolerance of most advanced organisms.

At a fine scale, the complexities of single chert samples become apparent. High-spatial-resolution data from single chert nodules show systematic antiphase periodicity in $\delta^{18}O$ and δD values (Sharp et al., 2002). The hydrogen and oxygen isotope variations cannot be explained in terms of diagenesis, because δD and $\delta^{18}O$ values change in opposite

---

[8]The temperature–$\delta^{18}O$ relationship of Line A is from experimental fractionation factors. The hydrogen–temperature dependence is empirical.

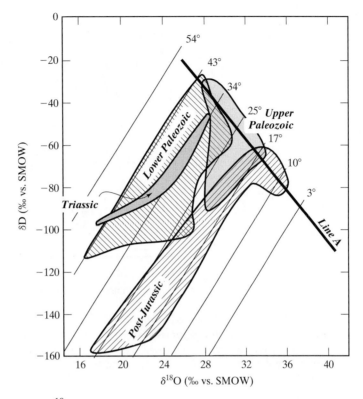

FIGURE 8.9: $\delta D - \delta^{18}O$ **plot of Phanerozoic cherts.** Samples in equilibrium with ocean water should plot on line A, an empirical fit to the highest delta values for each group. Lines of constant temperature are drawn with a $\delta D - \delta^{18}O$ slope of 8 and are based on the quartz–water fractionation curve. Each age group plots in a field with slope 8, indicating that the samples formed in equilibrium with some combination of ocean water and meteoric water.

directions (Fig. 8.10). Instead, the authors suggested that during the formation of the chert nodule in shallow sediments, temperatures periodically varied by up to ~10°C. This work illustrates both the complexity of single nodules and the resilience of the Jurassic-age sample to diagenesis.

8.3.3 Diagenesis

All of the preceding conclusions are valid only if one can evaluate and account for diagenetic effects. As we have seen, Knauth and Lowe recognized that significant diagenesis occurred in some Precambrian cherts, whereas others may not have changed by more than 3‰. Murata et al. (1977) measured $\delta^{18}O$ values of silica phases in the Miocene Monterey Shale formation in California, in the United States, and found that there were abrupt changes in crystallography with depth, related to a progression from biogenic opal (opal-A), to opal-CT, to microcrystalline quartz. For each silica polymorph, the $\delta^{18}O$ values are constant. However, a jump is observed across each group boundary (Fig. 8.11). Once a phase transformation occurred, the $\delta^{18}O$ values remain unchanged until the next polymorphic transition. If the samples had reequilibrated continuously, then there should be a

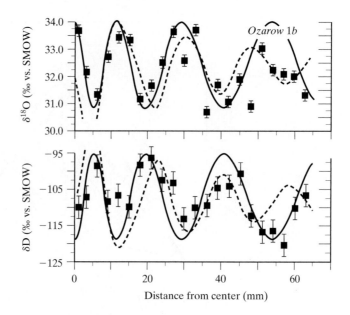

FIGURE 8.10: Variations in hydrogen and oxygen isotope ratios as a function of distance from center for chert nodules of Jurassic age. The antiphase periodicity of the nodules cannot be explained by diagenesis, which would lower both $\delta^{18}O$ and δD values in the same direction. Instead, the data are interpreted as reflecting temperature changes of ~10°C. (After Sharp et al., 2002.)

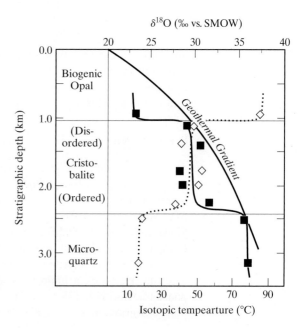

FIGURE 8.11: Variations in $\delta^{18}O$ values of chert vs. depth for the Monterey formation, California, United States. $\delta^{18}O$ values (diamonds) jump across each phase transition. The temperatures (filled squares) of each transition agree with the expected geothermal gradient. Complications associated with changing pore-water values should be considered. (After Murata et al., 1977.)

smooth variation in the $\delta^{18}O$ value with depth. Temperatures of transformation were estimated as 35–50°C from opal-A to opal-CT and 45–80°C from opal-CT to microcrystalline quartz. Inherent in these results is the assumption that the $\delta^{18}O$ values of the pore waters remained at 0‰. Monterey formation samples were reanalyzed with a stepwise fluorination technique developed to remove the effects of water contamination in the hydrous silica polymorphs. Results from the stepwise fluorination data give $\delta^{18}O$ values for the opal-CT that are considerably higher than earlier estimates, suggesting a temperature of 20°C for this transition (Matheney and Knauth, 1993).

8.3.4 Application to Recent Sediments

Fresh biogenic silica has a high abundance of organic-matter and high water content. Over millions of years, it recrystallizes to microcrystalline quartz. Water and organic matter are lost, so the silica phase becomes essentially pure quartz. Whether changes in $\delta^{18}O$ values occur during the recrystallization processes is open to debate, but there is no problem with analyzing ancient cherts via conventional fluorination after only a minimal amount of pretreatment. For recent, and especially for modern, samples, the complications of organic matter and high water content pose serious problems. Even purifying the silica phase is not a trivial task. Settling techniques are used to separate biogenic silica from detrital clays, and complicated chemical treatments are used to remove carbonate and iron oxide overgrowths. Organic matter is removed by treatment with NaOCl or hydrogen peroxide.

The most difficult task is to remove water from the highly hydrated silica, which may be up to 12 wt %. Several techniques have been employed to remove water without causing irreproducible shifts in the $\delta^{18}O$ value of the remaining dehydrated silica.

Heating opal at moderate temperatures in a vacuum will remove all water, but the resultant $\delta^{18}O$ values are not reproducible. Labeyrie (1974) showed that rapid sintering at 1,000°C in vacuum removes all water, with only a small $\delta^{18}O$ shift. The reproducibility of the method was later improved by first equilibrating the samples with water of known $\delta^{18}O$ value (Labeyrie and Juillet, 1982). Techniques have also been developed wherein repeated partial fluorination reactions strip away the hydrous component of the silica (Matheney and Knauth, 1989). The results from each method are similar, lending credence to their validity.

Given a method of analysis, modern and cultured samples could be analyzed in order to calibrate the silica–water thermometer. Diatoms[9] appear to form in equilibrium with ocean water, in accordance with the empirical equation (Leclerc and Labeyrie, 1987)

$$t = 17.2 - 2.4(\delta^{18}O_{si} - \delta^{18}O_w - 40) - 0.2(\delta^{18}O_{si} - \delta^{18}O_w - 40)^2 \quad (t \text{ in } °C). \quad (8.8)$$

with $\delta^{18}O_{si}$ and $\delta^{18}O_w$ on the SMOW scale

Sponges and Radiolaria apparently do not precipitate in oxygen isotope equilibrium with ocean water. An empirical estimate from amorphous *inorganically* crystallized silica over the temperature range 34–93°C (Kita et al., 1985) is

$$1000 \ln \alpha = \frac{3.52 \times 10^6}{T^2} \quad (T \text{ in K}), \quad (8.9)$$

---

[9]Diatoms are marine algae that deposit internal silica frustules. Because they are photosynthetic, they are only found in near-surface waters. Thus, their $\delta^{18}O$ value must reflect sea-surface temperatures.

which is in agreement with extrapolations of high-temperature experimental data, but *not* with empirical estimates from biogenic silica.

Remember that the fractionation between biogenic phosphate and carbonate is insensitive to temperature. As a result, carbonate–phosphate pairs cannot be used to determine temperature independently. The oxygen isotope fractionation between biogenic silica–carbonate does have a significant temperature dependence, so that it may be useful for temperature estimates that are free from assumptions about ocean water composition. Coexisting diatoms and (carbonate) foraminifera can be used both as a thermometer *and* as a way of calculating the $\delta^{18}O$ value of the ocean at the time the deposits were formed. By measuring $\delta^{18}O$ values of both silica and carbonate, Shemesh et al. (1992), using an equation slightly modified from equation (8.8), were able to constrain variations in $\delta^{18}O_{water}$ and temperatures from a Holocene marine core. The validity of this type of calculation, however, has been questioned (Clayton and Shemesh, 1992). Further study is needed to evaluate the utility of combined silica–carbonate thermometry.

The degree of precision required and the questions addressed vary considerably with the age of the material. When looking at Precambrian cherts, we are concerned with trends at a scale of 10‰ or more. Slight uncertainties in fractionation factors and minor diagenesis can safely be ignored. The questions addressed are rather general, in part because the time resolution is so poor. At the other end of the spectrum are recent samples, for which high analytical and temporal precision are needed in order to make meaningful interpretations of short-term and subtle climate variations. A detailed study of recent and modern diatoms highlights the difficulties in applying silica thermometry to high-precision paleoclimate reconstruction.

Schmidt et al. (1997) and Brandriss et al. (1998) measured the $\delta^{18}O$ values of cultured diatoms—those taken from the diatom blooms in the ocean surface water and recent samples taken from ocean sediment cores. Sediment cores from the Weddell Sea, Antarctica, and from the North Atlantic were measured with the use of stepwise fluorination and gave $\delta^{18}O$ values of 46.8 and 44.1‰, respectively. The values correspond to temperatures of $-4.6°C$ for the Antarctic sample and 3°C for the North Atlantic sample, assuming a $\delta^{18}O$ value of 0‰ for ocean water. Both temperature estimates are too low. Three explanations were proposed: (1) The samples underwent early diagenesis with pore waters with positive $\delta^{18}O$ values; (2) they formed out of equilibrium with the water; and (3) the temperature equation is not correct.

Cultured diatoms from the same study have $\delta^{18}O$ values that are 3 to 10‰ lower than what would be expected from the fractionation equation (8.8) or the modified version of Shemesh et al. (1992). The cultured data are in better agreement with the inorganic quartz–water fractionation equation (8.9), but not completely. Although the empirical fractionation equation could "work" for many recent diatoms, it may, in fact, be calibrated to diatoms equilibrated with pore waters that have a slightly positive $\delta^{18}O$ value.

8.3.5 Other Silica Applications

The $\delta^{18}O$ value of quartz has distinct ranges for different types of rock. The $\delta^{18}O$ values of igneous rocks range from 8 to 10‰, metamorphic rocks 10 to 16‰. Quartz from sandstones and beaches has a narrow range of 10 to 13‰. Shales are higher, ranging from 15 to 24‰, similar to authigenic quartz and quartz overgrowths. Cherts have the highest $\delta^{18}O$ values, ranging from 19 to 34‰. Blatt (1987) makes the very interesting case that the high $\delta^{18}O$ values of shales and mudstones cannot simply be reworked igneous and metamorphic rocks. Authigenic overgrowths formed during clay diagenesis and chert fragments are likely contributors, thereby explaining the high $\delta^{18}O$ values.

Burial diagenesis does not change the $\delta^{18}O$ value of quartz, unless it is accompanied by recrystallization. With this in mind, sources of eolian transported fine-grained quartz can be traced with oxygen isotope geochemistry. Mizota and Matsuhisa (1995) showed that quartz in the Canary Archipelago was sourced from the North African Sahara desert, on the basis of $\delta^{18}O$ values of quartz and $^{87}Sr/^{86}Sr$ ratios of mica. Sridhar et al. (1978) measured quartz from soils in the southwest United States and the Hawaiian Islands. Surprisingly, the $\delta^{18}O$ values of fine-grained materials from all of the locations examined ranged from 17 to 20‰. These values are neither those of metamorphic and igneous rocks (8 to 16‰) nor those of of cherts (~30‰). Sources could be shales or mixtures of light igneous material with heavier quartz of low-temperature origin. The quartz from Hawaii is probably sourced from Asian soils.

One last application is the use of silica phytoliths as a proxy for continental paleoclimate reconstruction. Phytoliths are the siliceous secondary cell wall found in many grasses, particularly those abundant in grasslands and steppes. In a series of papers by Webb and Longstaffe (e.g., 2003), the $\delta^{18}O$ value of phytolith silica was shown to be a function of the composition and temperature of soil water. Leaves were affected by humidity as well, due to transpiration. Empirical relationships between phytolith $\delta^{18}O$ values and temperatures were derived. Applications to ancient samples have not yet been made.

8.4 CLAY MINERALS

8.4.1 Early "Bulk" Sample Studies

Bulk $\delta^{18}O$ and δD values for clastic sedimentary rocks range from +8 to +25‰ and from −80 to −25‰, respectively, with some notable exceptions. The range of delta values reflects mixtures of material from a wide range of rock types and authigenic minerals formed in the sediment column. Of greatest interest to the petroleum geologist are the neoformed minerals, as they strongly control the permeability of potential petroleum reserves. Their isotopic composition is a function of the pore-water composition, formation mechanism, degree of reequilibration following formation, and temperature of formation.

The first detailed studies of fine-grained, noncarbonate sediments were made by Savin and Epstein (1970a, b, c), who hoped to answer the questions of whether clay minerals form in equilibrium with their surroundings and how easily they are altered after formation. In order to avoid complexities associated with mineral purification, these authors tried to measure nearly monomineralic samples, such as those from kaolinite and montmorillonite (bentonite) deposits. In analyzing shales, they made a small correction to the $\delta^{18}O$ values for clay minerals due to quartz contamination.

Savin and Epstein drew a number of basic conclusions. The $\delta^{18}O$ values of clay minerals range mostly from 16 to 26‰, higher than the values for igneous and most metamorphic rocks. Such high $\delta^{18}O$ values must have formed at low temperatures, where $\Delta^{18}O_{mineral-water}$ values are large. But the scatter in the data was large enough to indicate an "exotic" terrigenous component as well. The δD values are similar to those of igneous and metamorphic rocks, ranging from −90 to −40‰. Combined $\delta D-\delta^{18}O$ values of kaolinites formed a linear array that paralleled the meteoric water line. Savin and Epstein proposed that if the massive kaolinite deposits formed in mild and wet continental climates, then the oxygen and hydrogen isotope composition of the clays would be buffered by meteoric water, a concept that was later expanded on by Lawrence and Taylor (1971) (Fig. 8.12). On the basis of estimated meteoric water values and temperatures of kaolinitization, the $\alpha_{kaolinite-water}$ values for oxygen and

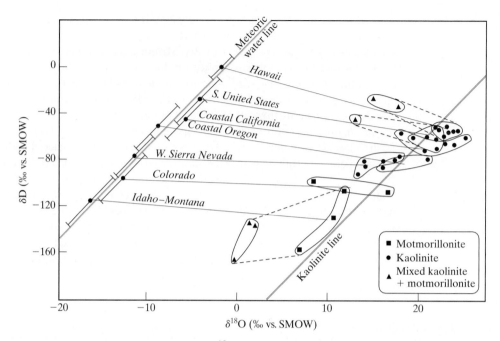

FIGURE 8.12: Correlation between $\delta^{18}O$ and δD values of kaolinite–montmorillonite with assumed local meteoric water. The offsets can be explained by mineral–water fractionations at ~25°C. (After Lawrence and Taylor, 1971.)

hydrogen were estimated as 1.027 and 0.970, respectively. Similar fractionation factors were calculated for montmorillonite and glauconite.

Savin and Epstein found that the $\delta^{18}O$ values of illites from ocean cores average 15.5‰, far lower than the equilibrium value for samples in equilibrium with seawater. Similarly, montmorillonites from ocean-core samples are approximately 17‰, also lower than the equilibrium value. The authors concluded that these minerals were of detrital origin and that they had not reequilibrated on the ocean floor, even over millions of years. Later, Yeh and Epstein (1978) demonstrated that hydrogen isotope exchange of detrital clays does not occur on the ocean floor in millions of years, except for the smallest size fraction (<0.1 μm).

8.4.2 Grain-Size Considerations

The next advance in studies of clay minerals occurred when people started picking apart the various phases in clastic deposits. The separation and purification of clay minerals is extremely laborious and requires a number of chemical and settling techniques (Sheppard and Gilg, 1996), but the information to be gained is substantial. The initial disequilibrium between clay minerals of different-size fractions and when and how reequilibration occurs as diagenesis increases are beautifully illustrated in a series of detailed isotopic studies of a sediment core from the Gulf coast in the southwestern United States (Yeh and Savin, 1977; Yeh, 1980). $\delta^{18}O$ and δD values were measured as a function of grain size and depth (Fig. 8.13). The data

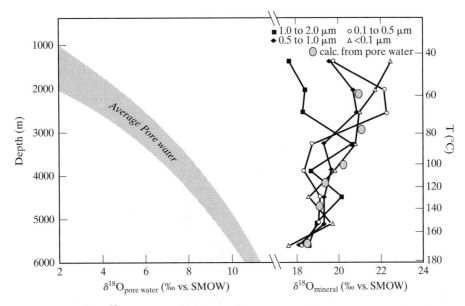

FIGURE 8.13: $\delta^{18}O$ values of illite–smectite from a Tertiary Gulf Coast shale sequence and $\delta^{18}O$ pore-water values calculated from temperature–fractionation relations and independent estimates from similar sequences (Milliken et al., 1981). The calculated $\delta^{18}O$ values of clays (assuming the smooth variation in pore waters) fits the 0.5- to 1.0-μm fraction remarkably well.

are plotted along with calculated porewater values. The smooth and large variation in pore-water values fits well the finer grain-size fractions for all but the shallowest levels.

The δD values also change with depth (Fig 8.14). Hydrogen equilibrium is attained for all grain sizes at ~70°C. Oxygen equilibrium may occur at slightly higher temperatures, a phenomenon almost certainly related to recrystallization due to the onset of active diagenesis and characterized by the reaction of aqueous fluids with unstable detrital components (Sharp, 1998). Below this temperature, the finer grain-size fractions are out of equilibrium with the pore fluid, indicating that they are detrital minerals of continental origin.

The combined evolution of δD–$\delta^{18}O$ values for the Gulf Coast waters is shown in Fig. 8.15. Pore waters gradually attain higher $\delta^{18}O$ and lower δD values with depth as exchange occurs with newly forming authigenic minerals. The large changes in both isotope ratios is evidence that fluid/rock ratios were low. Otherwise we would see the minerals changing their composition at least as much as the pore water does.

Longstaffe and Ayalon (1987) made a detailed petrographic–isotopic study of diagenetic minerals in a Cretaceous transitional marine-to-continental sequence in west-central Alberta, Canada. They were able to recognize an increase in temperature with only a slight increase in $\delta^{18}O$ values of pore waters, followed by a rapid decrease in $\delta^{18}O$ values of porewaters due to the infiltration of meteoric water and, finally, a reduction of the temperature to present-day conditions (Fig. 8.16). (For more information, see Longstaffe, 1987; Savin and Lee, 1988; and Longstaffe, 1989. A discussion of fractionation factors for hydrous phyllosilicates can be found in Savin and Lee, 1988.)

198 Chapter 8 Low-Temperature Minerals, Exclusive of Carbonates

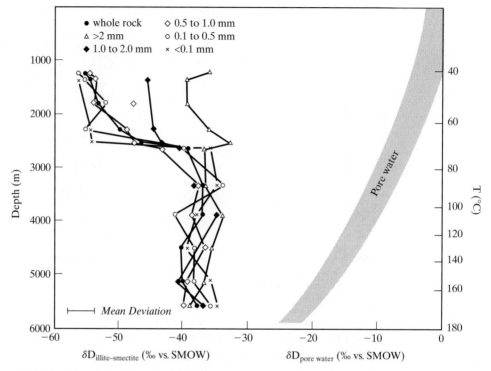

FIGURE 8.14: **δD values of mixed illite–smectite as a function of grain size.** As with oxygen, all grain sizes come to equilibrium with pore water at approximately 70°C, probably due to mineralogical recrystallization. (After Yeh, 1980.)

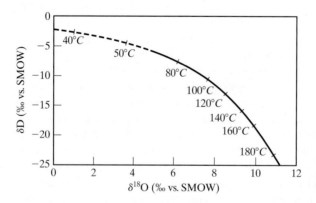

FIGURE 8.15: Combined $\delta^{18}O$–δD variations in pore waters from Gulf Coast sediments. Particularly large changes in $\delta^{18}O$ values can be explained by low fluid/rock ratios.

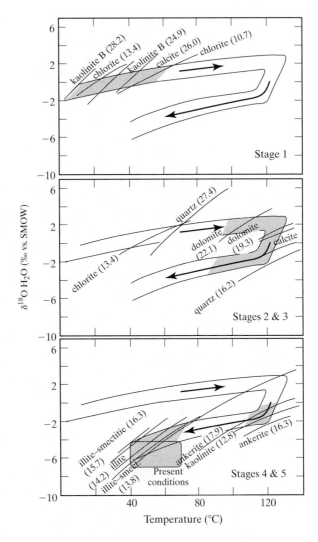

FIGURE 8.16: **Evolution of calculated oxygen isotope ratios of pore waters from a transitional marine–terrestrial sequence from the Cretaceous-age Alberta Basin.** Oxygen isotope ratios first increase due to exchange with newly formed authigenic minerals and then decrease as meteoric water infiltrates the system. (After Longstaffe and Ayalon, 1987.)

8.5 IRON OXIDES

The iron oxides hematite (Fe_2O_3) and goethite (α-FeOOH) are common low-temperature alteration phases in the terrestrial and marine environment. They are found in a bewildering array of environments when water and the oxidizing conditions of the near surface are encountered—from oceanic spreading centers, to soils, to precipitates on bones. Goethite has a small $Fe(CO_3)OH$ component. This means that oxygen, hydrogen, and carbon isotopes can all be measured from the same sample, providing multiple constraints on the conditions attending its formation. Yapp (2001) presents a nice review of the field.

As with many fine-grained materials, sample preparation is important. Selective dissolution methods are used to purify different ferric oxides. In addition, silicates are invariably intermixed with fine-grained ferric oxides, so that mass balance techniques are necessary to retrieve the pure ferric oxide $\delta^{18}O$ value (e.g., Yapp, 1998; Bao et al., 2000). The analyzed data must be interpreted in terms of known fractionation factors, in order to retrieve temperatures of formation. The discrepancy between calibrations for goethite–water is quite remarkable. (See Yapp, 2001, Fig. 2). Less discrepancy exists for hydrogen isotope fractionation, but that may be the result of only one experiment having been published!

The interpretation of goethite data can be complicated. The fractionation between goethites forming in modern soil and local mean meteoric water range from -1.5 to $6.3‰$ (Bao et al., 2000). Such a large spread of $\Delta^{18}O$ values cannot be explained by temperature variations alone. Instead, processes such as evaporation in the soils in which the goethites are forming causes dramatic shifts in the $\delta^{18}O$ values of soil water.

Yapp (1997) measured the $\delta^{18}O$ and δD values of goethites from a recent bog iron deposit and lateritic soil. His temperature estimates using oxygen isotope thermometry of goethites for the recent materials were in excellent agreement with modern (summer) values. Turning to ancient samples (Ordovician to Cretaceous), he predicted that $\delta^{18}O$ values of goethites from low latitudes had formed under conditions of high rainfall and that the meteoric water cycle affecting the late Cretaceous samples was different from that of today due to the effects of the Late Cretaceous seaway (Yapp, 1998).

Goethite geochemistry has also been used to calculate the partial pressure of $CO_2(pCO_2)$ in the atmosphere. CO_2 is formed in soil from the oxidation of organic matter. The pCO_2 in soils increases downward from the surface, while the $\delta^{13}C$ value decreases. Yapp and Poths (1992) measured the concentration and $\delta^{13}C$ value of the $Fe(CO_3)OH$ component of a goethite deposit of Late Ordovician age and found that the atmospheric pCO_2 during the goethite formation was ~ 16 times higher than present values although it should be noted that there are a considerable number of assumptions in this determination.

The foregoing examples of low-temperature materials used for reconstructing paleoclimate or conditions of diagenesis are by no means complete. Researchers have analyzed pretty much anything that exists near the Earth's surface, and more often than not, the information is particularly useful. Amber, eggshells, snails, coal, collagen, hackberries, hair, manganese coatings, and volcanic ash are only a few of the additional materials studied with stable isotope geochemistry to address surficial processes.

PROBLEM SET

1. a. From the kaolinite fractionation line in Figure 8.12, calculate the $\alpha_{\text{kaol-water}}$ value for both oxygen and hydrogen.
   b. On the basis of elementary fractionation relations, do mixed kaolinite–montmorillonite samples form at lower or higher temperatures than pure kaolinite and montmorillonite?

2. The box model for describing the $\delta^{18}O$ value of an animal's body water (Fig. 8.4) provides an explanation for the correlation between local meteoric water and drinking water having a slope less than unity. Construct a similar model for hydrogen isotopes. What slope would you predict?

3. When Kolodny et al. (1983) measured fish bones in Lake Baikal, they found that the $\delta^{18}O$ values of the bones varied from $6.3‰$ near the surface to $9.2‰$ in the deepest levels. Assuming that the $\delta^{18}O$ value of the lake is constant, what temperature change is there between the highest and lowest sampling levels in their study?

4. A kangaroo rat can derive all of its water by metabolizing food—so-called metabolic water. If the kangaroo rat oxidizes carbohydrates (general formula CH_2O), how will the $\delta^{18}O$ of its body water change as a function of the $\delta^{18}O$ of local meteoric water. Assume that the ingested carbohydrates have a $\delta^{18}O$ value similar to that of the local meteoric water.

5. In the Southern Arizona desert, kangaroo rats have $\delta^{13}C$ values ranging from -26 to $-23‰$, whereas other granivorous rodents have $\delta^{13}C$ values ranging from -26 to $-9‰$ (PDB) (Smith et al., 2002). What would you conclude about the diet of the kangaroo rats vs. those of other rodents?

6. Why would the D/H ratio of pore waters *decrease* with depth, while the $^{18}O^{16}O$ ratio for the same waters *increases* with depth?

ANSWERS

1. Answer for 1. From Figure 8.12, the $\delta^{18}O_{kaolinite} - \delta^{18}O_{water}$ is $\sim 25‰$. From the definition of $\alpha_{a-b} = \dfrac{1000 + \delta_a}{1000 + \delta_b}$, we have $\alpha \approx \dfrac{1000 + 25}{1000 + 0} = 1.025$. For hydrogen, the $\delta D_{kaol} - \delta D_w$ value is $\sim -30‰$. Substituting for the definition of α gives a value of $0.97‰$.

2. For hydrogen isotopes, there is no constant input value as there is for oxygen (i.e., atmospheric oxygen). Therefore the slope should be 1.

3. Using Longinelli's equation, we have $t(°C) = 111.4 - 4.3(\delta_p - \delta_w + 0.5)$. Because the $\delta^{18}O$ value of the water is constant, the effect of changing temperature on $t(°C)$ is given by $-4.3(\delta^{18}O_{phosphate})$. If we change $\delta^{18}O_{phosphate}$ by $2.9‰$, we change the temperature by $2.9 \times 4.3 = 12.5°C$.

4. For pigs and humans, the relationship is given by $\delta^{18}O_{bw} = 0.88\delta^{18}O_w + C$ and $\delta^{18}O_{bw} = 0.60\delta^{18}O_w + C$, respectively (equations (8.2) and (8.3), where C is a constant). Equation 8.7 gives the general relationship. If CH_2O combines with O_2 to give $CO_2 + H_2O$ in equilibrium with each other, then two-thirds of the oxygen comes from the air. The relationship is then $\delta^{18}O_{bw} = 0.33\delta^{18}O_w + C$.

5. The difference in delta values is related to the animals' different diets. Apparently, kangaroo rats eat only C_3 plants, whereas other rodents eat either C_3 or C_4 plants, depending upon availability. Most likely, kangaroo rats stash C_3 seeds from the winter months.

6. Because water concentrates the heavy isotope of hydrogen, but the light isotope of oxygen relative to clay minerals.

REFERENCES

Ayliffe, L. K., and Chivas, A. R. (1990). Oxygen isotope composition of the bone phosphate of Australian kangaroos: Potential as a palaeoenvironmental recorder. *Geochimica et Cosmochimica Acta* **54**, 2603–2609.

Ayliffe, L. K., Chivas, A. R., and Leakey, M. G. (1994). The retention of primary oxygen isotope compositions of fossil elephant skeletal phosphate. *Geochimica et Cosmochimica Acta* **58**, 5291–5298.

Ayliffe, L. K., Lister, A. M., and Chivas, A. R. (1992). The preservation of glacial–interglacial climatic signatures in the oxygen isotopes of elephant skeletal phosphate. *Palaeogeography, Palaeoclimatology, Palaeoecology* **99**, 179–191.

Balasse, M., Smith, A. B., Ambrose, S. H., and Leight, S. R. (2003). Determining sheep birth seasonality by analysis of tooth enamel oxygen isotope ratios: The late Stone Age site of Kasteelberg (South Africa). *Journal of Archaeological Science* **30**, 205–215.

Bao, H., Koch, P. L., and Thiemens, M. H. (2000). Oxygen isotopic composition of ferric oxides from recent soil, hydrologic, and marine environments. *Geochimica et Cosmochimica Acta* **64**, 2221–2231.

Barrick, R. E., and Showers, W. J. (1994). Thermophysiology of *Tyrannosaurus rex*; evidence from oxygen isotopes. *Science* **265**, 222–224.

Blake, R. E., O'Neil, J. R., and Garcia, G. A. (1997). Oxygen isotope systematics of biologically mediated reactions of phosphate: I. Microbial degradation of organophosphorus compounds. *Geochimica et Cosmochimica Acta* **61**, 4411–4422.

Blatt, H. (1987). Oxygen isotopes and the origin of quartz. *Journal of Sedimentary Petrology* **57**, 373–377.

Brandriss, M. E., O'Neil, J. R., Edlund, M. B., and Stoermer, E. F. (1998). Oxygen isotope fractionation between diatomaceous silica and water. *Geochimica et Cosmochimica Acta* **62**, 1119–1125.

Bryant, J. D., and Froelich, P. N. (1995). A model of oxygen isotope fractionation in body water of large mammals. *Geochimica et Cosmochimica Acta* **59**, 4523–4537.

Bryant, J. D., Froelich, P. N., Showers, W. J., and Genna, B. J. (1996). A tale of two quarries: Biologic and taphonomic signatures in the oxygen isotope composition of tooth enamel phosphate from modern and Miocene equids. *Palaios* **11**, 397–408.

Bryant, J. D., Koch, P. L., Froelich, P. N., Showers, W. J., and Genna, B. J. (1996). Oxygen isotope partitioning between phosphate and carbonate in mammalian apatite. *Geochimica et Cosmochimica Acta* **60**, 5145–5148.

Cerling, T. E., Ehleringer, J. R., and Harris, J. M. (1998). Carbon dioxide starvation, the development of C_4 ecosystems, and mammalian evolution. *Philosophical Transactions of the Royal Society of London* **3553**, 159–171.

Cerling, T. E., Wang, Y., and Quade, J. (1993). Expansion of C4 ecosystems as an indicator of global ecological change in the Late Miocene. *Nature* **361**, 344–345.

Clayton, R. N., and Shemesh, A. (1992). Silica–carbonate isotopic temperature calibration [discussion and reply]. *Science* **258**, 1162–1163.

Clutton-Brock, J., and Noe-Nygaard, N. (1990). New osteological and carbon-isotope evidence on Mesolithic dogs: Companions to hunters and fishers at Star Carr (Yorkshire, England, UK), Seamer Carr (Yorkshire, England, UK) and Kongemose (Sjaelland, Denmark). *Journal of Archaeological Science* **17**, 643–654.

Cormie, A. B., Schwarcz, H. P., and Gray, J. (1994). Relation between hydrogen isotopic ratios of bone collagen and rain. *Geochimica et Cosmochimica Acta* **58**, 377–391.

Crowson, R. A., Showers, W. J., Wright, E. K., and Hoering, T. C. (1991). Preparation of phosphate samples for oxygen isotope analysis. *Analytical Chemistry* **63**, 2397–2400.

Driessens, F. C. M. (1980). The mineral in bone, dentine and tooth enamel. *Bulletin des Sociétés Chimiques Belges* **89**, 663–689.

Fricke, H. C. (2003). Investigation of early Eocene water-vapor transport and paleoelevation using oxygen isotope data from geographically widespread mammal remains. *Geological Society of America Bulletin* **115**, 1088–1096.

Fricke, H. C., and O'Neil, J. R. (1996). Inter- and intra-tooth variation in the oxygen isotope composition of mammalian tooth enamel: Some implications for paleoclimatological and paleobiological research. *Palaeogeography, Palaeoclimatology, Palaeoecology* **126**, 91–99.

Fricke, H. C., and Rogers, R. R. (2000). Multiple taxon–multiple locality approach to providing oxygen isotope evidence for warm-blooded theropod dinosaurs. *Geology* **28**, 799–802.

Friedman, I., and O'Neil, J. R. (1977). Compilation of stable isotope fractionation factors of geochemical interest. In *U. S. Geological Survey Professional Paper*, Vol. 440-KK.

Huertas, A. D., Iacumin, P., Stenni, B., Chillón, B. S., and Longinelli, A. (1995). Oxygen isotope variations of phosphate in mammalian bone and tooth enamel. *Geochimica et Cosmochimica Acta* **59**, 4299–4305.

Kita, I., Taguchi, S., and Matsubaya, O. (1985). Oxygen isotope fractionation between amorphous silica and water at 34–93°C. *Nature* **314**, 83–84.

Knauth, L. P., and Epstein, S. (1976). Hydrogen and oxygen isotope ratios in nodular and bedded cherts. *Geochimica et Cosmochimica Acta.* **40**, 1095–1108.

Knauth, L. P., and Lowe, D. R. (1978). Oxygen isotope geochemistry of cherts from the Onverwacht Group (3.4 billion years), Transvaal, South Africa, with implications for secular variations in the isotopic composition of cherts. *Earth and Planetary Science Letters* **41**, 209–222.

Koch, P. L., Fisher, D. C., and Dettman, D. L. (1989). Oxygen isotopic variation in the tusks of extinct proboscideans; a measure of season of death and seasonality. *Geology* **17**, 515–519.

Koch, P. L., Heisinger, J., Moss, C., Carlson, R. W., Fogel, M. L., and Behrensmeyer, A. K. (1995). Isotopic tracking of change in diet and habitat use in African elephants. *Science* **267**, 1340–1343.

Kohn, M. J., Schoeninger, M. J., and Valley, J. W. (1996). Herbivore tooth oxygen isotope compositions; effects of diet and physiology. *Geochimica et Cosmochimica Acta* **60**, 3889–3896.

Kolodny, Y. and Luz, B. (1991). Oxygen isotopes in phosphates of fossil fish; Devonian to recent. In *Stable Isotope Geochemistry: A Tribute to Samuel Epstein*, Vol. 3 (ed. H. P. Taylor, Jr., J. R. O'Neil, and I. R. Kaplan), pp. 105–119. San Antonio: Lancaster Press, Inc.

Kolodny, Y., Luz, B., and Navon, O. (1983). Oxygen isotope variations in phosphate of biogenic apatites, I. Fish bone apatite—rechecking the rules of the game. *Earth and Planetary Science Letters* **64**, 393–404.

Labeyrie, L., and Juillet, A. (1982). Oxygen isotopic exchangeability of diatom valve silica; interpretation and consequences for paleoclimatic studies. *Geochimica et Cosmochimica Acta* **46**, 967–975.

Labeyrie, L. J. (1974). New approach to surface seawater palaeotemperatures using $^{18}O/^{16}O$ ratios in silica of diatom frustules. *Nature* **248**, 40–42.

Lawrence, J. R., and Taylor, H. P., Jr. (1971). Deuterium and oxygen-18 correlation: Clay minerals and hydroxides in Quaternary soils compared to meteoric waters. *Geochimica et Cosmochimica Acta* **35**, 993–1003.

Leclerc, A. J., and Labeyrie, L. (1987). Temperature dependence of oxygen isotopic fractionation between diatom silica and water. *Earth and Planetary Science Letters* **84**, 69–74.

Lécuyer, C., Grandjean, P., O'Neil, J. R., Cappetta, H., and Martineau, F. (1993). Thermal excursions in the ocean at the Cretaceous–Tertiary boundary (northern Morocco): $\delta^{18}O$ record of phosphatic fish debris. *Palaeogeography, Palaeoclimatology, Palaeoecology* **105**, 235–243.

Longinelli, A. (1965). Oxygen isotopic composition of orthophosphate from shells of living marine organisms. *Nature* **207**, 716–719.

Longinelli, A. (1973). Preliminary oxygen-isotope measurements of phosphate from mammal teeth and bones. *Colloque International du CNRS* **219**, 267–271.

Longinelli, A. (1984). Oxygen isotopes in mammal bone phosphate; a new tool for paleohydrological and paleoclimatological research? *Geochimica et Cosmochimica Acta* **48**, 385–390.

Longinelli, A., and Nuti, S. (1973). Revised phosphate–water isotopic temperature scale. *Earth and Planetary Science Letters* **19**, 373–376.

Longstaffe, F. J. (1987). Stable isotope studies of diagenetic processes. In *Short Course in Stable Isotope Geochemistry of Low Temperature Processes*, Vol. 13 (ed. T. K. Kyser), pp. 187–257. Saskatoon, SK, Canada: Mineralogical Society of Canada.

Longstaffe, F. J. (1989). Stable isotopes as tracers in clastic diagenesis. In *Short Course in Burial Diagenesis*, Vol. 15 (ed. I. E. Hutcheon), pp. 201–277. Montreal: Mineralogical Association of Canada.

Longstaffe, F. J., and Ayalon, A. (1987). Oxygen-isotope studies of clastic diagenesis in the Lower Cretaceous Viking Formation, Alberta; implications for the role of meteoric water. In *Diagenesis of Sedimentary Sequences*, Vol. 36 (ed. J. D. Marshall), pp. 277–296. Liverpool: Geological Society Special Publication.

Luz, B., Kolodny, Y., and Horowitz, M. (1984). Fractionation of oxygen isotopes between mammalian bone phosphate and environmental drinking water. *Geochimica et Cosmochimica Acta* **48**, 1689–1693.

Luz, B., Kolodny, Y., and Kovach, J. (1984). Oxygen isotope variations in phosphate of biogenic apatites; III, Conodonts. *Earth and Planetary Science Letters* **69**, 255–262.

Matheney, R. K., and Knauth, L. P. (1989). Oxygen-isotope fractionation between marine biogenic silica and seawater. *Geochimica et Cosmochimica Acta* **53**, 3207–3214.

Matheney, R. K., and Knauth, L. P. (1993). New isotopic temperature estimates for early silica diagenesis in bedded cherts. *Geology* **21**, 519–522.

Milliken, K. L., Land, L. S., and Loucks, R. G. (1981). History of burial diagenesis determined from isotopic geochemistry, Frio Formation, Brazoria County, Texas. *American Associate of Petroleum Geologists Bulletin* **65**, 1397–1413.

Mizota, C., and Matsuhisa, Y. (1995). Isotopic evidence for the eolian origin of quartz and mica in soils developed on volcanic materials in the Canary Archipelago. *Geoderma* **66**, 313–320.

Murata, K. J., Friedman I., and Gleason, J. D. (1977). Oxygen isotope relations between diagenetic silica minerals in Monterey Shale, Temblor Range, California. *American Journal of Science* **277**, 259–272.

O'Neil, J. R., Roe, L. J., Reinhard, E., and Blake, R. E. (1994). A rapid and precise method of oxygen isotope analysis of biogenic phosphates. *Israeli Journal of Earth Science* **43**, 203–212.

Perry, E. C., Ahmad, S. N., and Swilius, T. M. (1978). The oxygen isotope composition of 3,800 m.y. old metamorphosed chert and iron formation from Isukasai, West Greenland. *Journal of Geology* **86**, 223–239.

Perry, E. C., Jr. (1967). The oxygen isotope chemistry of ancient cherts. *Earth and Planetary Science Letters* **3**, 62–66.

Savin, S. M., and Epstein, S. (1970a). The oxygen and hydrogen isotope geochemistry of clay minerals. *Geochimica et Cosmochimica Acta* **34**, 35–42.

Savin, S. M., and Epstein, S. (1970b). The oxygen and hydrogen isotope geochemistry of ocean sediments and shales. *Geochimica et Cosmochimica Acta* **34**, 43–63.

Savin, S. M., and Epstein, S. (1970c). The oxygen isotopic compositions of coarse grained sedimentary rocks and minerals. *Geochimica et Cosmochimica Acta* **34**, 323–329.

Savin, S. M., and Lee, M. (1988). Isotopic studies of phyllosilicates. In *Hydrous Phyllosilicates*, Vol. 19 (ed. S. W. Bailey), pp. 189–223. Chelsea, MI: Mineralogical Society of America.

Schmidt, M., Botz, R., Stoffers, P., Anders, T., and Bohrmann, G. (1997). Oxygen isotopes in marine diatoms: A comparative study of analytical techniques and new results on the isotope composition of recent marine diatoms. *Geochimica et Cosmochimica Acta* **61**, 2275–2280.

Sharp, Z. D. (1998). Application of stable isotope geochemistry to low-grade metamorphic rocks. In *Low-Grade Metamorphism* (ed. M. Frey and D. Robinson), pp. 227–260. London: Blackwell.

Sharp, Z. D., Atudorei, V., and Furrer, H. (2000). The effect of diagenesis on oxygen isotope ratios of biogenic phosphates. *American Journal of Science* **300**, 222–237.

Sharp, Z. D., and Cerling, T. E. (1997). Fossil isotope records of seasonal climate and ecology: Straight from the horse's mouth. *Geology* **26**, 219–222.

Sharp, Z. D., Durakiewicz, T., Migaszewski, Z. M., and Atudorei, V. N. (2002). Antiphase hydrogen and oxygen isotope periodicity in chert nodules. *Geochimica et Cosmochimica Acta* **66**, 2865–2873.

Shemesh, A., Charles, C. D., and Fairbanks, R. G. (1992). Oxygen isotopes in biogenic silica; global changes in ocean temperature and isotopic composition. *Science* **256**, 1434–1436.

Shemesh, A., Kolodny, Y., and Luz, B. (1983). Oxygen isotope variations in phosphate of biogenic apatites: II. Phosphorite rocks. *Earth and Planetary Science Letters* **64**, 405–416.

Sheppard, S. M. F., and Gilg, H. A. (1996). Stable isotope geochemistry of clay minerals. *Clay Minerals* **31**, 1–24.

Smith, K. F., Sharp, Z. D., and Brown, J. H. (2002). Isotopic composition of carbon and oxygen in desert fauna: Investigations into the effects of diet, physiology, and seasonality. *Journal of Arid Environments* **52**, 419–430.

Sponheimer, M., and Lee-Thorp, J. A. (1999). Isotopic evidence for the diet of an early hominid, Australopithecus africanus. *Science* **283**, 368–370.

Sridhar, K., Jackson, M. L., Clayton, R. N., Gillette, D. A., and Hawley, J. W. (1978). Oxygen isotopic ratios of quartz from wind-erosive soils in the Southwestern United States in relation to aerosol dust. *Soil Science Society of America Journal* **42**, 158–162.

Thewissen, J. G. M., Roe, L. J., O'Neil, J. R., Hussain, S. T., Sahni, A., and Bajpai, S. (1996). Evolution of Cetacean osmoregulation. *Nature* **381**, 379–380.

Tudge, A. P. (1960). A method of analysis of oxygen isotopes in orthophosphate; its use in the measurement of paleotemperatures. *Geochimica et Cosmochimica Acta* **18**, 81–93.

Tuross, N., Behrensmeyer, A. K., Eanes, E. D., Fisher, L. W., and Hare, P. E. (1989). Molecular preservation and crystallographic alterations in a weathering sequence of wildebeest bones. *Applied Geochemistry* **4**, 261–270.

Vennemann, T. W., Fricke, H. C., Blake, R. E., O'Neil, J. R., and Colman, A. (2002). Oxygen isotope analysis of phosphates; a comparison of techniques for analysis of Ag_3PO_4. *Chemical Geology* **185**, 321–336.

Webb, E. A., and Longstaffe, F. J. (2003). The relationship between phytolith- and plant-water $\delta^{18}O$ values in grasses. *Geochimica et Cosmochimica Acta* **67**, 1437–1449.

Yapp, C. J. (1997). An assessment of isotopic equilibrium in goethites from a bog iron deposit and a lateritic regolith. *Chemical Geology* **135**, 159–171.

Yapp, C. J. (1998). Paleoenvironmental interpretations of oxygen isotope ratios in oolitic ironstones. *Geochimica et Cosmochimica Acta* **62**, 2409–2420.

Yapp, C. J. (2001). Rusty relics of Earth history: Iron(III) oxides, isotopes, and surficial environments. *Annual Review of Earth and Planetary Science* **29**, 165–199.

Yapp, C. J., and Poths, H. (1992). Ancient atmospheric CO_2 pressures inferred from natural goethites. *Nature* **355**, 342–344.

Yeh, H., and Savin, S. M. (1977). Mechanism of burial metamorphism of argillaceous sediments; 3, O-isotope evidence. *Geological Society of America Bulletin* **88**, 1321–1330.

Yeh, H. W. (1980). D/H ratios and late-stage dehydration of shales during burial. *Geochimica et Cosmochimica Acta.* **44**, 341–352.

Yeh, H. W., and Epstein, S. (1978). Hydrogen isotope exchange between clay minerals and sea water. *Geochimica et Cosmochimica Acta.* **42**, 140–143.

CHAPTER 9

NITROGEN

9.1 INTRODUCTION

Nitrogen is a trace phase in rocks and the major component of air. Nevertheless, 97.98% of all nitrogen occurs in crystalline rocks! Air is the source of 2%, and only 0.001% is tied up in organic matter. Of that 0.001%, 60% is dissolved in the ocean, 25% is in soil organic matter, and 1.5% is in living biomass (Hübner, 1986). In other words, only 70 ppb of the total nitrogen is contained in living matter. Although minor in abundance, organic nitrogen is of tremendous importance (other than for life!), because almost all nitrogen isotope fractionation occurs by metabolic or metabolically related processes. Over the eons, this has led to a range of nitrogen isotope compositions that span well over 100‰. Even in the mantle, the substantial range of $\delta^{15}N$ values has been attributed by some to the subduction of surficial material (Beaumont and Robert, 1999; Marty and Dauphas, 2003).

The two stable isotopes of nitrogen are ^{14}N and ^{15}N, with a $^{14}N/^{15}N$ ratio of 272 in air. Because the ratio is constant, air nitrogen is taken as our standard, given by[1]

$$\delta^{15}N(\text{‰ vs. AIR}) = \left(\frac{(^{15}N/^{14}N)_{\text{sample}}}{(^{15}N/^{14}N)_{\text{AIR}}} - 1 \right) 1000 \qquad (9.1)$$

A reference gas of N_2 from air is easily made by removing CO_2 and water cryogenically and removing O_2 by reaction with copper oxide. The remaining gas will be N_2 with a trace of Ar. Solid reference samples are also available from NIST and the IAEA (Appendix 1).

---

[1]In the agricultural literature, the $\delta^{15}N$ value is defined as $\delta^{15}N(\text{AIR ‰}) = \left(\frac{(\text{at\% }^{15}N)_{\text{sample}}}{(\text{at\% }^{15}N)_{\text{AIR}}} - 1 \right) 1000$, which is nearly, but not quite, identical to the definition given here.

Nitrogen is a trace element in rocks, and because nitrogen isotope ratios have traditionally been some of the most difficult to measure, nitrogen isotope geochemistry has not been thoroughly embraced by the geochemical community. Nitrogen is difficult to transfer in vacuum lines, because it cannot simply be frozen with liquid nitrogen. Instead, it needs to be *adsorbed* on zeolite-filled cold fingers. Also, at low nitrogen levels, even small leaks will compromise an analysis. Contamination with CO will have a drastic effect on measured $\delta^{15}N$ ratios, due to the interference at mass 29 ($^{13}C^{16}O$). (See Problem 7 at the end of the chapter for more information). For analyses of trace amounts of nitrogen in (especially mantle) minerals, static-source mass spectrometers are required.

Many analytical problems have been eliminated with the coupling of the elemental analyzer and mass spectrometer, which allows for the combustion and analysis of N-bearing compounds to be made in continuous-flow mode. (See Section 2.8.3.) Nitrogen analyses of many solids can now be made rapidly and with little effort (except at low concentrations). This capability has raised the status of nitrogen as an important isotopic tracer, especially for pollution studies and within the biological community. It must be stressed however, that for nitrogen dissolved in water, sophisticated wet-chemical procedures are generally required to convert the nitrogen-bearing ion ((NH_4^+), (NO_3^-), etc.) to a solid form suitable for analysis (Kendall, 1998), although exciting new methods that employ bacterial denitrification have drastically simplified the procedure (Sigman et al., 2001; Coplen et al., 2004). The discussion in this chapter is concerned mainly with nitrogen processes in the surficial environment. Nitrogen isotope geochemistry of the mantle is discussed in Section 11.2.3.

9.2 THE NITROGEN CYCLE

Nitrogen forms a number of oxidation states from $+5(NO_3^-)$ to $-3(NH_4^+)$. A simplified nitrogen cycle is shown in Fig. 9.1. Diatomic nitrogen is removed from air by microorganisms (particularly *Rhizobium* sp.) living symbiotically in higher plants or lichens. The process is called *nitrogen fixation*; a typical reaction is

$$N_{2(g)} + 3 H_2O_{(g)} \xrightarrow{\text{nitrogenase}} 2 NH_{3(g)} + \frac{3}{2}O_{2(g)} \qquad (9.2)$$

Nitrogen fixation is an energy-consuming process. The ammonia or ammonium ion is extremely important in fertilizer, to the point where $5-14 \times 10^{10}$ kg/yr of nitrogenous fertilizers are produced by industrial nitrogen fixation alone.

Assimilation or *immobilization* is the processes whereby NH_4^+ or NO_3^- is incorporated into living tissue. The reverse of these processes is the degradation of organic matter by heterotrophic bacteria and the release of NH_4^+ in a process called *mineralization* (also called *ammonification*).

Nitrification is the oxidation of ammonia to NO_3^- by nitrifying organisms (e.g., chemotrophic bacteria). Nitrification is an energy-releasing process and is used by organisms as an energy source. A typical reaction sequence is (Kaplan, 1983)

$$NH_3 + 3/2\, O_2 \longrightarrow HNO_2 + H_2O \qquad (9.3a)$$

$$KNO_2 + \tfrac{1}{2}O_2 \longrightarrow KNO_3. \qquad (9.3b)$$

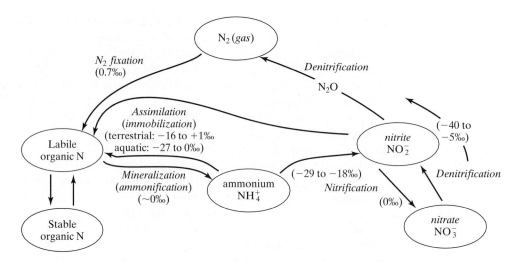

FIGURE 9.1: Simplified diagram of the nitrogen cycle. Note that nitrification is generally thought of as the conversion of ammonium to nitrate, with nitrite an intermediate step. Likewise, denitrification is the conversion of nitrate to N_2 and/or N_2O gas, but nitrite is an intermediate phase. Numbers in parentheses indicate average fractionations ($\delta^{15}N_{product} - \delta^{15}N_{source}$) associated with each process.

Another example of nitrification is the two-step process given by a first oxidation by *Nitrosomonas* ($NH_4^+ \to NO_2^-$) and a second step by *Nitrobacter* ($NO_2^- \to NO_3^-$). Oxygen comes from both H_2O and O_2 for this nitrification process.

Denitrification is the process whereby NO_3^- and NO_2^- are converted to N_2O and, ultimately, N_2 gas by anaerobic bacteria, some fungi, and aerobic bacteria. Denitrification accompanies the degradation of organic matter, as, for example, with glucose:

$$5\, C_6H_{12}O_6 + 24\, NO_3^- + 24\, H^+ \longrightarrow 30\, CO_2 + 42\, H_2O + 12\, N_2(g). \qquad (9.4)$$

Denitrification tends to occur in deeper layers of soil or in poorly aerated soils where the $p(O_2)$ is low. In the ocean, denitrification is most active in stagnant water masses and where the $p(O_2)$ is low. Correspondingly, denitrification increases with depth in the ocean. Atmospheric N_2 would be exhausted in 100 million years if it were not for denitrification processes.

9.3 NITROGEN ISOTOPE FRACTIONATION

The nitrogen isotope fractionations attending the various processes shown in Fig. 9.1 are difficult to quantify because most of the transformations are metabolically driven or kinetically controlled. *They are not equilibrium reactions.* As we have seen for biologically mediated carbon reduction and will see in Chapter 10 for sulfate reduction, the magnitude (and even sign) of fractionation can be highly variable, depending upon the reaction rates and availability of nutrients. For example, nitrification, given by the multistep transformation *organic nitrogen* $\to$ $NH_4^+ \to NO_2^- \to NO_3^-$, may have different isotopic fractionations associated with each step, and within each step the fractionations can be variable, depending on ambient conditions.

The δ^{15}N value of the product nitrate will be anywhere from -12 to $-29‰$ lighter than the ammonium from which it forms (Kendall, 1998).

In this chapter, we will use the standard notation $\Delta^{15}N_{product-reactant}$ to indicate the difference of the δ^{15}N values of a product phase (e.g., NO_3^- in the example in the previous paragraph) and the reactant or source (organic nitrogen in the example). In much of the nitrogen literature, the fractionation is described in terms of the *enrichment factor*

$$\varepsilon = (\alpha - 1)1000 = \left(\frac{\delta_p - \delta_r}{\delta_r + 1000}\right)1000, \tag{9.5}$$

where δ_p and δ_r are the delta values of the product and reactant, respectively. Except when δ_r values are very large (e.g., in ^{15}N-enriched tracer experiments), ε is nearly identical to $\delta_p - \delta_r$, or $\Delta^{15}N_{product-reactant}$.

9.3.1 Nitrogen Fixation

Nitrogen fixation is generally considered as a single process in terms of isotopic fractionation, because δ^{15}N values are measured on the product plant or bacterium, regardless of the pathway from $N_{2(g)}$ to organic matter. Nitrogen isotope fractionation associated with fixation is generally small. Hoering and Ford (1960) measured fractionations ($\delta^{15}N_{fixed} - \delta^{15}N_{air}$) ranging from -2.2 to $+3.7‰$ ($n = 4$) and considered the average fractionation between atmospheric N_2 and fixed nitrogen in organic matter to be near 0‰. A compilation by Fogel and Cifuentes (1993) gives ranges from -3 to $+1‰$; one by Hübner (1986) gives an average value of -0.7 ± 1.6 ‰; (Fig. 9.2). The scatter indicates, not some sort of analytical error or uncertainty, but rather that real variations in fractionation for this and all other pathways exist.

9.3.2 Mineralization

The fractionation associated with the breakdown of organic matter to soil ammonium is small ($\Delta = 0 \pm 1‰$). As pointed out by Kendall (1998), some researchers define mineralization as the breakdown of organic matter and the conversion to nitrate. Under such circumstances, fractionations can be large and variable, but the differences are due, not to the mineralization step itself, but rather the nitrification of ammonium to nitrate.

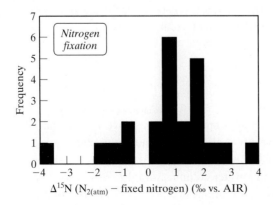

FIGURE 9.2: Fractionation associated with N_2 fixation. The average value is 0.72‰ (meaning that the organisms have δ^{15}N values less than 0‰). (Data from Hübner, 1986.)

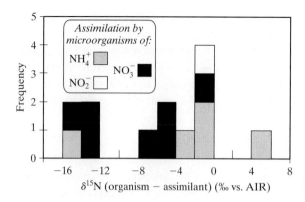

FIGURE 9.3: Nitrogen isotope fractionation during assimilation by microorganisms. On the basis of limited data, the fractionation is similar, regardless of whether ammonium, nitrate, or nitrite is the assimilant. (Data from Hübner, 1986.)

9.3.3 Assimilation

Assimilation by microorganisms causes a strong and variable discrimination, favoring ^{14}N (Fig. 9.3). There is no appreciable difference between assimilation of NH_4^+, NO_2^-, and NO_3^-. Higher plants show much smaller fractionations, averaging only $-0.25‰$ ($\delta^{15}N_{plant} - \delta^{15}N_{assimilant}$). A compilation of data for ammonium assimilation by aquatic algae spans a very large range from -27 to $0‰$ (Fogel and Cifuentes, 1993). The wide range of delta values can be modeled in terms of kinetic processes in which rates are controlled by (1) the availability of nitrogen, (2) enzymes responsible for NH_3 fixation, and (3) the diffusion of NH_3 through the cell walls. Velinsky et al. (1991) found that ammonium assimilation in anoxic waters was strongly dependent on NH_4^+ concentrations. In waters with NH_4^+ concentrations of 40 μM, the fractionation between particulate organic matter and NH_4^+ was modeled to be -20 to $-30‰$. In waters with concentrations of only 9 μM, fractionations were -5 to $-15‰$. In other words, discrimination is possible only when the concentration of the reactants is high.

9.3.4 Nitrification

Nitrification is a two-step process going from NH_4^+ through NO_2^- and, finally, NO_3^- (equation (9.3)). The second part of the reaction (equation (9.3b), $NO_2^- \rightarrow NO_3^-$) is quantitative, meaning that all nitrite is converted to nitrate, so there can be no nitrogen isotope fractionation associated with this step. Published estimates for the fractionation of ammonium to nitrite (equation 9.3(a)) range from -18 to $-29‰$, with the nitrite (and, ultimately, nitrate) having lower $\delta^{15}N$ values than the ammonium precursor. The fractionation depends on the proportions of ammonium and nitrate after reaction. Obviously, if all of the ammonium is converted to nitrate in a "closed system," then the $\delta^{15}N$ value of the nitrate will be identical to that of the original ammonium reservoir.

9.3.5 Denitrification

Laboratory experiments give a $\Delta^{15}N_{N_2\,gas-dissolved\,nitrate}$ value of -17 to $-20‰$. Measured fractionations from soil samples are often less, between -12 and $-14‰$ (e.g., Blackmer and Bremner, 1977). Mariotti et al. (1982) found the fractionation $\Delta^{15}N_{N_2O-NO_2^-}$ to range from $-33‰$ to $-11‰$. Cline and Kaplan (1975) measured the concentration and $\delta^{15}N$ values of dissolved nitrate in a water column from the eastern tropical North Pacific Ocean. They were able to model the variations in $\delta^{15}N$ values in terms of diffusion theory when the $\Delta^{15}N_{N_2\,gas-dissolved\,nitrate}$ value was $-40‰$ (Fig. 9.4).

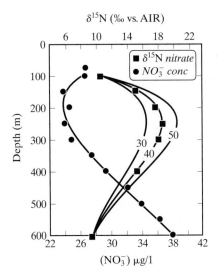

FIGURE 9.4: Dissolved nitrate concentration and $\delta^{15}N$ value in a depth profile from the eastern tropical North Pacific Ocean. As nitrate is converted to N_2, the nitrate concentration decreases and its $\delta^{15}N$ value increases, because the $\delta^{15}N$ value of the evolved N_2 gas is ~40‰ lighter. The data are modeled under the assumption that $\Delta^{15}N_{NO_3-N_2}$ = 30, 40, and 50‰, as shown by curves through isotope data. (After Cline and Kaplan, 1975.)

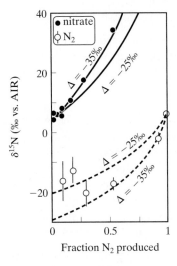

FIGURE 9.5: Variations in the $\delta^{15}N$ values of dissolved nitrate and N_2 gas from water wells from the Kalahari desert. The data can be modeled by a Rayleigh fractionation process with a $\Delta^{15}N$ value between N_2 gas (product) and dissolved nitrate (reactant) of −35‰ When only a small amount of nitrate remains, its $\delta^{15}N$ value becomes extremely high, while the $\delta^{15}N$ value of the total N_2 approaches the original nitrate value. Solid lines are the instantaneous nitrate value; dashed lines are the *total* N_2 value for a given proportion of N_2 produced. (After Heaton, 1986.)

Denitrification has large isotope fractionation effects due to the "distillation" of N_2 gas. In shallow aquifers, N_2 gas produced by denitrification can be lost by diffusion to the atmosphere. This is a classic Rayleigh fractionation process with a large coefficient of fractionation. If a large amount of N_2 is produced, the nitrogen isotope composition of the remaining nitrate can change significantly. Figure 9.5 illustrates the magnitude of this effect. It also shows how we can use nitrogen isotopes to evaluate the amount of nitrate that has been removed from a system. In a contaminated aquifer, this obviously is an important tool for water quality studies.

9.4 THE CHARACTERISTIC $\delta^{15}N$ VALUES OF VARIOUS MATERIALS

Now that we have the chemistry and fractionation factors for a number of chemical transformations involving nitrogen, it should be a relatively straightforward task to make sense of the variability and range of $\delta^{15}N$ values of the different reservoirs. For example, from Fig 9.1, it is clear that the fractionation between N_2 gas and labile organic nitrogen is close to zero. So we should expect that nitrogen-fixing plants[2] have $\delta^{15}N$ values close to zero, and indeed, this is the case. Especially when growing in nitrate-poor soil (so that the only source of nitrogen is N_2), nitrogen-fixing plants have $\delta^{15}N$ values that are within 2‰ of air (Shearer and Kohl, 1986). As shown next, we can use similar logic to explain the $\delta^{15}N$ values of a number of different reservoirs.

9.4.1 Plants and Soil

Nitrogen-fixing plants have $\delta^{15}N$ values close to zero. Other plants cannot fix N_2 and instead incorporate nitrogen by assimilating NH_4^+ or NO_3^- from soil. The $\delta^{15}N$ values of plants are strongly dependent on those of the soil, which are in part controlled by plants. In order to predict the $\delta^{15}N$ value of plants, it is necessary to know the range of $\delta^{15}N$ values of soil *and* the mechanisms of uptake from the soil. $\delta^{15}N$ values of soil range from -10 to $+15$‰, with most soils between $+2$ and $+5$‰ (Kendall, 1998). The positive values are loosely tied to a preferential loss of ^{14}N during the decomposition of particulate nitrogen sources. Unfortunately, specific factors controlling soil $\delta^{15}N$ values are complex and defy quantification. Even the sources of extracted nitrogen are variable: Tree roots preferentially assimilate soil nitrate, whereas microorganisms tend to incorporate soil ammonium (Nadelhoffer and Fry, 1988). Nevertheless, some general guidelines can be established:

- Denitrification is most intense in poorly drained or poorly oxidized soils, because nitrate-consuming organisms become active only when oxygen levels are low. The subsequent loss of N_2 gas—the product of denitrification—increases the $\delta^{15}N$ value of any remaining nitrate (Fig. 9.5).
- Soils with abundant leaf litter tend to have lower $\delta^{15}N$ values than surrounding regions with less foliage. An explanation for this trend is that the preferential uptake of ^{14}N by plants results in higher $\delta^{15}N$ values of the soil. In heavily vegetated areas, the ^{15}N-depleted plant material is returned to the soil as leaf litter.
- Anthropogenic activity can strongly affect the $\delta^{15}N$ of soil. In one study, the average $\delta^{15}N$ value of cultivated soils was 5.0 ± 3.5‰, compared with 6.8 ± 6.4‰ for uncultivated soils, due to the addition of nitrogenous fertilizers with low $\delta^{15}N$ values (Hübner, 1986). In some soils, there are variations with depth, while in others, no such correlation is found. Controls include drainage, the total N content of soil, and the precipitation rate.

In nitrogen-limited soils, the $\delta^{15}N$ value of plants is close to that of the soils, and no discrimination is possible. In nutrient-rich soils, the fractionation between plants and dissolved inorganic nitrogen can be several per mil. Trees tend to have slightly lower $\delta^{15}N$ values than soil, due to the negative fractionation during assimilation. Heterotrophic fungi, by contrast, may have $\delta^{15}N$ values higher than those of the soil (Högberg, 1997). Overall, tree leaves have $\delta^{15}N$ values that range from -8 to $+3$‰ (Peterson and Fry, 1987).

9.4.2 Other Terrestrial Reservoirs

Fertilizers. Fertilizers generally have a $\delta^{15}N$ range from -4 to $+4$‰, with the low values related to an atmospheric N_2 source. Organic fertilizers range from $+6$ to $+30$‰, related to the processes occurring in animal wastes (Kendall, 1998). The $\delta^{15}N$ value of animals increases by

---

[2] Nitrogen-fixing plants are those that are able to assimilate N_2 gas directly. In fact, nitrogen fixation represents a symbiotic relationship between these plants and bacteria that live on their roots, but the isotopic effect is the same nevertheless.

~3‰ at each higher trophic level. (See Section 9.5.) The most important factor in this increase is the excretion of isotopically light urine. Therefore, there is an enrichment in ^{15}N from plants to animals. Animal waste gets further enriched in ^{15}N by the subsequent volatilization of isotopically light ammonia.

Rain. The sources of nitrogen in rain are the volatilization of ammonia, nitrification and denitrification of soils, and anthropogenic sources. Hoering (1957) first measured the $\delta^{15}N$ values of NH_4^+ and NO_3^- in rain from the roof of the chemistry laboratory at the University of Arkansas. He found that, although there was significant variability in the $\delta^{15}N$ value of each component ($\delta^{15}N\ NH_4^+ = -0.1$ to 9.0‰; $\delta^{15}N\ NO_3^- = -7.2$ to $+3.4$), the fractionation between the two phases could be explained in terms of a kinetic fractionation between ammonium and nitrate. $\delta^{15}N$ values of nitrate in rain have since been found to range from ~ -10 to $+9$‰. Heaton (1986) gives an average of -5‰ (for South Africa), while a compilation by Kendall (1998) ranges from ~ -3 to $+9$‰, with a strong mode at 0 to $+2$‰. In general, ammonium is lighter than nitrate by several per mil. The measured fractionation between NH_4^+ and NO_3^- in rain is dependent on the concentrations of the ions in precipitation. Variations from site to site are substantial, because inputs can be so different. Pure air has $\delta^{15}N$ values for NH_3 and NO_2 of -10.0 ± 2.6‰ and -9.3 ± 3.5‰, respectively (Hübner, 1986). The effects of mixing different reservoirs are clear when just a few "end-member" sources are considered. Barnyard NH_3 has a $\delta^{15}N$ value of $+25$‰, NO_x from automobile exhaust is $\sim +3.7$‰, and fumaroles from southern Kamchatka have $\delta^{15}N$ values of $(NH_4)_2SO_4$ as low as -31‰ (Hübner, 1986). Freyer (1978) measured $\delta^{15}N$ values of -12.0 ± 1.9‰ for NH_4^+ in rainwater from Jülich, Germany. Other published values of NH_4^+ in rainwater range from -9.7 to $+6.9$. Not surprisingly, variations can be large between storms and even within individual storms, because the sources of nitrogen—fossil-fuel burning, ocean denitrification, etc.—themselves have a large range of $\delta^{15}N$ values. Peterson and Fry's (1987) average estimates for precipitation are -18 to $+8$‰ for NH_4^+ and -15 to $+3$‰ for NO_3^-.

Fossil fuels. Peat and coal average $+0.8$‰, with a standard deviation of 1.6‰, ranging up to 6.3‰ (Hoering, 1955; Wada et al., 1975; Stiehl and Lehmann, 1980), similar to values obtained for modern peats and bogs. Crude oils are generally in the range from $+1.0$ to $+6.7$‰, while natural gas has far more variation (-10.5 to 14.4‰; Hoering and Moore, 1958). The $\delta^{15}N$ value of natural gas changes quite drastically with the distance of migration from its source. An example from north Germany shows a systematic increase from -8.7 to $+18.0$‰ as migration distances increase, likely due to Rayleigh fractionation attending denitrification (Stahl, 1977).

9.4.3 Nitrogen in the Oceans

The fact that the $\delta^{15}N$ values of most ocean materials are positive is easily explained in terms of the nitrogen cycle in the ocean. Nitrogen is one of the most important limiting nutrients in the ocean, so that productivity is limited by the availability of useable nitrogen.[3] The major inputs are river runoff, rain, and the fixation of molecular N_2 by marine blue-green algae (Fig. 9.6). Outputs or sinks of nitrogen in the ocean include burial in sediment and denitrification.

---

[3] Nitrogen and phosphorus are strongly correlated with a nitrogen/phosphate ratio of ~15. Both nutrients become exhausted at the same time. The constant ratio is probably tied to a biochemical feedback mechanism. If dissolved nitrate levels become low, nitrogen-fixing blue green algae will produce nitrate, restoring the biochemical ratio. If nitrate values become high, nonnitrifying organisms would have an advantage, consuming nitrate disproportionately and again driving the ratio back to its balanced state.

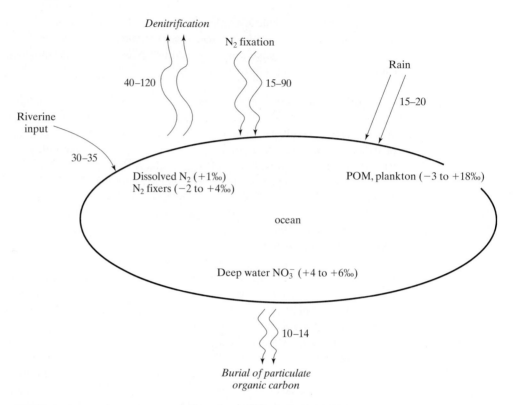

FIGURE 9.6: **Nitrogen system for the oceans.** Sources and sinks (and their fluxes in 10^{12} gm-N_2/yr) are shown as roman and italicized text, respectively. $\delta^{15}N$ values of ocean materials are in boldface. Data are from the following sources: Kaplan (1983); Macko et al. (1984); Berner and Berner (1987); Peterson and Fry (1987); and Fogel and Cifuentes (1993).

We assume that the nitrogen cycle is balanced, with inputs equaling outputs, but how variable this balance is over time is not known. The fluxes of each exchange path are shown in Fig. 9.6. It is clear that the average values are not known well enough to quantify. Even more intractable is an attempt to quantify isotopic mass balance, because the $\delta^{15}N$ values of each source are quite variable, as are the fractionations accompanying any transfer from one reservoir to another.

In spite of the uncertainties, several gross features are apparent. The major fractionation in the global ocean cycle occurs during denitrification, with N_2 being strongly depleted in ^{15}N relative to its source. All other sources and sinks are associated with rather small fractionation effects. (See Fig. 9.1.) Clearly, a loss of light N_2 back to the atmosphere will result in a positive average $\delta^{15}N$ value of the ocean. The positive $\delta^{15}N$ value of organic material in sediments is retained during subduction, as is seen both in rocks (Bebout and Fogel, 1992) and in volcanic fumaroles sourced in oceanic sediments (Fischer et al., 2002).

There is significant spatial variation in the $\delta^{15}N$ value of dissolved nitrate. In the Eastern North Pacific Ocean, $\delta^{15}N$ values range from +6.5‰ in the Antarctic intermediate water mass (at depth) up to +18.8‰ in the active denitrification zone (Cline and Kaplan, 1975).

Average $\delta^{15}N$ values of various reservoirs, compiled from many sources, are shown in Fig. 9.7. There are always unusual samples that have higher or lower values (Coplen et al., 2002), but the figure should serve as a guide for average ranges that are commonly found for each material.

Section 9.4: The Characteristic $\delta^{15}N$ Values of Various Materials

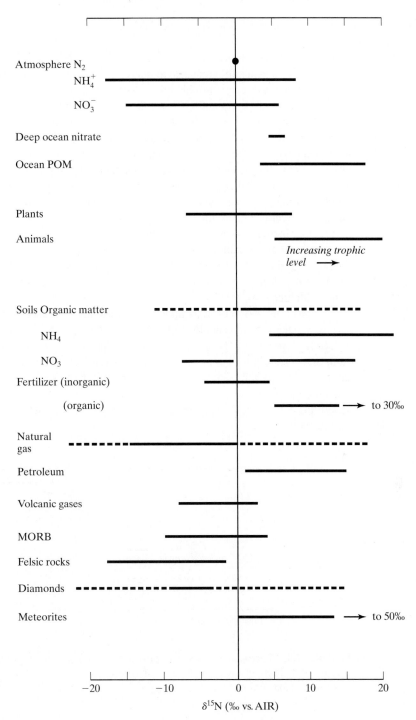

FIGURE 9.7: Average $\delta^{15}N$ values for common materials.

9.5 NITROGEN ISOTOPE RATIOS IN ANIMALS

The $\delta^{15}N$ values of animals are related to their diet (DeNiro and Epstein, 1981). The $\delta^{15}N$ value of an animal is generally greater than that of the food it eats, and the $\delta^{15}N$ values increase by 3–4‰ for each successive trophic level.[4] Stable nitrogen isotope ratios are therefore an important ecological tool for quantifying trophic position and for reconstructing dietary preferences.

Not all tissues in a body have the same $\delta^{15}N$ value. Milk, blood, and muscle tend to have $\delta^{15}N$ values 1–3‰ heavier than the diet. Urinary urea and bile have $\delta^{15}N$ values 2 to 4‰ more negative than the diet (Ambrose, 1991). The loss of ^{15}N-depleted urine is probably the primary cause for the elevated overall $\delta^{15}N$ value of animals relative to their diet, although it has been shown that there can be preferential uptake of ^{15}N relative to diet. At each successive trophic level, the $\delta^{15}N$ value of the food source increases, and the animals' $\delta^{15}N$ values follow suit. The effect is most regular and intense in marine communities. As seen in Fig. 9.8, there is a regular increase of ~3‰ per trophic level. The effect on terrestrial communities is not as large and is more variable, controlled by numerous factors. For example, several authors have found a correlation between the $\delta^{15}N$ value of animals and the relative annual rainfall (Sealy et al., 1987). It appears that animals which are more water stressed excrete a concentrated urine with higher $\delta^{15}N$ values.

The combination of $\delta^{15}N$ and $\delta^{13}C$ values has been used in well over 100 publications to investigate the trophic ecology of birds and mammals (Kelly, 2000). Combined $\delta^{15}N$–$\delta^{13}C$ values from prehistoric bone collagen are a valuable tool for distinguishing different populations and constraining a community's diet. To a first degree, we can state that the $\delta^{15}N$ values are controlled by the trophic level of diet and the $\delta^{13}C$ values are controlled by the relative dietary proportions of C_3 and C_4 plants. Communities subsisting mainly on an animal diet will inherit the $\delta^{13}C$ value of their prey, perhaps with a subtle offset towards higher values (DeNiro and Epstein, 1978).

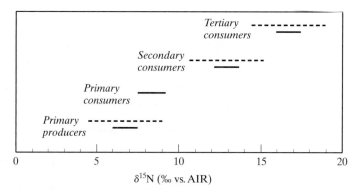

FIGURE 9.8: Nitrogen isotope compositions of marine plants and animals as a function of trophic level. There is a regular increase of approximately 3‰ per trophic level. (Data from Minagawa and Wada, 1984 (solid); Schoeninger and DeNiro, 1984 (dashed).)

---

[4]A trophic level is the position a group of organisms occupies in the food chain. Each successive trophic level consumes the one below it. Hence, trophic level 1 consists of the plants, trophic level 2 the herbivores, trophic level 3 the carnivores, etc.

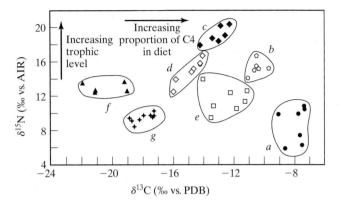

FIGURE 9.9: $\delta^{13}C$ and $\delta^{15}N$ values of bone collagen from Native American communities:
a. Western Anasazi, Southern Utah (A.D. 1–1300)
b. Nantucket Island, Massachusetts (A.D. 1000–1600)
c. Northwest coast Haida and Tlingit salmon fishers (historic)
d. Southern California coastal (A.D. 1400–1800)
e. Plains Arikara, South Dakota (A.D. 1650–1733)
f. Southern Ontario, Middle Woodland period (A.D. 1–400)
g. Modern Chicago, United States.
See text for details. Compilation and references from Martin (1999) and Schoeller et al. (1986).

Figure 9.9 shows a compilation for a number of North American Native American communities. The combined carbon and nitrogen isotope values are easily explained in terms of the communities' assumed diets, and they place constraints on the diet in cases where ambiguities exist. For example, the western Anasazi have the lowest $\delta^{15}N$ values and highest $\delta^{13}C$ values. It can be concluded that they had a maize-based diet[5] with only minor animal consumption (Martin, 1999). The southern Ontario communities (Schwarcz et al., 1985) had a diet consisting of C_3 plants and animals that consumed C_3 plants. The elevated $\delta^{15}N$ values relative to the Anasazi community indicate at least a partial animal diet. The highest $\delta^{15}N$ values are found in the northwest coastal communities that subsisted in large part on salmon (Schoeninger and Moore, 1992). Salmon are at a high trophic level, and this fact is reflected in the isotope data.

PROBLEM SET

1. Use the equations given in Chapter 4 to generate the Rayleigh fractionation curves shown in Fig. 9.5.
2. The fractionation during ammonium assimilation by aquatic algae spans a very large range, with $\delta^{15}N$ ranging from -27 to $0‰$. How would the availability of NH_3 ultimately affect the $\delta^{15}N$ value of the algae?

---

[5]Maize is a C_4 plant. (See Chapter 7 for a more thorough discussion of C_3 and C_4 plants.)

218 Chapter 9 Nitrogen

3. In downhole hydrology tests of an aquifer "downstream" from an agricultural center, dissolved nitrate content and $\delta^{15}N$ values of nitrate in each sample were measured. Assume that the concentration and $\delta^{15}N$ value of the water were initially the same and that nitrogen is lost *via* denitrification only by Rayleigh loss of N_2 gas. Consider the following table:

Sample	NO_3^- Concentration	NO_3^- $\delta^{15}N$
1	37.4	7.36
2	18.4	32.4
3	4.7	83.0
4	28.9	16.4
5	20.6	28.5
6	39.7	5.25

What is the average denitrification enrichment factor ($\Delta^{15}N$ between product N_2 gas and dissolved nitrate)? (*Hint*: Graph the Rayleigh fractionation equation for different values of α, and see what fits.)

4. A local grocer claims that the apples he sells are "organic"—that only natural fertilizers are used. You don't trust him and decide to measure the nitrogen isotope ratio of the apple to see if he's telling the truth. The $\delta^{15}N$ value of your apple is $+19\text{‰}$. What do you conclude?

5. If the $\delta^{15}N$ value of air is 0‰, but the discrimination associated with denitrification was $+20$ (instead of negative), what would the average $\delta^{15}N$ value of the ocean be?

6. On the basis of Fig. 9.9, where on the diagram would you expect an Eskimo community to plot? Assume a diet of seal and occasional polar bear.

7. Does the contamination of an N_2 sample by CO cause an apparent increase or decrease in the measured $\delta^{15}N$ value. (*Hint*: Check out Table 1.4.)

ANSWERS

1. From equation 4.16; $\delta^{15}N_{(Nitrate)} = (\delta^{15}N_{nitrate,\ initial} + 1000)F^{(\alpha-1)} - 1000$. For $\Delta_{(N_2\ gas-nitrate)} = -35$, $\alpha = 0.965$. For an initial $\delta^{15}N_{nitrate} = 5\text{‰}$ (from graph), the above equation becomes:

   $\delta^{15}N_{(Nitrate)} = (1005)F^{(-0.035)} - 1000$. The $\delta^{15}N_{(N_2\ gas)}$ will be 35‰ less than the dissolved nitrate.

2. As with carbon, availability affects the fractionation during the conversion from one form to another. If there is very little ammonium available, there can be very little discrimination. The algae takes "whatever it can get." If there is excess ammonium, then the algae can discriminate, in which case they preferentially take the light isotopologues. In general, more food available implies more discrimination.

3. Use equation (4.16) to answer the question. We have $\delta_{nitrate} = [\delta_{nitrate,i} + 1000]F^{(\alpha-1)} - 1000$. α is the fractionation between product (N_2) and reactant (nitrate). Start with a $\delta_{nitrate,initial}$ value of 5.25 (the highest value) and $F = \text{conc}/39.7$ (where 39.7 is the highest value). Then test for different values of α to see what fits the data best. The simple Rayleigh distillation equation can also be linearized, so that a simple linear regression can be applied to the data. See Sharp et al. (2003) for details.

4. Keep buying apples! A $+16\text{‰}$ $\delta^{15}N$ value is consistent with an organic fertilizer nitrogen source. See Fig. 9.7.

5. Here is a classic problem of steady state. The assimilation of nitrogen in the ocean has no fractionation, but the fractionation of N_2 leaving the ocean has a positive value of +20. That is, the $\delta^{15}N$ of the N_2 leaving the ocean is 20‰ heavier than the $\delta^{15}N$ of the ocean itself. In order to have a steady-state balance, the flux $\times$ $\delta^{15}N$ value entering the ocean must equal the flux $\times$ $\delta^{15}N$ value leaving the ocean. Otherwise the ocean $\delta^{15}N$ value would change with time. If the fractionation is 20‰, then the ocean must have an average fractionation value of -20‰ to have the same flux $\times$ $\delta^{15}N$ value.

6. An Eskimo community should have very high $\delta^{15}N$ values and intermediate $\delta^{13}C$ values. The diet comes from a very high trophic level (seal, etc.), making the $\delta^{15}N$ value high, whereas carbon is of marine origin, intermediate between C_3 and C_4 plants.

7. Big increase. The $^{13}C/^{12}C$ ratio of CO is 1.11/98.89 vs. 0.37/99.759 for N_2. Calculate the apparent increase on $\delta^{13}C$ for a 5% contamination by N_2 of typical "air" value.

REFERENCES

Ambrose, S. H. (1991). Effects of diet, climate and physiology on nitrogen isotope abundances in terrestrial foodwebs. *Journal of Archaeological Science* **18**, 293–317.

Beaumont, V., and Robert, F. (1999). Nitrogen isotope ratios of kerogens in Precambrian cherts: A record of the evolution of atmosphere chemistry? *Precambrian Research* **96**, 63–82.

Bebout, G. E., and Fogel, M. L. (1992). Nitrogen-isotope compositions of metasedimentary rocks in the Catalina Schist, California: Implications for metamorphic devolatilization history. *Geochimica et Cosmochimica Acta* **56**, 2839–2849.

Berner, E. K., and Berner, R. A. (1987). *The Global Water Cycle*. Englewood Cliffs, NJ: Prentice-Hall.

Blackmer, A. M., and Bremner, J. M. (1977). N-isotope discrimination in denitrification of nitrate in soils. *Soil Biology & Biochemistry* **9**, 73–77.

Cline, J. D., and Kaplan, I. R. (1975). Isotopic fractionation of dissolved nitrate during denitrification in the eastern tropical North Pacific Ocean. *Marine Chemistry* **3**, 271–299.

Coplen, T. B., Bohlke, J. K., and Casciotti, K. L. (2004). Using dual-bacterial denitrification to improve $\delta^{15}N$ determinations of nitrates containing mass-independent ^{17}O. *Rapid Communications in Mass Spectrometry* **18**, 245–250.

Coplen, T. B., Hopple, J. A., Böhlke, J. K., Peiser, H. S., Rieder, S. E., Krouse, H. R., Rosman, K. J. R., Ding, T., Vocke, R. D. J., Révész, K. M., Lamberty, A., Taylor, P., and DeBièvre, P. (2002). *Compilation of Minimum and Maximum Isotope Ratios of Selected Elements in Naturally Occurring Terrestrial Materials and Reagents*. Reston, VA: United States Geological Survey.

DeNiro, M. J., and Epstein, S. (1978). Influence of diet on the distribution of carbon isotopes in animals. *Geochimica et Cosmochimica Acta* **42**, 495–506.

DeNiro, M. J., and Epstein, S. (1981). Influence of diet on the distribution of nitrogen isotopes in animals. *Geochimica et Cosmochimica Acta* **45**, 341–351.

Fischer, T. P., Hilton, D. R., Zimmer, M. M., Shaw, A. M., Sharp, Z. D., and Walker, J. A. (2002). Contrasting nitrogen isotope behavior along the Central America margin: Implications for the nitrogen balance of the Earth. *Science* **297**, 1154–1157.

Fogel, M. L., and Cifuentes, L. A. (1993). Isotope fractionation during primary production. In *Organic Geochemistry* (ed. M. H. Engel and S. A. Macko), pp. 73–98. New York: Plenum Press.

Freyer, H. D. (1978). Seasonal trends in NH_4^+ and NO_3^- nitrogen isotope composition in rain collected at Jülich, Germany. *Tellus* **30**, 83–92.

Heaton, T. H. E. (1986). Isotopic studies of nitrogen pollution in the hydrosphere and atmosphere; a review. *Chemical Geology* **59**, 87–102.

Hoering, T. (1957). The isotopic composition of the ammonia and the nitrate ion in rain. *Geochimica et Cosmochimica Acta* **12**, 97–102.

Hoering, T. C. (1955). Variations in N-15 abundance in naturally occurring substances. *Science* **122**, 1233–1234.

Hoering, T. C., and Ford, T. H. (1960). The isotope effect in the fixation of nitrogen by *Azotobacter*. *Journal of the American Chemical Society* **82**, 376–378.

Hoering, T. C., and Moore, H. E. (1958). The isotopic composition of the nitrogen in natural gases and associated crude oils. *Geochimica et Cosmochimica Acta* **13**, 225–232.

Högberg, P. (1997). ^{15}N natural abundance in soil–plant systems. *New Phytologist* **137**, 179–203.

Hübner, H. (1986). Isotope effects on nitrogen in the soil and biosphere. In *Handbook of Environmental Isotope Geochemistry*, Vol. 2 (ed. P. Fritz and J. C. Fontes), pp. 361–425. Amsterdam: Elsevier.

Kaplan, I. R. (1983). Stable isotopes of sulfur, nitrogen and deuterium in Recent marine environments. In *Stable Isotopes in Sedimentary Geology*, Vol. 10 (ed. M. A. Arthur, T. F. Anderson, I. R. Kaplan, J. Veizer, and L. S. Land), pp. 2-1–2-108. Columbia, SC: SEMP Short course.

Kelly, J. F. (2000). Stable isotopes of carbon and nitrogen in the study of avian and mammalian trophic ecology. *Canadian Journal of Zoology* **78**, 1–27.

Kendall, C. (1998). Tracing nitrogen sources and cycling in catchments. In *Isotope Tracers in Catchment Hydrology* (ed. C. Kendall and J. J. McDonnell), pp. 519–576. Amsterdam: Elsevier.

Macko, S. A., Entzeroth, L., and Parker, P. L. (1984). Regional differences in nitrogen and carbon isotopes on the continental shelf of the Gulf of Mexico. *Naturwissenschaften* **71**, 374–375.

Mariotti, A., Germon, J. C., Leclerc, A., Catroux, G., and Létolle, R. (1982). Experimental determination of kinetic isotope fractionation of nitrogen isotopes during denitrification. *Stable Isotopes, Proceedings of the 4th International Conference*. Amsterdam: Elsevier, 459–464.

Martin, S. L. (1999). Virgin Anasazi diet as demonstrated through the analysis of stable carbon and nitrogen isotopes. *Kiva* **64**, 495–513.

Marty, B., and Dauphas, N. (2003). The nitrogen record of crust–mantle interaction and mantle convection from Archean to Present. *Earth and Planetary Science Letters* **206**, 397–410.

Minagawa, M., and Wada, E. (1984). Stepwise enrichment of ^{15}N along the food chains: Further evidence and the relation between δ^{15}N and animal age. *Geochimica et Cosmochimica Acta* **48**, 1135–1140.

Nadelhoffer, K. J., and Fry, B. (1988). Controls on natural nitrogen-15 and carbon-13 abundances in forest soil organic matter. *Soil Science Society of America Journal* **52**, 1633–1640.

Peterson, B. J., and Fry, B. (1987). Stable isotopes in ecosystem studies. *Annual Review of Ecology and Systematics* **18**, 293–320.

Schoeller, D. A., Minagawa, M., Slater, R., and Kaplan, I. R. (1986). Stable isotopes of carbon, nitrogen and hydrogen in the contemporary North American human food web. *Ecology of Food and Nutrition* **18**, 159–170.

Schoeninger, M. J., and DeNiro, M. J. (1984). Nitrogen and carbon isotopic composition of bone collagen from marine and terrestrial animals. *Geochimica et Cosmochimica Acta* **48**, 625–639.

Schoeninger, M. J., and Moore, K. (1992). Bone stable isotope studies in archaeology. *Journal of World Prehistory* **6**, 247–296.

Schwarcz, H. P., Melbye, J., Katzenberg, M. A., and Knyf, M. (1985). Stable isotopes in human skeletons of southern Ontario; reconstructing palaeodiet. *Journal of Archaeological Science* **12**, 187–206.

Sealy, J. C., van der Merwe, N. J., Thorp, J. A. L., and Lanham, J. L. (1987). Nitrogen isotopic ecology in southern Africa: Implications for environmental and dietary tracing. *Geochimica et Cosmochimica Acta* **51**, 2707–2718.

Sharp, Z. D., Papike, J. J., and Durakiewicz, T. (2003). The effect of thermal decarbonation on stable isotope compositions of carbonates. *American Mineralogist* **88**, 87–92.

Shearer, G., and Kohl, D. H. (1986). N_2-fixation in field settings: Estimations based on natural ^{15}N abundance. *Australian Journal of Plant Physiology* **13**, 699–757.

Sigman, D. M., Casciotti, K. L., Andreani, M., Barford, C., Galanter, M., and Bohlke, J. K. (2001). A bacterial method for the nitrogen isotopic analysis of nitrate in seawater and freshwater. *Analytical Chemistry* **73**, 4145–4153.

Stahl, W. J. (1977). Carbon and nitrogen isotopes in hydrocarbon research and exploration. *Chemical Geology* **20**, 121–149.

Stiehl, G., and Lehmann, M. (1980). Isotopenvariationen des Stickstoffs humoser und bituminöser natürlicher organischer Substanzen. *Geochimica et Cosmochimica Acta* **44**, 1737–1746.

Velinsky, D. J., Fogel, M. L., Todd, J. F., and Tebo, B. M. (1991). Isotopic fractionation of dissolved ammonium at the oxygen–hydrogen sulfide interface in anoxic waters. *Geophysical Research Letters* **18**, 649–652.

Wada, E., Kadonaga, T., and Matsuo, S. (1975). ^{15}N abundance in nitrogen of naturally occurring substances and global assessment of denitrification from isotopic viewpoint. *Geochemical Journal* **9**, 139–148.

CHAPTER 10

SULFUR

10.1 INTRODUCTION

Sulfur isotope (bio)geochemistry has broad applications to geological, biological, and environmental studies. Sulfur is an important constituent of Earth's lithosphere, biosphere, hydrosphere, and atmosphere and occurs as a major constituent in evaporites and ore deposits. Many of the characteristics of sulfur isotope geochemistry are analogous to those of nitrogen and, especially, carbon, because all three elements occur in reduced and oxidized forms[1] and undergo a change in oxidation state as a result of biological processes.

There are four stable isotopes of sulfur (Table 10.1). The relative ratios vary in constant proportions in both terrestrial and extraterrestrial samples, although important $\delta^{33}S$ anomalies have recently been found in meteorites, ancient sulfate, and some modern deposits. In general, however, there is no need to analyze more than any two of the four stable isotopes of sulfur, the others being easily computed from simple relationships. The isotopes that are commonly measured are ^{34}S and ^{32}S, simply because these are the two most abundant of the four. All $\delta^{34}S$ values are reported relative to the Cañon Diablo Troilite (CDT) standard (Ault and Jensen, 1963) by the standard equation defining delta:

$$\delta^{34}S_{sample} = \frac{\left(\frac{^{34}S}{^{32}S}\right)_{Sample} - \left(\frac{^{34}S}{^{32}S}\right)_{CDT}}{\left(\frac{^{34}S}{^{32}S}\right)_{CDT}} \times 1000. \quad (10.1)$$

---

[1]The valence of sulfur ranges from -2 to $+6$. In natural inorganic compounds, intermediate valence compounds are generally not stable.

TABLE 10.1: Abundance and mass (amu) of the different isotopes of sulfur. Quantity in parentheses represents uncertainty in the last two decimal places.

Isotope	Abundance	Mass
^{32}S	95.03957(90)	31.97207
^{33}S	0.74865(12)	32.97146
^{34}S	4.19719(87)	33.96786
^{36}S	0.01459(21)	35.96708

(Coplen et al., 2002)

Cañon Diablo is the meteorite found at Meteor Crater, Arizona, in the United States. The isotopic composition of troilite[2] in iron meteorites has a very restricted range, from 0.0 to 0.6‰ (Kaplan and Hulston, 1966), and is similar to that of the bulk Earth, making it an ideal reference material. However, recent high-spatial-resolution analyses of CDT have shown it to be slightly heterogeneous (Beaudoin et al., 1994). Secondary synthetic argentite (Ag_2S) and other sulfur-bearing standards have since been developed, with $\delta^{34}S$ values defined relative to the accepted CTD value of 0‰ (Appendix 1). The primary reference is IAEA-S-1, a synthetic silver sulfide with a $\delta^{34}S$ value of -0.3‰ relative to that of CDT (Robinson, 1995). Secondary standards with high and low $\delta^{34}S$ values have been synthesized so that both the absolute value on the CDT scale and laboratory-based corrections related to the "stretching factor" can be determined.

10.2 ANALYTICAL TECHNIQUES

A discussion of the analytical procedures for measuring sulfur isotope ratios is necessary because different isotope ratios may be obtained with different methods. This apparently serious problem is mitigated by analyzing samples and internationally accepted standards using the same methods. A correction procedure can then be applied (see Appendix 2 for a sample calculation) to bring measured data into agreement with the officially recognized scale.

Part of the reason that sulfur has been the "outcast" isotope is the problems associated with making mass spectrometric determinations. Two gases—SO_2 and SF_6—can be used for sulfur isotope determinations. Both have advantages and disadvantages. SO_2 is a highly polar molecule, making it "sticky" in the source of a mass spectrometer and resulting in a memory between samples and hence reduced accuracy. The problem is compounded when trace amounts of water are in the vacuum line, leading to the formation of sulfuric acid, a nonvolatile, corrosive acid. Extraspecial care, including heating the source and frequent cleaning, must be taken to keep the mass spectrometer clean. As a result, some laboratories have been reluctant to put SO_2 into their mass spectrometers. Those which make routine measurements of SO_2 often have a dedicated SO_2 mass spectrometer or configure their mass spectrometer for SO_2 analysis for a block of time, followed by a thorough cleaning. Recent continuous-flow techniques have changed the playing field for SO_2 (Giesemann et al., 1994). Only a very small amount of SO_2 enters the mass spectrometer, and the continuous He stream keeps the source clean. In addition, analyses are very rapid and can be made on a small amount of material.

---

[2]Troilite is a sulfide with mineral composition FeS. It differs from the common pyrrhotite, which has an approximate composition of Fe_7S_8, in that troilite is stoichiometric. Troilite is not found in terrestrial materials.

The second problem with SO_2 measurements is that there is an unavoidable mass spectrometer uncertainty. The 34/32 ratio is determined by measuring the mass ratio 66/64, given by $^{34}S^{16}O^{16}O/^{32}S^{16}O^{16}O$. However, $^{32}S^{18}O^{16}O$ also has mass 66, and the two are uncorrelated. There is no way to separate or distinguish $^{34}S^{16}O^{16}O$ from $^{32}S^{18}O^{16}O$ by measuring only the mass 64 and 66 intensities. Generally, a correction is made on the basis of measured $\delta^{34}S$ values of standard materials (Box 10.1). SO_2 is generated from sulfides by oxidation at high temperatures with either O_2 gas or an oxidized species such as copper oxide. If oxygen in the newly formed SO_2 has a constant $\delta^{18}O$ value, then the correction based on the interference of ^{18}O with the mass 66 signal can easily be applied. However, different sulfides may react with oxygen at different rates, and the $\delta^{18}O$ value of the SO_2 generated could vary among sulfides. The problem is minimized by converting all samples to Ag_2S (an easily oxidized species). This conversion step, however, is being abandoned (for convenience) by most laboratories that use continuous-flow techniques, so there may be a small bias in comparing $\delta^{34}S$ values of coexisting sulfides. The problem is more serious when one is comparing coexisting sulfides and sulfates. Oxygen may be inherited from the sulfate during thermal decrepitation to SO_2. Thus, the oxygen in SO_2 gas generated from sulfates and sulfides could have quite different $\delta^{18}O$ values. The problem is alleviated by comparing results from sulfate standards with sulfate samples and results from sulfide standards with sulfide samples, although this assumes that the $\delta^{18}O$ values of sulfate samples and standards are similar.

In contrast to SO_2, SF_6 is a clean, inert gas. Because fluorine is monoisotopic, the different masses measured in the mass spectrometer for the ionized species SF_5^+ (127, 128, 129, 131) are all uniquely related to a single isotope of sulfur. SF_6 has a further advantage over SO_2 in that it has a very high ionization potential, 90% ionization compared with ~50% (Halas, 1985). The drawbacks to measuring SF_6 gas are twofold: First, SF_5^+ is a very heavy ion and can be measured only on large-radius mass spectrometers configured to analyze heavy masses. Second, the extraction technique involves fluorination and is therefore significantly more elaborate and dangerous than SO_2 combustion methods (although recent developments in laser fluorination make the SF_6 method more straightforward than earlier procedures (Beaudoin and Taylor, 1994).

BOX 10.1 The ^{18}O problem with SO_2.

$^{13}C/^{12}C$ ratios are measured on CO_2 gas. The 45/44 ratio is equal to $(^{12}C^{16}O^{17}O + ^{12}C^{17}O^{16}O + ^{13}C^{16}O^{16}O)/^{12}C^{16}O^{16}O$. The contribution to mass 45 by ^{17}O (the ^{17}O correction) can be made for terrestrial samples because the $\delta^{17}O$ value is half that of the $\delta^{18}O$ value. By measuring both 46/44 and 45/44 ratios of a gas, it is possible to extract the exact $\delta^{13}C$ value of a sample, regardless of the $\delta^{18}O$ value of the CO_2 (Craig, 1957).

The $\delta^{34}S$ value of SO_2 gas is determined from the 66/64 ratios. Mass 64 is entirely $^{32}S^{16}O^{16}O$, but mass 66 has contributions from $^{34}S^{16}O^{16}O$, $^{32}S^{18}O^{16}O$, $^{32}S^{16}O^{18}O$, and $H_2SO_2^+$ (Halas, 1985). Unlike the situation with CO_2, there is no way to correct for the ^{18}O contribution to mass 66, because the $^{34}S/^{32}S$ and $^{18}O/^{16}O$ ratios are uncorrelated. If the $\delta^{18}O$ value of the gas could be independently determined, then a correction could be made if both standard and sample have the same $\delta^{18}O$ value. Alternatively, the ^{18}O correction is determined with samples that have been measured independently by the SF_6 method. (For a further discussion on SO_2 and ^{18}O correction, see Rees, 1978.)

A number of investigators have found that the relative difference between samples with very different $\delta^{34}S$ values is larger when measured as SF_6, compared with SO_2. It is generally assumed that most analytical artifacts tend to shrink the differences among samples with significantly different isotope ratios. The analytical technique with the least amount of compression of the isotope scale is generally considered to be the most accurate. In a comparative study, Rees (1978) found that the spread of values determined with SF_6 extraction methods was larger than that obtained by using SO_2 by a factor of 1.034. (For example, if a $\delta^{34}S$ value of a sample was 20.3‰ (CDT) with the use of SO_2, a value of 21.0‰ would be obtained with SF_6.) Overall, SF_6 is probably more accurate and precise. However, the rapidity, simplicity, and safety of SO_2-based extraction techniques is a major factor for researchers who would like to have large numbers of analyses and who do not require the very high-precision and multiple-isotope data that can be obtained only by measuring SF_6.

10.3 EQUILIBRIUM FRACTIONATIONS AND GEOTHERMOMETRY

Equilibrium fractionations for selected sulfur-bearing compounds are shown in Fig. 10.1. The sulfur isotope fractionation between sulfates and sulfides is enormous and highly temperature sensitive. Unfortunately, sulfur isotope equilibrium between sulfates and sulfides occurs only at very high temperatures. (See Rye, 2005, for a comprehensive review.) Sulfides and sulfates, commonly found to coexist in ore deposits, are rarely in equilibrium, even when temperatures

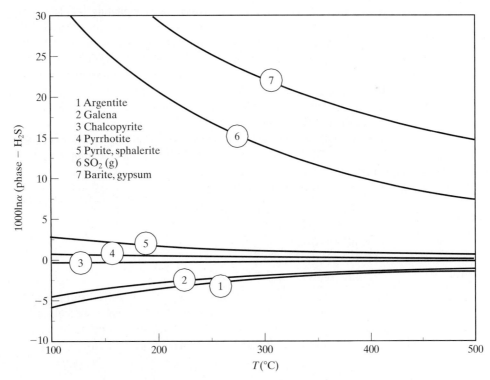

FIGURE 10.1: Sulfur isotope fractionations between selected phases. Among the sulfides, galena–pyrite and galena–sphalerite have the largest fractionations.

in excess of 500°C had been reached. If equilibrium exchange between sulfate and sulfides *could* occur at low temperatures, the fractionations would be huge. (The expected equilibrium gypsum–galena fractionation is over 70‰ at 25°C!)

Equilibrium between oxidized and reduced sulfur species is most commonly observed in (a) coexisting sulfides and sulfates in magmatic and hydrothermal systems formed above 250°C, (b) H_2S and SO_2 gas from volcanic fumaroles, and (c) dissolved sulfide and sulfide minerals, as well as coprecipitating sulfide minerals in hydrothermal fluids (Nielsen, 1979).

Applications of sulfur isotope thermometry have focused mainly on coexisting sulfide pairs. One would expect that mineral pairs with the largest fractionations should be most suitable for thermometry, and indeed, sphalerite–galena pairs work very well as a geothermometer. In most cases, the calculated isotope temperature agrees well with other independent estimates, especially above 300°C. Reasonable temperature estimates from the sphalerite–galena thermometer may be obtained in ore deposits formed as low as 120°C if the two minerals precipitated simultaneously. Below this temperature, the $\Delta^{34}S$ (sphalerite–galena) values are smaller than expected for equilibrium; that is, they correspond to temperatures that are too high.

Because of its large fractionation with other sulfides, pyrite should be an excellent mineral for thermometry, but it is not. In the majority of cases, pyrite–sulfide pairs (and chalcopyrite–sulfide pairs) give erroneous temperatures, the fractionations often being reversed. The reasons for this are not well understood, but are probably related to different sources of sulfur for pyrite and coexisting minerals.

Chemical processes relating to the precipitation of sulfides differ fundamentally from those responsible for silicate formation in that dissolved sulfur can exist in more than one oxidation state. Therefore, the fractionation between the precipitating sulfide and the solution is a function of both temperature *and* the dominant aqueous sulfide species (Sakai, 1968; Ohmoto, 1972).

The four principal aqueous sulfur-bearing species above 250°C are ΣSO_4^{2-}, H_2S, HS^- and S^{2-}. (ΣSO_4^{2-} is the sum of SO_4^{2-}, HSO_4^{2-}, KSO_4^{2-}, etc). Which of these species dominates in a fluid depends on the pH, $f(O_2)$, and temperature of the fluid (Fig. 10.2). At low pH and $f(O_2)$ conditions, H_2S makes up the vast majority of the sulfur species, with ΣSO_4^{2-} and S^{2-} existing in vanishingly small quantities. S^{2-} dominates at very high pH conditions (lower right of the figure), and ΣSO_4^{2-} is the principal aqueous sulfur species at high $f(O_2)$. In between these fields, where the contours are drawn, proportions of one of the minor species is at least 1%. The isotopic composition of minerals precipitating from the sulfur-bearing solution depends on which of the aqueous sulfur species is most abundant.

Consider the following example:

- The bulk $\delta^{34}S$ value of the fluid ($\delta^{34}S_\Sigma$) is 0‰.
- $T = 250°C$.
- pH = 4.5.
- $\log f(O_2) = -38$.
- total sulfur = 1.0 mole/kg H_2O.

Under these conditions, more than 99% of the dissolved sulfur will be in the form of H_2S (point *a* in Fig. 10.2). The $\delta^{34}S$ value of H_2S is equal to the bulk value ($\delta^{34}S_\Sigma$) of 0‰. For phases precipitating in equilibrium with H_2S, such as pyrite, sphalerite, galena, and the dissolved species SO_4^{2-} and S^{2-}, the following fractionations apply:

$$\Delta^{34}S\ (H_2S - \text{pyrite}) = -1\ ‰;$$
$$\Delta^{34}S\ (H_2S - \text{sphalerite}) = 0‰$$
$$\Delta^{34}S\ (H_2S - \text{galena}) = 2‰$$

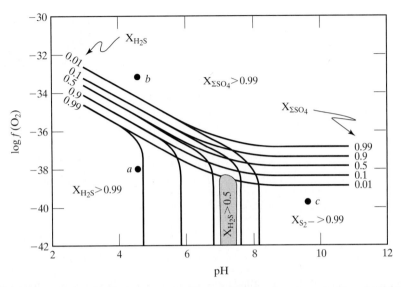

FIGURE 10.2: Stability field of aqueous sulfur species as a function of pH and $f(O_2)$. At low pH and $f(O_2)$, H_2S is the dominant dissolved sulfur species (point *a*); at high pH and low $f(O_2)$, S_2 is the dominant species (point *c*); and at high $f(O_2)$, ΣSO_4 is the dominant species (point *b*). The three-phase field represents an oversimplification of the system, because HS^- is a significant component at intermediate pH and low $f(O_2)$, as shown by the shaded region, where $X_{H_2S} > 0.5$. (Modified from Ohmoto, 1972.)

$$\Delta^{34}S\,(H_2S - S_2^-) = 4‰$$
$$\Delta^{34}S\,\left(H_2S - \Sigma SO_4^{2-}\right) = -25‰$$

The *initial* $\delta^{34}S$ values of sulfides precipitating from the H_2S-rich fluid will be 1‰ (pyrite), 0‰ (sphalerite), and −2‰ (galena) (Fig. 10.3). If all of the sulfur in solution is converted to solid sulfides, then the $\delta^{34}S$ value of the sulfides will be determined by the relation

$$\delta^{34}S_\Sigma = X_{pyrite} \cdot \delta^{34}S_{pyrite} + X_{sphalerite} \cdot \delta^{34}S_{sphalerite} + X_{galena} \cdot \delta^{34}S_{galena}, \quad (10.2)$$

where X is the mole fraction of sulfur in each of the sulfide phases and $\delta^{34}S_\Sigma$ is the $\delta^{34}S$ value of the total system. The $\delta^{34}S$ value of each phase can be determined by substituting the fractionation relationships $\Delta^{34}S(pyrite - sphalerite)$, $\Delta^{34}S$ (pyrite − galena) into equation (10.2).

If the same fluid were to be oxidized, then the dominant sulfur species would be ΣSO_4^{2-} (point *b* in Fig. 10.2). In this case, sulfides precipitating from the fluid would have *initial* $\delta^{34}S$ values of −24‰ (pyrite), −25‰ (sphalerite), and −27‰ (galena) (Fig. 10.3). If all of the aqueous sulfur is precipitated as sulfides, the *final* $\delta^{34}S$ values will the determined by equation (10.2), as before. But if only a small fraction of the dissolved sulfur is converted to sulfides, the $\delta^{34}S$ values of these minerals will be very different from the case where H_2S is the dominant ionic sulfur-bearing species. The third case is when the $f(O_2)$ remains low, but the pH increases, such as would happen if the fluid entered a graphitic carbonate rock (point *c* in Fig. 10.2). Now the dominant dissolved sulfur species is S_2^-, with a $\delta^{34}S$ value of 0. The *initial* $\delta^{34}S$ values of crystallizing sulfides in this case will be 5‰ (pyrite), 4‰ (sphalerite), and 2‰ (galena) (Fig. 10.3).

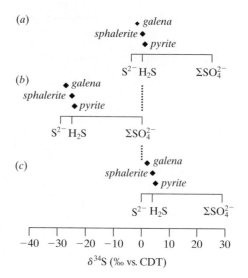

FIGURE 10.3: Illustration of the initial $\delta^{34}S$ values of minerals forming from a fluid with a $\delta^{34}S_\Sigma$ value of 0‰. (a) H_2S as dominant fluid species; (b) ΣSO_4^{2-} as dominant fluid species; (c) S_2^- as dominant fluid species. (Modified from Nielsen, 1979.)

10.4 SULFATE AND SULFIDE FORMATION AT LOW TEMPERATURES: THE SEDIMENTARY SULFUR CYCLE

Essentially all dissolved sulfur in the ocean is in the oxidized form. The residence time of reduced sulfur evolved from black smokers, etc., is only on the order of minutes.[3] The residence time of sulfate in seawater is 2×10^7 y, so sulfur is well mixed and the $\delta^{34}S$ value of the modern ocean has a constant value of 21.0‰ ± 0.25‰ (Rees et al., 1978).[4] Sulfur is removed from the ocean by the precipitation of sulfate and sulfide minerals and is returned to the ocean through the erosion of sediments.

The sulfur cycle has a number of similarities to the carbon cycle. In both cases, the element forms solids in both oxidized and reduced states. The oxidized forms are carbonates and sulfates; the reduced forms are organic matter and sulfides for carbon and sulfur, respectively. The isotopic fractionation between seawater and the oxidized forms is small and nearly constant, so that the delta value of the oxidized form can be used as a proxy for the ocean value at the time of mineral formation. The near-surface reduction of carbon and sulfur from dissolved carbonate and sulfate, respectively, is a biologically mediated process. Both are kinetically controlled, and the product phases—organic matter and sulfide—have much lighter (and variable) delta values than those of the ocean.[5] The long-term carbon and sulfur

---

[3]In Archean times, reduced sulfur would have been the stable ionic species in the ocean.
[4]The value of 21.0‰ was determined with the SF_6 method. Earlier estimates obtained with the SO_2 method are 20.0‰, and a large number of publications have used (and continue to use) this 20.0‰ value.
[5]Note, however, that fractionation during the assimilation of sulfur is small (2–3‰), whereas fractionation during the assimilation of carbon is large (Kaplan, 1983, page 2–4).

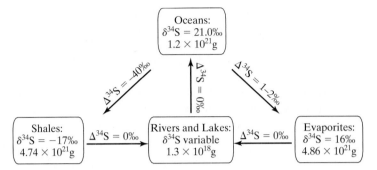

FIGURE 10.4: Box model of the exospheric (outer-crust) sulfur cycle. The $\delta^{34}S$ value of the ocean is a function of transfer to and from sulfide (shales box); the *modern* value is 21.0‰. (After Claypool et al., 1980. Abundances of each reservoir are modern values after Holser et al., 1988.) Note that published estimates for reduced sulfur (total sulfur) range from 0.29 to 0.55.

delta values of the ocean are controlled by the amount of organic matter (carbon) and sulfide (sulfur) stored in sediments (Fig. 10.4). Note that the average crustal $\delta^{34}S$ value is given by $\frac{\sum m_i \times \delta^{34}S_i}{\sum m_i}$, where m_i is the mass of each reservoir. Using the data in Fig. 10.4 gives a value of 2‰, close to the bulk Earth value of 0‰.

The sulfur isotope fractionation between evaporitic sulfate minerals and dissolved sulfate is approximately 1–2‰ (Thode and Monster, 1965). Holser and Kaplan (1966) compiled the fractionations from experiments and analyses of modern evaporites. The fractionation obtained from experiments is 1.1 ± 0.9‰. Recent evaporites were, on average, 0.4 ± 1.2‰ heavier than dissolved ocean sulfate. Modern barites measured by the SF_6 method averaged 0.2‰ heavier than dissolved ocean sulfate (Paytan et al., 1998). The similarity between the $\delta^{34}S$ value of sulfate minerals and dissolved sulfate means that ancient sulfates can be used as a proxy for the $\delta^{34}S$ value of the ocean at the time the minerals formed.

Sulfides are produced mostly as a by-product of bacterial sulfate reduction. *Assimilatory reduction* occurs in autotrophic organisms when sulfur is incorporated into proteins, particularly as S^{2-} in amino acids. Assimilatory reduction involves a change in valence from +6 to −2. The bonding of the product sulfur is similar to that the dissolved sulfate ion, and fractionations are small −4.5 to +0.5‰ (Kaplan, 1983). The $\delta^{34}S$ value of organic sulfur in extant marine organisms incorporated by assimilatory processes is generally depleted in ^{34}S by 0 to 5‰ relative to the ocean.

The second reduction mechanism—*dissimilatory reduction*—is performed by heterotrophic organsisms, particularly *Desulfovibrio desulfuricans*. The overall reaction can be represented as

$$2\,CH_2O + SO_4^{2-} \longrightarrow 2\,HCO_3^- + H_2S. \tag{10.3}$$

Dissimilatory reduction is also called bacterial sulfate reduction and is extremely important in terms of the Earth's sulfur budget. It is the major surficial process for changing sulfate to sulfide. Bacterial sulfate reduction is an energy-yielding, anaerobic process that occurs only in reducing environments, such as organic-rich ocean sediments. The bacteria use sulfate as an electron acceptor (oxidant). H_2S is given off as a by-product and is quickly consumed in the formation of metal sulfides. The sulfate reduction pathway consists of four steps (Goldhaber and Kaplan, 1974). The rates of these steps are quite different, as is the fractionation associated with each step. The slower the step, the larger is the fractionation. In a dissimilatory reduction reaction limited by the reactant sulfate, essentially all available sulfur will be consumed, and sulfur isotope fractionation between reactant and product will be very small. When there is excess sulfate and reaction rates are slow, isotopic fractionation will be much larger. Reaction rates are most rapid near neutral pH conditions and are independent of sulfate concentrations above ~0.5 mM. (See Canfield, 2001, for an extensive review.) Measured fractionations under experimental conditions range from −20 to −46‰ at low rates of sulfate reduction to −10‰ at high reduction rates. The $\delta^{34}S$ value of sulfides of modern marine sediments is typically around −40‰; however, a wide range, from ~−40‰ to ~+3‰, is observed (Fig. 10.5). In general, and for the purposes of modeling, a fractionation of ~40 ± 10‰ between sulfate and sulfide is often assumed (Claypool et al., 1980).

If, during dissimilatory sulfate reduction, all of the pore-water sulfate were converted to sulfides—the "closed-system" case—then the $\delta^{34}S$ of the bulk sulfides would necessarily equal the dissolved seawater sulfate $\delta^{34}S$ value of +21‰ (modern value). The amount of sulfide precipitated (in wt %) would also necessarily equal the amount of sulfur initially present in the pore-water sulfate. What is observed in most oceanic sediment, however, are very low $\delta^{34}S$ values of the newly formed sulfide (Fig. 10.5) and far more precipitated sulfide than is expected from the amount of original dissolved sulfate in pore water alone (Fig. 10.6). The isotope and concentration data can be explained only if ocean water circulates freely through the upper sediment, via "open-system" behavior. Sulfate reduction continues due to the influx of fresh sulfate from the overlying ocean water. Some ^{34}S-enriched sulfate must be returned to the ocean, thereby explaining the low $\delta^{34}S$ values of the sulfide mineralization. Bioturbation strongly aids the mixing process, which is most intense in the upper 6 cm of sediment.

Sulfate reduction and sulfide precipitation continue only as long as (1) organic matter is available for sulfate-reducing bacteria, (2) reactive iron is present to react with H_2S, and (3) sulfate is available as a reactant. The exchange of pore-water sulfate with ocean water decreases with depth, as the porosity is reduced due to compaction of the sediments. Sulfate reduction will quickly halt if the H_2S concentration is allowed to build up. As long as there

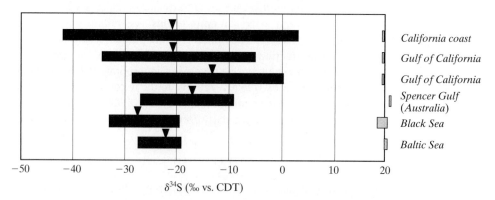

FIGURE 10.5: Sulfur isotope composition of pyrite precipitated in modern sediments (solid black bars) and coexisting ocean-water sulfate (gray boxes). Averages for pyrite analyses are given by arrowheads. In general, pyrite formed from bacterial sulfate reduction is 40‰ lighter than seawater sulfate. (After Schidlowski et al., 1977; 1983.)

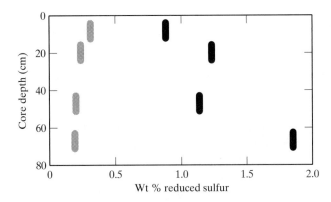

FIGURE 10.6: Expected (gray) and measured (black) reduced sulfur abundance from sediment cores from the Gulf of California. The expected amount is calculated from the pore-water volumes and sulfate concentrations, which decrease with depth. The higher measured abundance of reduced sulfur is evidence for a continued influx of seawater sulfate after initial burial. (After Kaplan, 1983.)

are metal cations present to react with the evolving H_2S, however, the sulfate reduction reactions will continue and sulfides will precipitate. Some of the H_2S by-product will also leave the system by simple degassing at shallow levels. In the marine environment, neither sulfate nor ferric iron generally limits the reaction. Instead, it is the abundance of easily metabolized carbon that controls the extent of sulfate reduction reactions.

10.5 SECULAR VARIATIONS IN SULFUR

10.5.1 Long-Term Variations

The $\delta^{34}S$ values of marine evaporites reflect the isotopic composition of dissolved ocean sulfate at the time of their formation. Because the fractionation between sulfate minerals and dissolved sulfate is close to zero, neither the creation nor the destruction of evaporites will have much effect on the $\delta^{34}S$ value of the oceans. By contrast, the large fractionation between dissolved sulfate and bacterially produced sulfide minerals means that the production or destruction of sulfides will strongly influence the $\delta^{34}S$ value of the ocean. During periods of intense weathering of argillaceous sediments—notably, shales—the $\delta^{34}S$ value of the ocean decreases, while during periods of vigorous bacterial sulfate reduction and the preferential removal of ^{32}S, the $\delta^{34}S$ value of ocean sulfate increases.

The secular curve for the $\delta^{34}S$ value of sulfates is shown in Fig. 10.7. The $\delta^{34}S$ values range from less than 10‰ to over 35‰. The gross characteristics of the secular curve can be explained in terms of the removal or addition of reduced sulfur to the oceans. Note that the incorporation of old evaporitic sulfate with a very different composition from that of the ocean (e.g., 35‰ sulfate into a 10‰ ocean) could have a measurable effect.

- Periods of high biological activity (by sulfate-reducing bacteria) increase the $\delta^{34}S$ value of the ocean (and evaporites) due to the removal of ^{34}S-depleted sulfur as sedimentary sulfides. Favorable paleogeographic conditions, including extensive shallow marine settings, and abundant organic matter (as a food source), are necessary for the extensive production of sulfide. Periods of high $\delta^{34}S$ values (e.g., the Cambrian and Devonian) also have widespread carbon-rich sediments.
- Intense weathering lowers the $\delta^{34}S$ value of the ocean. Low $\delta^{34}S$ values of evaporites due to the extensive erosion of shales require elevated orogenic activity, such as the Devonian-to-Permian change associated with the Variscan orogeny (Fig. 10.7).

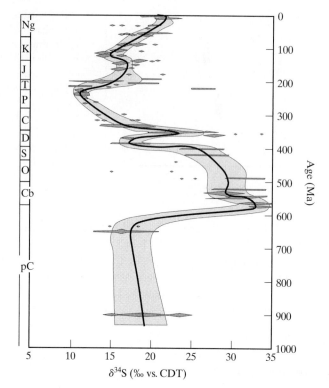

FIGURE 10.7: Secular variations in the $\delta^{34}S$ value of evaporates. (After Claypool et al., 1980.)

The significance of the sulfate–sulfide ratio can be appreciated when one recognizes that 50% of the total photosynthetically produced O_2 is locked up in evaporite sulfate (Schidlowski et al., 1977). To put this in context, the shift from $\delta^{34}S$ values of 10‰ in the late Permian to 20‰ in the Tertiary requires a shift of 5×10^{20} grams of sulfur to the shale reservoir (Holser and Kaplan, 1966), releasing an equivalent amount of oxygen. 10^{20} grams of oxygen is equal to all of the diatomic oxygen in the modern atmosphere.

10.5.2 Alternative Approaches: Barite and Trace Carbonates

Relative to carbon and strontium, the secular sulfur isotope curve is poorly defined, because evaporites were deposited only at certain periods in the Earth's past and are difficult to date with precision. Special conditions are required for their deposition and they are easily eroded. Other sulfur species, especially barite and trace sulfate in carbonates, have been used to expand and refine the sulfur curve. The $\delta^{34}S$ values of barites are strongly dependent on the type of deposit. Thick stratiform barite deposits faithfully reproduce the evaporite curve, while thin beds or nodular barites have $\delta^{34}S$ values that are too high and most likely form in diagenetic settings within sulfate-reducing sediments (Cecile et al., 1983; Goodfellow and Jonasson, 1984). Massive barite deposits require unusual geological conditions, such that the $\delta^{34}S$ values may have a high contribution from a hydrothermal source. A recent sulfur isotope stratigraphic curve was generated for deep-sea marine barites of Cretaceous and Cenozoic age with unprecedented temporal resolution (Paytan et al., 2004). The marine barites precipitate directly in the seawater column and have $\delta^{34}S$ values that are virtually identical to that of

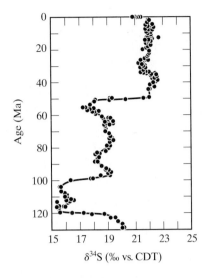

FIGURE 10.8: $\delta^{34}S$ values of marine barite covering the Cenozoic. The high-resolution curve clearly illustrates rapid changes in the sulfur isotope composition of the ocean. (After Paytan et al., 2004.)

dissolved sulfate. The high-resolution curve shows some very rapid changes, such as a 5‰ change between 125 and 120 Ma (Fig. 10.8). Such large rapid shifts require huge changes in input and output fluxes. The rapid shifts may also occur in part because sulfate concentrations in the ocean were low at that time. These "second-order" features (on a scale of 10^7 yr) are similar to other, more poorly defined equivalents in older evaporites (Holser, 1977) and suggest that the seawater sulfate curve may not be as smooth as is drawn in Fig. 10.7.

Sulfur is a ubiquitous trace element in sedimentary carbonates. Concentrations range from several tens of ppm in inorganic carbonates to several thousand ppm in some biogenic carbonates. As long as no fractionation accompanies the incorporation of sulfur into the carbonate and no diagentic alteration takes place, carbonates offer an attractive method for refining the secular sulfur curve. Carbonates may provide a more representative $\delta^{34}S$ value of the ocean, because they precipitate in more open conditions as opposed to restricted marine basins, (Strauss, 1997). Burdett et al. (1989) constructed a high-resolution secular sulfur curve for the past 25 Ma using foraminifera. Their results agree well with Fig. 10.8, but have much more scatter, which the authors attribute to possible diagenesis. More recently, a global secular $\delta^{34}S$ curve for the Paleozoic and Mesozoic has been generated from nearly 300 measurements of structurally substituted sulfate in the calcite lattice of carbonates (Kampschulte and Strauss, 2004). The curve matches earlier evaporite data, but has far higher temporal resolution.

10.5.3 Time Boundaries

Strauss (1997) reviewed secular variations across time boundaries characterized by profound biological or geological changes. Due to the paucity of evaporite data, all time-boundary studies have been made on sedimentary sulfides. Carbon isotope excursions can be expected in response to major extinction and radiation events, as the response time to carbon is relatively rapid. Enhanced sulfate reduction requires abundant organic matter as a food source, as is evidenced by a correlation between the abundance of organic matter and sedimentary pyrite. During a catastrophic event, in which productivity plunges, the $\delta^{34}S$ values of the oceans should decrease. The subsequent biological radiations should have the opposite effect. The $\delta^{34}S$ values of the oceans should first decrease across a time boundary associated with a catastrophic extinction and then increase during the period of recovery. The magnitude of the effect is related to the intensity of the extinction event, the rate of recovery, and the size of the oceanic sulfur reservoir.

Four extinction events have been studied (see Strauss, 1997, for references): the Precambrian–Cambrian, the Frasnian–Famennian, the Permian–Triassic, and the Cretaceous–Tertiary boundaries. Of these, only the Permian–Triassic event shows the expected sulfur trend. Fluctuations occur at other boundaries, but no secular variations have been observed. Part of the reason for the irregular results between sections may be related to the problems that are inherent in analyzing sulfides instead of sulfates. In addition, local effects may mask any global sulfur variations.

10.5.4 Archean Sulfates: Clues to the Early Atmosphere

Oxidation weathering of sulfides is the major source of sulfate. Virtually all sulfate oxygen in post-Archean sediments is photosynthetically derived. Prior to the development of an oxygenated ocean and atmosphere, photolithotrophic (green and purple) sulfur bacteria would have been the principal source of oxygen for the production of sulfate (Monster et al., 1979). Sulfur isotope fractionation during photolithotrophic oxidation is small, so that in the absence of extensive bacterial sulfate reduction, all sulfide and sulfate minerals would have had $\delta^{34}S$ values close to the whole-earth value of 0‰. If photosynthetic algae first became plentiful in the oceans, there should have been a buildup of sulfate in the oceans prior to the ultimate development of free oxygen in the atmosphere. Abundant sulfate-reducing bacterial activity could occur only when sulfate became abundant and the ocean had been oxygenated in at least some locations. Once the sulfate-reducing bacteria became abundant, the dispersion of $\delta^{34}S$ values in sediments (due to isotope partitioning between sulfide and sulfate) would increase toward its modern range that reaches greater than 70‰. The advent of sulfate-reducing bacteria was a major evolutionary step that preceded the oxygenation of the atmosphere.

A number of authors have measured the $\delta^{34}S$ values of ancient sulfides and sulfates, although Archean sulfate mineralization is rare.[6] Monster et al. (1979) showed that the divergence of sulfur isotope ratios in sediments occurred some time between 3.1 and 2.8 billion years in the past (Fig. 10.9). Earlier sediments and igneous sulfide samples have an extremely limited range of $\delta^{34}S$ values of $+0.5 \pm 1.0$‰. By the middle Archean, the $\delta^{34}S$ values ranged from -20 to $+20$‰. On the basis of these data, it was concluded that low ocean sulfate concentrations persisted until ~2.8 Ga. Kakegawa and Ohmoto (1999) measured the $\delta^{34}S$ values of pyrite grains hosted in sedimentary pyretic shales of the 3.4- to 3.2-Ga Fig Tree Group of the Barberton Greenstone Belt, South Africa. The $\delta^{34}S$ values ranged from -0.8 to $+4.4$‰, suggesting that the sulfides were a product of the microbial reduction of seawater sulfate. Such a conclusion itself suggests that seawater sulfate was an appreciable component of the Archean ocean, at least locally.

---

[6]No Archean evaporites exist. Instead, gypsum and anhydrite have been replaced by barite. There is some question as to whether primary $\delta^{34}S$ values were preserved during diagenetic alteration to barite.

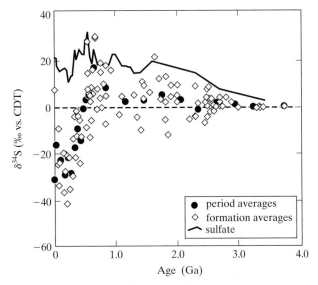

FIGURE 10.9: Sulfur isotope compositions of sedimentary sulfides and sulfates through time. The delta values begin to spread out between 3.1 and 2.8 Ga as sulfate-reducing bacteria become more abundant. (Reprinted from Canfield, 2004, with permission from American Journal of Science.)

10.5.5 Sulfur Isotope Anomalies: Mass-Independent Fractionation

Further constraints on Archean surficial conditions can be made from multiple isotope measurements of sulfides. Under almost all geological conditions, the four isotopes of sulfur follow a predictable mass-dependent fractionation in which $\delta^{33}S = 0.515 \times \delta^{34}S$, and $\delta^{36}S = 1.90 \times \delta^{34}S$ (Hulston and Thode, 1965). Mass-independent fractionations occur in a number of gas-phase reactions. The isotopic composition of a mineral produced by mass-independent reactions does not follow the predicted relationships between the $\delta^{33}S$, $\delta^{34}S$, and $\delta^{36}S$ ratios. The $\Delta^{33}S$ values are defined as the deviation from the expected isotope ratios by the equation

$$\Delta^{33}S = 1000 \times [(1 + \delta^{33}S/1000) - (1 + \delta^{34}S\,1000)^{0.518} - 1] \qquad (10.4)$$

Farquhar et al. (2000) measured the $\delta^{33}S$, $\delta^{34}S$, and $\delta^{36}S$ of sulfates and sulfides covering a wide age range and found $\Delta^{33}S$ anomalies in samples older than about 2.1 Ga and, especially, older than 2.45 Ga (Fig. 10.10). They attributed these anomalies to different atmospheric conditions in the early atmosphere. Ozone would have been lacking in a diatomic oxygen-free atmosphere, so ultraviolet rays could penetrate farther into the atmosphere, resulting in mass-independent reactions. Alternatively, the photochemical oxidation of reduced sulfide species could have produced sulfate species that did not follow normal mass-dependent fractionations. Regardless of the actual mechanism, the $\Delta^{33}S$ anomalies in Archean sulfates clearly indicate very different conditions in the early atmosphere.

Mass-independent sulfur isotope anomalies have also been found in sulfuric acid layers in South Pole ice cores (Savarino, Bekki, Cole-Dai, and Thiemens, 2003; Savarino,

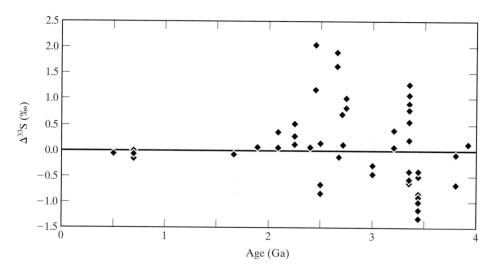

FIGURE 10.10: $\Delta^{33}S$ values as a function of age for sulfur-bearing samples. At ~2.1 Ga, $\Delta^{33}S$ anomalies are observed; at ~2.5 Ga, they become significantly larger. The anomalous samples are believed to be related to an oxygen-free atmosphere in the Archean. (After Farquhar et al., 2000.)

Romero, Cole-Dai, Bekki, and Thiemens, 2003). The proposed explanation is that SO_2 expelled into the upper atmosphere during volcanic explosions has undergone photoxidation by UV radiation. Sulfate from the relatively weak 1991 Cerro Hudson event had no isotope anomaly, because the ejected SO_2 never left the troposphere. In contrast, the Pinatubo eruption has a strong sulfur isotope anomaly, explained by the SO_2 ejecta reaching well into the stratosphere, where photolysis in the 190- to 220-nm spectral range would bring about the anomaly. Oxygen isotope anomalies in massive sulfates have also been measured and are presumably related to sulfur oxidation reactions in the upper atmosphere (e.g., Bao et al., 2000).

10.6 SULFUR ISOTOPE RATIOS IN THE TERRESTRIAL ENVIRONMENT

Variations in the terrestrial realm are far less predictable than in the oceanic environment. The largest natural flux of sulfur in the atmosphere is sulfate aerosols derived from sea spray, with a $\delta^{34}S$ value close to that of the ocean. H_2S has a very short residence time in the atmosphere and will be converted to SO_2 (and sulfuric acid) without sulfur isotope fractionation. In arid climates, the $\delta^{34}S$ value of soils and water is controlled by atmospheric input from aerosols. In humid climates, where rock weathering is extensive, input from the dissolution of sulfur-bearing rocks overwhelms the contribution from atmospheric precipitates. The $\delta^{34}S$ values of rivers is controlled by the local rock type. Hoefs (1987) compiled results from a number of river waters. Not surprisingly, the $\delta^{34}S$ values of rivers cover the same range as those of the rocks themselves. The $\delta^{34}S$ values of the Mackenzie River drainage system in Canada span a wide range, from -20 to $+20$‰, clearly indicating that sulfur is supplied from both marine evaporites and shales. The Amazon river has a surprisingly narrow range of $\delta^{34}S$ values, around 8‰. Longinelli and Edmond (1983) attributed the near-constant values to a strong Andean source of Permian evaporites. Sulfur isotope studies of individual drainage systems is most valuable for determining the sulfur source for the particular area being examined.

Soil δ^{34}S values are generally locally derived. In arid regions, especially near the ocean, δ^{34}S values are high, due to addition of sulfate from sea spray aerosols. Other high δ^{34}S values can be traced to the dissolution of evaporites or travertines (Krouse, 1989). Other sulfur sources include organic matter and, in recent sediments, fossil-fuel burning or other anthropogenic sources. It is now common for sulfur from anthropogenic sources to overwhelm all other sources. Krouse and Case (1983) measured the δ^{34}S profile of soils near a sour-gas-processing plant in Alberta, Canada. Soil profiles of soluble sulfur were measured 2 km and 10 km from the processing plant. The 2-km profiles have δ^{34}S values of 20‰ at the surface, identical to those emitted from the processing plant. The δ^{34}S values decrease with depth down to approximately 0‰. The 10-km profiles decrease steadily from ~5‰ at the surface to a plateau of −6 to −10‰ with depth. These lowest values are thought to represent preindustrial sulfur values. The use of sulfur isotopes as a tracer of anthropogenic sources is somewhat limited, however, given the huge number of possible sources and extensive mixing between the different sources.

10.7 OXYGEN ISOTOPE VARIATIONS IN SULFATES

Early measurements of the δ^{18}O values of dissolved sulfate in seawater range from 9.5 to 9.9‰ (Longinelli and Craig, 1967; Rafter and Mizutani, 1967; Cortecci, 1975). Later authors obtained a value of 8.6‰ (Holser et al., 1979; Zak et al., 1980). The cause of the discrepancy is not known (Longinelli, 1989). Nevertheless, the measured values are far from those predicted on the basis of equilibrium fractionation between dissolved sulfate and water, for which $\Delta^{18}O(SO_4^{2-}{}_{aq} - H_2O)$ is 38‰ (Lloyd, 1967; 1968). The reason for the disequilibrium is the dynamic conditions of input, output, and partial reequilibration that occur in the sulfate oxygen cycle (Holser et al., 1979). The main inputs and outputs for sulfate oxygen can be modeled along the lines for sulfur (Fig. 10.4). The oxygen source for the surficial oxidation of sulfides to sulfate is a combination of meteoric water and atmospheric O_2. Although the processes of oxidation and the importance of bacterial oxidation are not well known, average global δ^{18}O values for oxidized sulfate are around −2‰ (Holser et al., 1979). Using an average δ^{18}O value of +12‰ for dissolved sulfate derived from weathered evaporites, Holser et al. (1979) estimate that the δ^{18}O value of sulfate flowing into the ocean is ~+5‰. Dissolved sulfate will also fractionate oxygen during bacterial reduction to sulfide. Again, the magnitude of the effect is not well known. Holser et al. (1979) estimate the fractionation to be equal to −10‰. In other words, dissolved sulfate with low $^{18}O/^{16}O$ ratios will preferentially be reduced, leaving the remaining sulfate with a higher δ^{18}O value. The final fractionation occurs during the precipitation of sulfate minerals from dissolved sulfate. Results from various lines of evidence regarding this fractionation are often conflicting, ranging from 2 to nearly 10‰. An average $\Delta^{18}O_{evaporite-dissolved\ sulfate}$ value of 3.5 to 3.6‰ has been proposed (Holser et al., 1979; Claypool et al., 1980).

When all the data are taken together, a mass-balance equation for oxygen in sulfate can be derived. The modern δ^{18}O value of 8.6‰ is obtained if the proportions of input from sulfide and sulfate are about equal. The secular sulfate oxygen isotope curve shows variations on the order of 7‰, although there is a great deal of uncertainty in its construction (Claypool et al., 1980). Covariations between oxygen and sulfur are poor. Nevertheless, some of the features seen in the secular sulfur curve are seen for oxygen as well—notably, the lowering of delta values during the Permian, followed by a rapid rise at the Permian–Triassic boundary. Late Devonian and Carboniferous samples have the highest δ^{18}O values of 17‰.

A stratigraphic curve for δ^{18}O in barite has been constructed for the last 10 million years using Deep Sea Drill Cores (Turchyn and Schrag, 2004). From a nearly constant value of 9‰ in the Miocene, there is a monotonic increase of 5‰ between 6 and 3 Ma and then a decrease back to modern values of 7.9‰ in the last 3 million years. The slightly elevated $\delta^{18}O_{sulfate}$ values in the Miocene are interpreted to represent the oxidation of sulfide to a sulfate value between 10 and 14‰. The elevated $\delta^{18}O$ values may be related to changing intensities of bacterial disproportionation of sulfur compounds. The change in $\delta^{18}O$ values of 5‰ in several million years is far in excess of what is seen for oxygen in carbonates. The difference is clearly related to the fact that oxygen isotope equilibrium is attained between water and carbonate, which is not the case for sulfates.

PROBLEM SET

1. Comment on the relationship between the secular carbon and sulfur curves. Remember that the bacterial reduction of sulfate releases oxygen and the destruction of organic matter consumes oxygen.
2. The mass difference between ^{36}S and ^{32}S is larger by a factor of 2 than that between ^{34}S and ^{32}S. So why aren't the 36/32 ratios measured instead of the 34/32 ratios?
3. If SF_6^+ were the ionized species measured in a mass spectrometer for $^{34}S/^{32}S$ ratio determinations, what masses would be analyzed for normal samples?
4. Coal and petroleum incorporate sulfur by assimilatory sulfate reduction. What ranges of $\delta^{34}S$ values would be expected in these materials? Pyrite, formed by dissimilatory sulfate reduction, is often found in coal deposits. How would its $\delta^{34}S$ value compare?
5. Figure 10.3 shows the isotopic composition of minerals in equilibrium with a sulfur-rich fluid with an initial $\delta^{34}S$ value of 0‰. Comment on what the zoning profile of the newly formed minerals would look like if all aqueous sulfur were converted to sulfides (e.g., in a closed-system exchange). Assume that the cores formed first and the rims formed later. This might best be considered as a Rayleigh-like fractionation. (See McKibben and Eldridge, 1990, for further ideas.)
6. Shales are found at all stratigraphic intervals. Why aren't $\delta^{34}S$ values measured in sulfides hosted in shales in order to reconstruct the secular curve for seawater sulfate?
7. How would the $\delta^{34}S$ value of evaporites change during a period of intense evaporite formation? How about during a period of intense shale formation?

ANSWERS

2. The abundance of ^{36}S is only 1/40 that of ^{34}S, so that the low counting statistics dramatically decrease the precision and outweigh the benefit of the larger fractionations. (Table 10.1).
3. The two major isotopes of sulfur are 32 and 34. Fluorine has a mass of 19. $19 \times 6 + 32 = 146$ and $19 \times 6 + 34 = 148$
4. Assimilatory sulfate reduction has a small fractionation; dissimilatory sulfate reduction has a large fractionation. (See Section 10.4.)
5. As light sulfur is removed from the system as sulfide minerals, the remaining sulfur in solution gets heavier with time. Thus the rims of sulfide grains should be very heavy.
6. Unlike sulfates, which precipitate with nearly the same $\delta^{34}S$ value as the dissolved sulfate, sulfides precipitate with a kinetic effect associated with biogenic sulfur reduction. As with carbon, the kinetic effect is not constant.
7. In the first case, there would be very little effect. The fractionation between evaporites and seawater is small, so that the removal of sulfur in the form of evaporites would have little effect on the

remaining dissolved sulfate in the ocean. Shale formation includes the precipitation of sulfides, which strongly incorporate the light isotopes of sulfur. Therefore, sulfate formation during periods in which large shale deposits formed should be characterized by high $\delta^{34}S$ values.

REFERENCES

Ault, W. U., and Jensen, M. L. (1963). Summary of sulfur isotope standards. *Biogeochemistry of Sulfur Isotopes*, Proceedings of the National Science Foundation Symposium at Yale University. New Haven, CT: Yale University Press, pp. 16–29.

Bao, H., Thiemens, M. H., Farquhar, J., Campbell, D. A., Lee, C. C.-W., Heine, K., and Loope, D. B. (2000). Anomalous ^{17}O compositions in massive sulphate deposits on the Earth. *Nature* **406**, 176–178.

Beaudoin, G., and Taylor, B. E. (1994). High precision and spatial resolution sulfur isotope analysis using MILES laser microprobe. *Geochimica et Cosmochimica Acta* **58**, 5055–5063.

Beaudoin, G., Taylor, B. E., Rumble, D. I., and Thiemens, M. (1994). Variations in the sulfur isotope composition of troilite from the Cañon Diablo iron meteorite. *Geochimica et Cosmochimica Acta* **58**, 4253–4255.

Burdett, J. W., Arthur, M. A., and Richardson, M. (1989). A Neogene seawater sulfur isotope age curve from calcareous pelagic microfossils. *Earth and Planetary Science Letters* **94**, 189–198.

Canfield, D. E. (2001). Biogeochemistry of sulfur isotopes. In *Stable Isotope Geochemistry*, Vol. 43 (ed. J. W. Valley and D. R. Cole), pp. 607–636. Washington, DC: Reviews in Mineralogy and Geochemistry.

Canfield, D. E. (2004). The evolution of the Earth surface sulfur reservoir. *American Journal of Science* **304**, 839–861.

Cecile, M. P., Shakur, M. A., and Krouse, H. R. (1983). The isotopic composition of western Canadian barites and the possible derivation of oceanic sulphate $\delta^{34}S$ and $\delta^{18}O$ age curves. *Canadian Journal of Earth Sciences* **20**, 1528–1535.

Claypool, G. E., Holser, W. T., Kaplan, I. R., Sakai, H., and Zak, I. (1980). The age curves of sulfur and oxygen isotopes in marine sulfate and their mutual interpretation. *Chemical Geology* **28**, 199–260.

Coplen, T. B., Hopple, J. A., Böhlke, J. K., Peiser, H. S., Rieder, S. E., Krouse, H. R., Rosman, K. J. R., Ding, T., Vocke, R. D. J., Révész, K. M., Lamberty, A., Taylor, P., and DeBièvre, P. (2002). *Compilation of Minimum and Maximum Isotope Ratios of Selected Elements in Naturally Occurring Terrestrial Materials and Reagents*. Reston, VA: United States Geological Survey.

Cortecci, G. (1975). Isotopic analysis of sulfate in a South Pacific core. *Marine Geology* **19**, 69–74.

Craig, H. (1957). Isotopic standards for carbon and oxygen and correction factors for mass-spectrometric analysis of carbon dioxide. *Geochimica et Cosmochimica Acta* **12**, 133–149.

Farquhar, J., Bao, H., and Thiemens, M. (2000). Atmospheric influence of Earth's earliest sulfur cycle. *Science* **289**, 756–758.

Giesemann, A., Jager, H.-J., Norman, A. L., Krouse, H. R., and Brand, W. A. (1994). On-line sulfur-isotope determination using an elemental analyzer coupled to a mass spectrometer. *Analytical Chemistry* **66**, 2816–2819.

Goldhaber, M. B., and Kaplan, I. R. (1974). The sulfur cycle. In *The Sea*, Vol. 5 (ed. E. D. Goldberg), pp. 569–655. New York: John Wiley and Sons.

Goodfellow, W. D., and Jonasson, I. R. (1984). Ocean stagnation and ventilation defined by $\delta^{34}S$ secular trends in pyrite and barite, Selwyn Basin, Yukon. *Geology* **12**, 583–586.

Halas, S. (1985). On bias in $^{34}S/^{32}S$ data obtained using SO_2 gas in mass spectrometry. *Studies on Sulphur Isotope Variations in Nature*. Vienna: International Atomic Energy Agency, pp. 105–111.

Hoefs, J. (1987). *Stable Isotope Geochemistry*. Berlin: Springer-Verlag.

Holser, W. T. (1977). Catastrophic chemical events in the history of the ocean. *Nature* **267**, 403–408.

Holser, W. T., and Kaplan, I. R. (1966). Isotope geochemistry of sedimentary sulfates. *Chemical Geology* **1**, 93–135.

Holser, W. T., Kaplan, I. R., Sakai, H., and Zak, I. (1979). Isotope geochemistry of oxygen in the sedimentary sulfate cycle. *Chemical Geology* **25**, 1–17.

Holser, W. T., Schidlowski, M., Mackenzie, F. T., and Maynard, J. B. (1988). Geochemical cycles of carbon and sulfur. In *Chemical Cycles in the Evolution of the Earth* (ed. C. B. Gregor, R. M. Garrels, F. T. Mackenzie, and J. B. Maynard), pp. 105–173. New York: John Wiley & Sons.

Hulston, J. R., and Thode, H. G. (1965). Variations in the S^{33}, S^{34}, and S^{36} contents of meteorites and their relation to chemical and nuclear effects. *Journal of Geophysical Research* **70**, 3475–3484.

Kakegawa, T., and Ohmoto, H. (1999). Sulfur isotope evidence for the origin of 3.4 to 3.1 Ga pyrite at the Princeton gold mine, Barberton Greenstone Belt, South Africa. *Precambrian Research* **96**, 209–224.

Kampschulte, A., and Strauss, H. (2004). The sulfur isotopic evolution of Phanerozoic seawater based on the analysis of structurally substituted sulfate in carbonates. *Chemical Geology* **204**, 255–286.

Kaplan, I. R. (1983). Stable isotopes of sulfur, nitrogen and deuterium in Recent marine environments. In *Stable Isotopes in Sedimentary Geology*, Vol. 10 (ed. M. A. Arthur, T. F. Anderson, I. R. Kaplan, J. Veizer, and L. S. Land), pp. 2-1–2-108. Columbia, SC: SEMP Short course.

Kaplan, I. R., and Hulston, J. R. (1966). The isotopic abundance and content of sulfur in meteorites. *Geochimica et Cosmochimica Acta* **30**, 479–496.

Krouse, H. R. (1989). Sulfur isotope studies of the pedosphere and biosphere. In *Stable Isotopes in Ecological Research*. Ecological Studies, Vol. 68 (ed. P. W. Rundel, J. R. Ehleringer, and K. A. Nagy). New York: Springer-Verlag, pp. 424–444.

Krouse, H. R., and Case, J. W. (1983). Sulphur isotope abundances in the environment and their relation to long term sour gas flaring near Valleyview, Alberta. *RMD Report 83/18 to Research Management and Pollution Control Divisions of Alberta Environment* 110.

Lloyd, R. M. (1967). Oxygen-18 composition of oceanic sulfate. *Science* **156**, 1228–1231.

Lloyd R. M. (1968). Oxygen isotope behavior in the sulfate–water system. *Journal of Geophysical Research.* **73**, 6099–6110.

Longinelli, A. (1989). Oxygen-18 and sulphur-34 in dissolved oceanic sulphate and phosphate. In *The Marine Environment*, Vol. 3 (ed. P. Fritz and J. C. Fontes), pp. 219–255. Amsterdam: Elsevier.

Longinelli, A., and Craig, H. (1967). Oxygen-18 variations in sulfate ions in sea-water and saline lakes. *Science* **146**, 56–59.

Longinelli, A., and Edmond, J. M. (1983). Isotope geochemistry of the Amazon Basin; a reconnaissance. *Journal of Geophysical Research. C. Oceans and Atmospheres* **88**, 3703–3717.

McKibben, M. A., and Eldridge, C. S. (1990). Radical sulfur isotope zonation of pyrite accompanying boiling and epithermal gold deposition; a SHRIMP study of the Valles Caldera, New Mexico. *Economic Geology* **85**, 1917–1925.

Monster, J., Appel, P. W. U., Thode, H. G., Schidlowski, M., Carmichael, C. M., and Bridgwater, D. (1979). Sulfur isotope studies in early Archaean sediments from Isua, West Greenland; implications for the antiquity of bacterial sulfate reduction. *Geochimica et Cosmochimica Acta.* **43**, 405–413.

Nielsen, H. (1979). Sulfur isotopes. In *Lectures in Isotope Geology* (ed. E. Jager and J. C. Hunziker), pp. 283–312. Berlin: Springer.

Ohmoto, H. (1972). Systematics of sulfur and carbon isotopes in hydrothermal ore deposits. *Economic Geology* **67**, 551–578.

Paytan, A., Kastner, M., Campbell, D., and Thiemens, M. H. (2004). Seawater sulfur isotope fluctuations in the Cretaceous. *Science* **304**, 1663–1665.

Paytan, A., Kastner, M., Campbell, D. R., and Thiemens, M. H. (1998). Sulfur isotopic composition of Cenozoic seawater sulfate. *Science* **282**, 1459–1462.

Rafter, T. A., and Mizutani, Y. (1967). Oxygen isotopic composition of sulphates, 2. Preliminary results on oxygen isotopic variation in sulphates and relationship to their environment and to their delta (34)S values. *New Zealand Journal of Science* **10**, 816–840.

Rees, C. E. (1978). Sulfur isotope measurements using SO_2 and SF_6. *Geochimica et Cosmochimica Acta.* **42**, 383–390.

Rees, C. E., Jenkins, W. J., and Monster, J. (1978). The sulphur isotopic composition of ocean water sulphate. *Geochimica et Cosmochimica Acta.* **42**, 377–382.

Robinson, B. W. (1995). Sulphur isotope standards. In *Reference and Intercomparison Materials for Stable Isotopes of Light Elements*, Vol. TECDOC-825 (ed. IAEA), pp. 39–46. Vienna: International Atomic Energy Agency.

Rye, R. O. (2005). A review of the stable-isotope geochemistry of sulfate minerals in selected igneous environments and related hydrothermal systems. *Chemical Geology* **215**, 5–36.

Sakai, H. (1968). Isotopic properties of sulfur compounds in hydrothermal processes. *Geochemical Journal* **2**, 29–49.

Savarino, J., Bekki, S., Cole-Dai, J., and Thiemens, M. H. (2003). Evidence from sulfate mass independent oxygen isotopic compositions of dramatic changes in atmospheric oxidation following massive volcanic eruptions. *Journal of Geophysical Research, D, Atmospheres* **108**, doi:10.1029/2003JD003737

Savarino, J., Romero, A., Cole-Dai, J., Bekki, S., and Thiemens, M. H. (2003). UV induced mass-independent sulfur isotope fractionation in stratospheric volcanic sulfate. *Geophysical Research Letters* **30**, doi:10.1029/2003GL018134

Schidlowski, M., Hayes, J. M., and Kaplan, I. R. (1983). Isotopic inferences of ancient biochemestries: Carbon, sulfur, hydrogen, and nitrogen. In *Earth's Earliest Biosphere: Its Origin and Evolution* (ed. W. F. Schopf), pp. 149–186. Princeton, NJ: Princeton University Press.

Schidlowski, M., Junge, C. E., and Pietrek, H. (1977). Sulfur isotope variations in marine sulfate evaporites and the Phanerozoic oxygen budget. *Journal of Geophysical Research* **82**, 2557–2565.

Strauss, H. (1997). The isotopic composition of sedimentary sulfur through time. *Palaeogeography, Palaeoclimatology, Palaeoecology* **132**, 97–118.

Thode, H. G., and Monster, J. (1965). Sulfur-isotope geochemistry of petroleum, evaporites, and ancient seas. *American Association of Petroleum Geologists, Memoir* **4**, 367–377.

Turchyn, A. V., and Schrag, D. P. (2004). Oxygen isotope constraints on the sulfur cycle over the past 10 million years. *Science* **303**, 2004–2007.

Zak, I., Sakai, H., and Kaplan, I. R. (1980). Factors controlling the $^{18}O/^{16}O$ and $^{34}S/^{32}S$ isotope ratios of ocean sulfates, evaporites and interstitial sulfates from modern deep sea sediments. In *Isotope Marine Chemistry* (ed. E. D. Goldberg, Y. Horibe, and K. Saruhashi), pp. 339–373. Tokyo: Rokakuho.

CHAPTER 11

IGNEOUS PETROLOGY

11.1 INTRODUCTION

Igneous rocks make up far and away the majority of the Earth's crust and, if we include the mantle, overwhelm all other rock types. Inevitably, every rock at the Earth's surface owes its origin to igneous rocks. Processes that occur in the near-surface environment can ultimately be best understood if we have a firm understanding of the composition and compositional range of different isotopic reservoirs, particularly those of the mantle. Igneous rocks can undergo a variety of modifications on their way to the surface. From a pristine mantle-derived magma, isotopic changes occur during fractional crystallization, the assimilation of country rock, magma mixing, degassing, and hydrothermal alteration. Alteration can also occur at depth by subduction-related contamination and mantle metasomatism. All of these processes have been addressed with stable isotope geochemistry. This chapter is organized from the inside out. That is to say, we start from the mantle and work towards the surface.

Oxygen is the major component of the mantle. The oxygen isotope composition of the oceans and of all meteoric waters is ultimately controlled by hydrothermal interaction with oceanic igneous rocks of mantle origin. Carbon, sulfur, nitrogen, and hydrogen are all trace components in the mantle. Nevertheless, due to the immensity of the mantle, even relatively small amounts of degassing of these elements can affect, or has affected, their isotopic compositions in the terrestrial environment. Although there are important exceptions, the stable isotope compositional range of mantle materials is generally limited. The much larger isotopic variability seen in crustal igneous rocks is ultimately due to low-temperature, upper-crustal processes.[1] Many igneous rocks preserve an isotopic signature characteristic of a low-temperature alteration event. Where in their genesis these rocks inherited such a signature is an important aspect of stable isotope geochemisty as applied to igneous rocks.

---

[1] For oxygen and hydrogen, large variations are controlled by processes of fractionation related to the meteoric water cycle; for carbon, nitrogen, and sulfur, biogenic processes are far more important.

11.2 THE MANTLE

The mantle is heterogeneous no matter how it is viewed. A seismologist sees distinct heterogeneous zones and layers. This perspective does not necessarily correlate with a geochemist's view of the mantle. A seismic discontinuity related to density variations may not manifest itself chemically. From a geochemical standpoint, the mantle is heterogeneous with respect to elemental concentrations, radiogenic isotope geochemistry, and stable isotope geochemistry.

The chemical composition of the mantle probably *was* homogeneous early in its history, following major mixing events associated with core formation and degassing. The chemical composition of the mantle at that time would have been equal to that of the bulk silicate Earth, or BSE. Later formation of the continental crust led to the development of a depleted mantle component, so that in the simplest case we have three components: the BSE, the depleted mantle, and the continental crust.

The mantle has also been contaminated by the reintroduction of crustal material during subduction. On the basis of radiogenic isotope, rare-gas, and trace element data, a number of mantle subdivisions have been proposed at one time or another in an effort to explain multi-component chemical trends. (See van Keken et al., 2002, for a review.) On the basis of fundamental chemical differences between Mid Ocean Ridge Basalts (MORBs) and Ocean Island Basalts (OIBs), the mantle has been divided into a primitive mantle, whose chemical composition is similar to that of the BSE, and even a depleted mantle, manifest at midocean ridges (DMM—depleted mid-ocean ridge MORB mantle). Further subdivisions include a subcontinental mantle, which has been affected by subduction and may have been isolated from other regions of the mantle for extended periods, and more refined subdivisions have even been proposed, primarily on the basis of radiogenic isotope and rare-gas compositions. High ^{3}He/^{4}He ratios of OIB have been interpreted in terms of a primitive, undegassed component, but the low gas concentrations complicate the picture considerably: An undegassed component implies that the gas concentrations should be high! An enriched mantle source, defined by Rb/Sr, Sm/Nd, and (U + Th)/Pb ratios that are higher than that of the primitive mantle, is sampled at several hot spots (e.g., Hawaii, Pitcairn Island, and Samoa). Zindler and Hart (1986) further divided the enriched mantle into three components[2] and proposed that all oceanic mantle compositions could be derived from a mixing of these components and a depleted MORB mantle (DMM). Hart et al. (1992) added a lower mantle component, presumably transported by plumes. Given the chemical complexity of the mantle, it is now unclear whether the simple primitive mantle, representative of the BSE, even exists today!

11.2.1 Oxygen

In the absence of contamination by crustal material, the oxygen isotope composition of the mantle should have a very limited range, because fractionation between the major minerals and melts in the mantle are small. Excepting spinel and CO_2, the fractionation between olivine and other phases, such as pyroxene and silicate melts, will be far less than 1‰ at mantle temperatures. Fractional crystallization, which is the process whereby crystals are removed from a melt by gravitational settling, will have only a minor effect on the $\delta^{18}O$ value of the

---

[2] These components are HIMU (high (U + Th)/Pb without high Rb/Sr) and EM1 and EM2 (enriched mantle 1 and 2), which may contain recycled continental material or may have undergone mantle metasomatism.

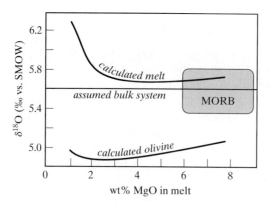

FIGURE 11.1: Calculated $\delta^{18}O$ values of melt and olivine crystals during progressive fractionation crystallization (toward lower MgO contents). Only when the residual melt reaches a very high SiO_2 content (low wt% MgO in melt) does the $\delta^{18}O$ value change appreciably. (Reprinted from Eiler, 2001, with permission of the Mineralogical Society of America.)

magma until the melt evolves beyond the composition of basalt into the andesite or dacite field, with the concomitant precipitation of quartz (Fig. 11.1). The removal of a CO_2-rich phase could have a greater effect on the isotopic composition of the remaining melt, but only if CO_2 concentrations were extremely high, a situation that is probably not realized in mantle free of contamination by subducted material.

A pristine sample of mantle, not contaminated by subducted material, would provide us with the BSE value. If this value were known, deviations from it could be used to evaluate subtle degrees of contamination. Sampling of primary mantle obviously poses a great problem. We can sample only at (or very near to) the Earth's surface, and materials that were once at depth have had abundant "opportunities" to be modified during their ascent by processes such as mixing, segregation, degassing, polymorphic transformation, assimilation, and the rapid surficial alteration of unstable phases. Therefore, one of the most difficult tasks in mantle studies is seeing through the modifications that can and often do occur. Three approaches have been used. The first is to use meteorite and lunar data as a proxy for the mantle, the second is to analyze mafic lavas, and the third is to analyze xenoliths or phenocrysts hosted in the lavas themselves.

Meteorites and lunar basalts. Meteorites and lunar basalts can be used as a proxy for the bulk Earth $\delta^{18}O$ value. The total range of $\delta^{18}O$ and $\delta^{17}O$ values of all meteorites is very large (Clayton et al., 1976). Enstatite chondrites, which represent the bulk Earth most closely, have a narrow range of $\delta^{18}O$ values centered around 5 to 6‰. The Moon lies exactly on the $\delta^{17}O$–$\delta^{18}O$ terrestrial fractionation line, strongly supporting a common origin for these two bodies (see Chapter 13 for more details). The $\delta^{18}O$ values of lunar samples have a very restricted range centered on 5.7‰. If the Earth–Moon system separated prior to any segregation and isotopic fractionation, then the $\delta^{18}O$ value of 5.7‰ should also apply to the bulk Earth. (See Chapter 13 for more on extraterrestrial samples.)

Mafic lavas. Mafic lavas and their nodules are the material at the Earth's surface closest in composition to the mantle. However, there is a wealth of evidence indicating that the chemical compositions of most mafic lavas at the surface have been modified from what they were at depth. Mafic magmas are often very fine grained and glassy. Surface alteration, degassing, and hydration occur very quickly. Phenocrysts tend to be less altered than the groundmass and generally have a more restricted range of $\delta^{18}O$ values than does bulk basalt (Kyser, 1986; Eiler et al., 1995). There is a strong correlation between Fe_2O_3/FeO or water content and $\delta^{18}O$ values of young submarine basalts (Fig. 5.7). Samples with the least degree of alteration are those with low water content and low Fe_2O_3/FeO ratios (Kyser et al., 1982).

The oxygen isotope composition of MORB is quite homogeneous. Ito et al. (1987) determined that the $\delta^{18}O$ value of fresh MORB glasses ranged from 5.3 to 6.2‰, averaging 5.7 ± 0.2‰. Slight variations correlating with Sr, Nd, and Pb isotope data were interpreted as contamination by a small percentage of recycled crustal material. Analyses by laser fluorination also have a limited range of $\delta^{18}O$ values that average 5.5‰, and similar conclusions have been made regarding contamination by subducted material (Eiler, 2001).

Ocean Island Basalts range from 4.6 to 7.5‰ (with a mean of 5.5 ± 0.5, 1σ) and continental basalts range from 4.5 to 8.1‰ (mean 6.1 ± 0.7) (Harmon and Hoefs, 1995). On the basis of the wide $\delta^{18}O$ range for "unmodified, primary mantle partial melts" (Mg number = 0.68–0.75)[3] of 5 to 7‰, Harmon and Hoefs concluded that the upper mantle is clearly heterogeneous with respect to oxygen isotope ratios (Fig. 11.2).

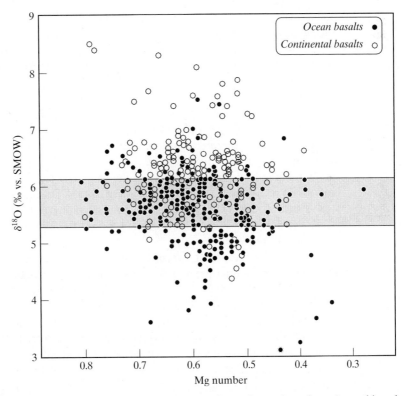

FIGURE 11.2: Compilation of oxygen isotope values of oceanic and continental basalts vs. Mg number. The wide range of $\delta^{18}O$ values for "pristine basalts" (Mg number = 0.68–0.75) has been used as evidence that the mantle is heterogeneous. Shaded area shows value for "pristine" mantle. (After Harmon and Hoefs, 1995.)

---

[3]The Mg number is defined as the molar ratio Mg/(Mg + Fe).

However, several lines of evidence indicate that the basalt data may not accurately reflect the isotopic composition of the mantle. A number of studies illustrate subtle alteration features that could easily have been overlooked. In samples that show little chemical evidence of alteration or contamination by crustal material during their ascent, unusual hydrogen isotope compositions or large variations in the isotopic composition of xenocrystic material may indicate cryptic alteration (Dobson and O'Neil, 1987; Baldridge et al., 1996; Feldstein et al., 1996). Clear evidence of basalt glass alteration can be seen in a comparison of the glass data with data from coexisting phenocrystic material. For example, the oxygen isotope ratios of fresh, recently erupted submarine volcanic glasses from the Pitcairn seamounts range from 5.8 to 7.4‰ (Woodhead et al., 1993). In contrast, the $\delta^{18}O$ values of olivine phenocrysts from basalts from Pitcairn Island are homogeneous at 5.2‰, a value indistinguishable from typical mantle values (Eiler et al., 1995). For these reasons, recent studies tend to concentrate more on phenocrysts or xenoliths than basalts themselves.

Phenocrysts and xenoliths. Olivine and pyroxene are the principal minerals making up mantle xenoliths. Very early on in the isotope game it became clear that olivine was notoriously difficult to fluorinate. Taylor and Epstein (1963) found that only 60 and 80‰ of olivine reacted with fluorine. One lone analysis gave an oxygen yield of 88%,[4] which they considered reliable. Several researchers (Reuter et al., 1965; Garlick, 1966) circumvented the problem by first fusing olivine with quartz of a known composition to make a pyroxene glass, which could easily be fluorinated. Clearly, the problems associated with analyzing olivine were well known.

In 1980, Javoy analyzed a number of olivine samples from peridotite massifs and obtained $\delta^{18}O$ values of 5.2 ± 0.08‰. Later work by a number of researchers showed that $\delta^{18}O$ values of olivine spanned a considerable range, from 4.5 to 7.5‰ (Mattey et al., 1994). The spread of data was interpreted in terms of heterogeneous mantle reservoirs (Kyser, 1986) or mantle metasomatism (Gregory and Criss, 1986). It now appears, however, that much of the variability may have been an artifact of the analytical technique. Fluorination of olivine by laser heating results in a far smaller range of $\delta^{18}O$ values, with a mantle average of 5.18 ± 0.28‰ (2σ) (Mattey et al., 1994) and a total range from 4.8 to 5.5‰. The $\Delta^{18}O_{cpx-ol}$ values average 0.4‰, consistent with a constant (temperature-insensitive) fractionation at mantle temperatures.

The results of the laser fluorination data indicate that the $\delta^{18}O$ value of the mantle is far less heterogeneous than previously thought. The large variations in oxygen isotope values found in OIB basalts therefore need to be considered in terms of alteration immediately prior to, during or post eruption. The mantle value of 5.2‰ has been used as a benchmark from which to estimate the degree of crustal contamination. For example, olivine phenocrysts from Pitcairn basalts are indistinguishable from the mantle average value of 5.2‰ (Eiler et al., 1995), whereas the basalts themselves are quite scattered. On the basis of "mantle" oxygen values, Eiler et al. concluded that the incorporation of subducted sediment is not the explaination for the unusual Sr, Nd, and Pb isotope ratios exhibited by these EM1 ocean island basalts. If the mantle is indeed as homogeneous as is presently thought, then even slight variations of a few tenths of a per mil—what would be considered "noise" in most studies—have geological

---

[4]The oxygen yield is the relationship between the amount of O_2 gas produced (determined manometrically) and the theoretical yield based on sample weight and stoichiometry. For example, quartz has 16.64 μmol O_2/mg quartz, such that 10 mg of quartz should evolve 166.4 μmol of O_2 during the fluorination procedure.

significance. The story may be quite different for sub-continental mantle. Olivines from the western United States have a considerable range of $\delta^{18}O$ values that tend to be higher than accepted for typical mantle (see Perkins et al., 2006).

Mantle eclogites. The discovery of oxygen isotope anomalies in mantle eclogites provided the earth science community with direct geochemical evidence for the subduction and recycling of crustal material. Originally misinterpreted as being a function of fractional crystallization, the large variations in $\delta^{18}O$ values of mantle eclogites were correctly identified by MacGregor and Manton (1986) and Ongley et al. (1987) as metamorphosed subducted oceanic crust. Unlike peridotites, which have a very narrow range of $\delta^{18}O$ values, mantle eclogites span a range from 2 to 8‰ (Fig. 11.3). The large variation in $\delta^{18}O$ values cannot be explained in terms of any known mantle processes. Instead, the spread in $\delta^{18}O$ values is remarkably similar to those seen in altered oceanic crust (Figs. 5.8, 5.9). The carbon and sulfur isotope data presented in the rest of this section support the presubduction near-surface alteration signature for eclogites.

11.2.2 Carbon

A considerable effort has been made towards gleaning an understanding of the carbon isotope systematics of the mantle. It is now quite clear that the range of carbon isotope values in mantle phases far exceeds that of oxygen. A number of mechanisms have been proposed for the large range, and compelling arguments have been made for each mechanism, but unfortunately, the geological community has yet to reach consensus. One important difference between carbon and oxygen isotope systematics is that carbon isotope fractionation between

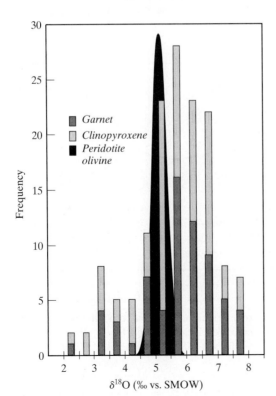

FIGURE 11.3: Oxygen isotope values of garnet and clinopyroxene in mantle eclogites. Also shown is the distribution of peridotites (not at same frequency scale, $n > 100$). (Data from MacGregor and Manton, 1986; Ongley et al., 1987; Shervais et al., 1988; Caporuscio, 1990; Mattey et al., 1994; Snyder et al., 1995; Viljoen et al., 1996.)

relevant mantle phases is large even at high temperatures (Fig. 11.4), primarily because carbon exists in different oxidation states. Contrast this behavior with that of oxygen, whereby fractionations are too small to have any significant consequence at mantle temperatures. A recent review of carbon in the mantle can be found in Deines (2002).

Carbon composition inferred from crustal reservoir. The crustal carbon abundance is a function of the carbon fluxes to and from the mantle. The huge carbon flux along midocean ridges is sufficient to generate all terrestrial carbon in 300 to 700 million years (DesMarais, 1985).[5] If all carbon in the crustal reservoir is derived from the mantle without appreciable isotopic fractionation, then the $\delta^{13}C$ value of the overall mantle can be determined from the bulk crustal value using simple mass-balance equations, as long as we know the sizes and $\delta^{13}C$ values of the two major crustal reservoirs, namely, carbonate and organic carbon. Assuming $\delta^{13}C$ values of 0‰ and -25‰ for carbonates and organic matter, respectively, with an abundance ratio of 4 to 1, gives a $\delta^{13}C$ value of -5‰ for the average crust. Other published estimates of this kind range from -4.5 to -7‰ (e.g., Hoefs, 1973).

Diamonds and graphite. Diamonds are arguably the best phase for addressing the $\delta^{13}C$ values of the mantle. They are unambiguously of mantle origin and, once formed, are certain to retain their carbon isotope ratio. $\delta^{13}C$ values of diamonds have a huge range, from -34 to $+5$‰ (Fig. 11.5). A number of observations are apparent from the compiled data: (1) There is a very pronounced mode at -6 to -5‰. (2) The data are skewed towards negative values, down to less than -30‰. (3) There is a sharp upper limit to the data at approximately -1‰, with a few rare exceptions at higher $\delta^{13}C$ values.

Several mechanisms have been proposed to explain the data. Those considered most probable are the following:

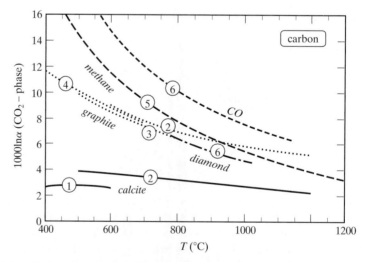

FIGURE 11.4: Carbon isotope fractionations between phases at moderate to high temperatures. (References: 1 Bottinga, 1969; 2 Scheele and Hoefs, 1992; 3 Chacko et al., 1991, and Scheele and Hoefs, 1992; 4 Bottinga, 1968; 5 Richet et al., 1977; 6 Deines, 1980b.)

---

[5]Presumably there is a return flux to the mantle *via* subduction of equivalent magnitude.

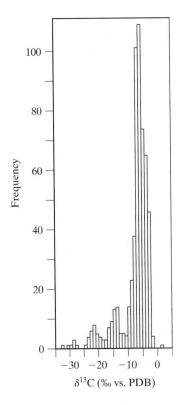

FIGURE 11.5: Carbon isotope values of diamonds. The mode is between −6 and −5‰ skewed towards low values. (Reprinted from Deines, 1980a, with permission from Elsevier.)

1. The $\delta^{13}C$ value of the mantle is between −6 and −5‰, and samples with higher, but especially lower, $\delta^{13}C$ values, are contaminated with subducted material. This argument is strongly supported by correlation of the $\delta^{13}C$ values of diamonds with their occurrence. Diamonds hosted in peridotites or with peridotite mineral inclusions are called P-type diamonds. The range of $\delta^{13}C$ values of P-type diamonds is rather limited between −7 and −5‰, with rare outliers as low as −18.9‰. In contrast, diamonds hosted by or having inclusions of eclogite minerals (E-type) have a similar range, but a mode at lower values from −15 to −13‰ (Deines, 1980a) (Fig. 11.6). As we saw earlier, the $\delta^{18}O$ values of eclogites are anomalous compared with that of the rest of the mantle and have been explained in terms of subducted material. As a logical extension, the low $\delta^{13}C$ values of the E-type diamonds could be explained by a subducted organic source. The explanation seems simple and convincing, but unfortunately, nothing really is so simple, and other, equally compelling explanations for the range of $\delta^{13}C$ values in eclogitic diamonds have been proposed.

2. Diamonds are formed by a partial oxidation of methane rising from deeper mantle levels. One argument against subducted organic material to explain the large range of $\delta^{13}C$ values is that similar ranges are seen in iron meteorites and carbonaceous chondrites (Deines and Wickman, 1975; Kerridge, 1985). Deines (1980a) considered a number of oxidation–reduction reactions in terms of P–T, $f(O_2)$, and $a(H_2O)$ that might explain the diamond variation. None of his models was able to explain all of the observations of mantle carbon chemistry, but he nevertheless concluded that if only a small fraction of methane were oxidized and precipitated as diamond, a large

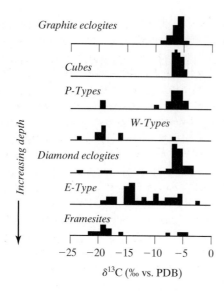

FIGURE 11.6: $\delta^{13}C$ values for diamonds of different types, expressed as a function of depth. The $\delta^{13}C$ values of the diamonds are tightly clustered at -7 to $-5‰$ for cubic diamonds at the highest level and spread out to lower values at greater depths for the E-Type diamonds. (See text and Deines, 1980a, for details of diamond type).

range of $\delta^{13}C$ values could be generated. Correlations with depth, morphology, and nitrogen content (Fig. 11.6) were ultimately interpreted in terms of primary heterogeneity in mantle carbon values.

3. The difference in $\delta^{13}C$ values of P-Type and E-Type diamonds is due to Rayleigh fractionation associated with a loss of CO_2. Cartigny et al. (1998) measured both $\delta^{13}C$ and $\delta^{15}N$ values of P-type and E-type diamonds. The E-type diamonds have a narrow range of $\delta^{15}N$ values compared with the P-Type. The combined carbon and nitrogen data, argued Cartigny and colleagues, could not be explained either by contamination of subducted material or by reduction of methane. Instead, they proposed that diamonds are formed by asthenospheric-derived carbonatitic melts that move into a more reduced lithosphere. CO_2 generated by the oxidation of the melts is enriched in ^{13}C, leaving a carbonatitic residue with low $\delta^{13}C$ values. Reduction of the carbonatitic residue ultimately produces diamonds in the eclogites. The peridotites differ from the eclogites in that they contain olivine, which is not stable with CO_2 at mantle conditions. Any CO_2 formed by oxidation of the carbonatitic liquid would immediately react with the host olivine to form magnesite, thereby preventing the escape of the volatile phase. As a result, diamonds in peridotites cannot undergo significant CO_2 loss and therefore do not share the large range of $\delta^{13}C$ values seen in eclogites. Variations of up to 7‰ have been found in a single diamond (Swart et al., 1983), supporting the idea that Rayleigh fractionation mechanisms cause the carbon isotope variability found in diamonds. All fractionation mechanisms (as opposed to crustal recycling) require that only a very small portion of the carbon pool be converted to diamonds.

Carbonatites and kimberlites. Carbonates in kimberlites and carbonatites are of deep-seated origin and can be used to constrain the $\delta^{13}C$ value of the mantle. There are advantages and disadvantages to using carbonates compared with diamonds. Carbonatites are large masses of carbon-bearing material and cannot have undergone considerable carbon isotope fractionation during their formation or ascent. In contrast to carbonatites, diamonds may have undergone considerable fractionation from their ultimate mantle source, as explained before. The disadvantage of analyzing carbonates is that their $\delta^{13}C$ values can be modified during or following eruption. Degassing of CO_2, surficial meteoric water alteration, diagenesis, and mixing with biogenic carbonates can all affect the $\delta^{13}C$ values of carbonatites. By contrast, none of these processes will influence the $\delta^{13}C$ values of diamonds.

Taylor et al. (1967) measured carbon and oxygen isotope ratios of carbonatites and were able to distinguish the most pristine samples from those which had undergone significant exchange on the basis of C–O trends. The lowest isotope ratios were used to define the "carbonatite box," with $\delta^{18}O$ and $\delta^{13}C$ values of 6 to 8‰ and -8 to -5‰, respectively. Other carbonatite studies have used additional criteria, such as radiogenic isotope and trace element ratios, as well as depth of emplacement, to subtract the effects of late alteration or degassing. Deines and Gold (1973) concluded that the average $\delta^{13}C$ value of carbonates is -5.1 ± 1.4‰, with statistically significant variation between carbonatite complexes. Carbonates in kimberlites are extremely fine grained. The $\delta^{13}C$ value of kimberlites is statistically indistinguishable from that of carbonatites, but $\delta^{18}O$ values are invariably higher due to hydrothermal exchange during or after emplacement.

Basaltic glass. Carbon is present in basaltic glass, trapped as fluid inclusions, along grain boundaries, and as a dissolved component in silicate minerals. Early studies found a wide range of $\delta^{13}C$ values, which, it is now apparent, was due to analytical problems. There are two distinct populations of carbon that are released when samples are step-heated. At temperatures below 600°C, degassing in an oxygen atmosphere generates a low-$\delta^{13}C$ CO_2 gas (~ -26‰). Upon further heating (in excess of 1,000°C), CO_2 gas is liberated with a $\delta^{13}C$ value of ~ -6.6‰. The CO_2 that is evolved from low-temperature heating is thought to be due to either organic contamination or fractionation during late degassing associated with emplacement. MORB glasses have a consistent $\delta^{13}C$ value of -6.3‰; other basalt types show slight variations from this value (Exley et al., 1986).

11.2.3 Nitrogen

The paucity of nitrogen isotope measurements in deep-seated rocks is in part related to analytical difficulties which only recently have been overcome. Nitrogen is only present in trace quantities, and even minor atmospheric contamination will compromise an analysis. The low concentration of nitrogen in the mantle is an analytical problem, but it is also a benefit. Even small amounts of recycled material (crustal contamination), for example, could significantly alter the $\delta^{15}N$ of a sample. Thus, nitrogen isotope ratios are highly sensitive indicators of mantle heterogeneities and mixing between reservoirs. In the last 10 years, our understanding of nitrogen isotope systematics in the mantle has improved dramatically; nevertheless, a complete characterization of mantle isotope reservoirs and end members is far from having been achieved.

A thorough review of nitrogen isotope systematics in mantle materials can be found in Marty and Dauphas (2003). The average $\delta^{15}N$ value of the mantle is -5‰ relative to AIR, although clear heterogeneities exist[6] (Fig. 11.7). The $\delta^{15}N$ value of gases extracted from MORB glass vesicles by crushing range from -5 to $+5$‰. Samples with high $^{40}Ar/^{36}Ar$ ratios (indicating minimal air contamination) have $\delta^{15}N$ values of -5 to -3‰ (Marty and Humbert, 1997). Negative $\delta^{15}N$ values of N_2 gases are found in volcanic gases (Sano et al., 2001; Fischer et al., 2002). In conjunction with other rare-gas geochemistry, they require a mantle component that is far less than -5‰, possibly as low as -15‰ (Mohapatra and

---

[6]One cannot turn to meteorites to constrain the nitrogen isotope ratio of the mantle: $\delta^{15}N$ values of chondrites range from -40 to $+50$‰ (Kung and Clayton, 1978).

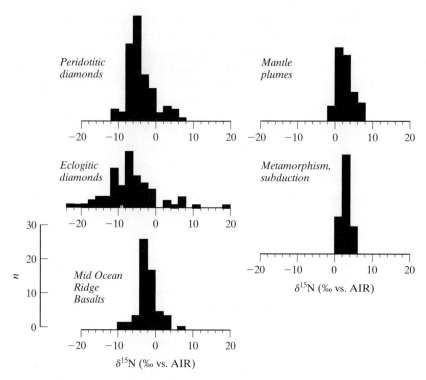

FIGURE 11.7: $\delta^{15}N$ values of different mantle sources. (After Marty and Dauphas, 2003.)

Murty, 2004). The $\delta^{15}N$ value of diamonds has a strong mode at $-5‰$ as well, although there is considerable scatter, with values ranging from -25 to $+15‰$.

In contrast to MORB, deep-mantle material, as sampled by mantle plumes, has $\delta^{15}N$ values that are typically positive, ranging from -2 to $8‰$. A number of explanations for the variations have been presented. Javoy (1997) suggested that heterogeneities in the mantle are remnants of heterogeneities of early accretion. Marty and Dauphas (2003) proposed that the difference between plume-related samples and MORB can be explained by secular variations in subduction. Archean sediments have $\delta^{15}N$ values as low as $-6‰$ (Beaumont and Robert, 1999), in comparison to modern sediments, which are almost always positive. Archean sediments with negative $\delta^{15}N$ values would have been subducted only to shallow levels (due to the high temperatures that existed early in the Earth's history), whereas younger material with positive $\delta^{15}N$ values could be subducted to much greater depths. In this scenario, the MORB data represent early subduction of material with negative $\delta^{15}N$ values, whereas mantle plumes tap a nitrogen reservoir related to much younger (and deeper) subduction. The reason that nitrogen (as opposed to other isotope systems) may trace crustal subduction so clearly is simply that it has a very low concentration in unaltered mantle material. Even slight additions to the mantle are therefore evident.

11.2.4 Hydrogen

Determining the hydrogen isotope composition of pristine mantle is a daunting endeavor. Almost any modification that occurs during the material's ascent will alter the hydrogen isotopic composition of whatever phase is hosting it. The very processes that bring a sample from the mantle to the surface are themselves aberrations from normal mantle conditions. For example, the eruption of a xenolith-bearing kimberlitic magma is initiated by infiltration of a metasomatic fluid, which could alter the δD value of a mineral in a xenolith. Hydrogen isotope measurements of bulk mantle minerals, fluid inclusions, and fresh glass from submarine basalts range from −120 to +13‰ (e.g., Sheppard and Epstein, 1970; Sheppard and Dawson, 1975; Kuroda et al., 1977; Boettcher and O'Neil, 1980; Kyser and O'Neil, 1984). In situ spot analyses from single amphiboles span a range of more than 50‰ (Deloule et al., 1991; Xia et al., 2002). There are a number of processes that can modify the hydrogen isotope composition of mantle phases during ascent. Exchange with water can either raise or lower the δD value of a glass or hydrous mineral. Simple degassing will lower the δD value, whereas dehydrogenation, or the removal of molecular H_2, will *raise* the δD value of a hydrous phase (Feeley and Sharp, 1996).

Fresh submarine basalts encompass a wide range of both δD values and water content (Kyser and O'Neil, 1984). Kyser and O'Neil were able to explain δD value vs. water content trends by degassing and the addition of seawater (Fig. 11.8). Extrapolating these trends back to the "unmodified" δD values of the original basalts, they concluded that the mantle has a "surprisingly constant" δD value of ∼−80‰. According to Kyser (1986, p. 160), the hydrogen isotope composition of the mantle is "surprisingly uniform and best explained by the presence of a homogeneous reservoir of hydrogen that has existed in the mantle since the very early history of the earth."

Boettcher and O'Neil (1980) analyzed a number of phlogopites and amphiboles from kimberlites and proposed a range of deep-seated H_2O from −79 to −58‰. In contrast to the δD values of the phlogopites, which had a relatively restricted range, the δD values of amphibole megacrysts were extremely scattered, ranging from −113 to +8‰. The highly variable δD values (combined with other chemical compositions) led the authors to conclude that there are complicated, near-surface processes that involve a discrete aqueous fluid. Late degassing behavior is supported by high-resolution ion-microprobe studies, in which variations of 50‰ over hundreds of microns are found.

The δD value of the mantle above subduction zones is inferred from materials associated with these environments. Poreda (1985) found that basalts from the Mariana Trough had a number of geochemical characteristics clearly indicating a subducted component, including elevated water contents and $δ^{18}O$ values, as well as low $^3He/^4He$ ratios. The δD values for

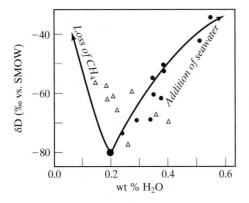

FIGURE 11.8: Correlation of δD value and water content for Kilauea (circles) and Mid-Atlantic Ridge (MAR) samples (triangles). The Kilauea samples can be explained by the addition of water, while the MAR samples follow a degassing-of-CH_4 trend. Both trends converge at an unmodified value of ∼−80‰ Trends are schematic. (After Kyser and O'Neil, 1984.)

these rocks range from −46 to −32, consistent with an aqueous component from a low-T altered seafloor basalt. High δD values are also found in fresh boninite glasses, supporting a subducted component to these materials (Dobson and O'Neil, 1987). Giggenbach (1992) compiled data for volcanic and geothermal discharges around the Pacific rim and proposed a magmatic "andesite" water component with a δD value of −20 ± 10‰. Basaltic water (δD = −60‰) is brought to the mantle from hydrated basalt, whereas the andesitic water is a mixture of oceanic crust, pore water, and clay minerals accumulated in marine sediments. The distinctly different δD values for basaltic and andesitic water are due to dehydration of the different components occurring at different levels in the subduction zone.

11.2.5 Sulfur

Hulston and Thode (1965) determined that the $\delta^{34}S$ values of meteorites are near 0‰, chondrites 0 ± 0.7‰, carbonaceous chondrites 0.6 ± 1.1‰, and achondrites and irons 0.1 ± 0.1‰.[7] The $\delta^{34}S$ values of MORB are quite constant at +0.3 ± 0.5‰ (Sakai et al., 1984), with a fractionation of 7.4‰ between sulfate and sulfide. The $\Delta^{34}S$(sulfate–sulfide) value of 7.4 is considered the equilibrium fractionation between sulfate and sulfide at magmatic temperatures.[8] What appeared at first glance to be a "well-behaved" system got more complicated as more data were collected (Fig. 11.9). In situ analyses of sulfide minerals in E-type diamonds range from +2.3 to +8.2‰ (Chaussidon et al., 1987), which is interpreted as evidence of subducted sulfur of terrestrial origin. Evidence for long-term recycling of oceanic and crustal sulfur was also found in metasomatized xenoliths from Dish Hill, California, where $\delta^{34}S$ values are near +7‰, sourced from subducted crustal sulfur (Wilson et al., 1996). Evidence for a subducted sulfur component is most definitive in a recent study by Farquhar et al. (2002), who found mass-independent $\Delta^{33}S$ anomalies in sulfides in diamonds from Orapa kimberlite. Such anomalies are normally found in Archean sediments and are known only from samples formed at the Earth's surface (Chapter 10). Their discovery in deep-mantle xenoliths is compelling evidence for recycling of an Archean sulfur component.

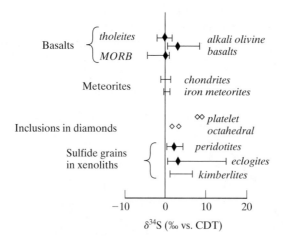

FIGURE 11.9: Sulfur isotope composition of sulfides from various sources. Most data are near zero, with the notable expection of eclogites and inclusions in diamonds from eclogites. These data, supported by $\delta^{33}S$ anomalies, indicate a near-surface origin of sulfur in eclogites. (After Chaussidon et al., 1987.)

---

[7]An analysis of sulfate was anomalously low at −5.7‰, a number confirmed by later studies.
[8]This fractionation can change by 1 to 3‰ over a temperature range of 700–1200°C.

11.3 EMPLACEMENT OF PLUTONIC ROCKS: INTERACTIONS WITH THE CRUST AND HYDROSPHERE

Magmas undergo a number of physio-chemical processes that change their isotopic composition on their perilous journey towards the Earth's surface. These include assimilation of crustal material, physical segregation by fractional crystallization, degassing, and hydrothermal alteration. Interaction of submarine basalts with seawater has already been discussed (Chapter 5); $\delta^{18}O$ values of rocks can be either raised or lowered, depending upon the temperature of interaction. The more general case of emplacement of plutonic rocks has been nicely outlined by Taylor in a number of publications, notably (Taylor, 1978) for plutonic granitic rocks. Compiled whole-rock oxygen isotope data for a number of igneous rocks (Fig. 11.10) show a number of trends and patterns that can adequately be explained by well-understood processes.

Stable isotopes, particularly those of oxygen and hydrogen, are well suited to the study of igneous rocks for the following reasons: (1) The oxygen and hydrogen isotope composition of the source region—the mantle—is well known and fairly uniform. (2) As has been discussed in regard to the mantle, simple fractional crystallization does not affect the $\delta^{18}O$ value of a magma appreciably. Therefore, deviations from mantle values are evidence of open-system behavior at some time during or after emplacement. (3) The $\delta^{18}O$ values of sedimentary rocks are far higher than those of igneous rocks. Combined with other isotopic systems, such as $^{87}Sr/^{86}Sr$, they can be used as a sensitive monitor for sedimentary contamination. (4) The oxygen and hydrogen isotope compositions of meteoric and ocean water are unique and far from equilibrium with deep-seated igneous rocks. Combined hydrogen and oxygen isotope

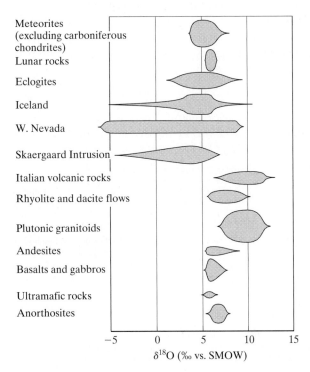

FIGURE 11.10: Compilation of oxygen isotope compositions of various igneous rocks. The $\delta^{18}O$ values of lunar, meteoritic, and most mafic and ultramafic rocks span a narrow range equal to the $\delta^{18}O$ values of the mantle. Plutonic granitoids in general range from 7 to 12‰, with values higher than 9‰ found in S-type granites. A large number of altered rocks plot both below and above the unaltered granitoids values. Volcanic andesite, rhyolite, and dacite rocks tend to have $\delta^{18}O$ values that are several per mil lower than those of their intrusive equivalents. Thickness of each shape roughly corresponds to number of analyses. (After Taylor, 1974.)

data can be used to identify and sometimes quantify fluid–rock interaction in cases where other isotope or chemical systems fail.

11.3.1 Normal Igneous Rocks

Unaltered igneous rocks have several characteristics in common. Whole-rock $\delta^{18}O$ values range from 6 to 10‰.[9] The $\delta^{18}O$ values of coexisting minerals follow the expected order of ^{18}O enrichment: Magnetite–biotite–hornblende–muscovite–plagioclase–potassium feldspar–quartz. The mineral fractionations correspond to high temperatures, but generally less than those expected for assumed granite emplacement. The explanation for the discrepancy is postsolidus exchange.[10] Typical $\Delta^{18}O$ (quartz − plagioclase) values are 1.5–2.5‰; $\Delta^{18}O$ (quartz − K − spar) values are 1.0–1.5‰. Volcanic rocks have $\delta^{18}O$ values that are generally 1–2‰ lower than chemically equivalent intrusive rocks. Finally, there is a general increase in $\delta^{18}O$ with increasing wt % Si. The increase, due in part simply to a higher modal abundance of quartz, is almost certainly related to the incorporation of crustal material at some point (Grunder, 1987). Taylor (1978) classifies granites (and, more generally, plutonic rocks) into normal, low-, and high-$\delta^{18}O$ groups (Table 11.1).

11.3.2 Shallow-Level Hydrothermal Alteration by Meteoric Water: Low-$\delta^{18}O$ Plutonic Rocks

Igneous bodies emplaced at shallow levels in the crust commonly undergo intense hydrothermal alteration. The high temperatures associated with intrusions set up hydrothermal convection cells that can drive significant amounts of aqueous fluids through a pluton and surrounding wall rock. Interaction with meteoric waters was recognized as early as 1963 (Taylor and Epstein), but the magnitude of water–rock interaction in shallow-level plutons was not realized for another 20 years. Meteoric water infiltration may have no effect on the chemical composition of a pluton other than on its stable isotope ratios (e.g., Criss and Taylor, 1983), simply because meteoric water contains no elements other than oxygen and hydrogen. Interaction with brines, on the other hand, can lead to intense chemical metasomatism and is a common mechanism for the generation of ore deposits.

During shallow-level emplacement of a pluton, the pluton is the heat engine driving convection. The effect is largest on rocks emplaced at shallow levels in permeable host rocks

TABLE 11.1: Characterization of granitoids.

Type	$\delta^{18}O$	δD	Comments
Normal	6 to 10‰	−85 to −50‰	This represents the vast majority of all granites, granodiorites, and quartz monzonites.
Low $\delta^{18}O$	<6‰	<−150 to −85	The δD values can be higher. Hydrothermally altered granites.
High $\delta^{18}O$	>10‰		S-type granites and low-T hydrothermal interaction.

(Taylor, 1978.)

---

[9]Values up to 14‰ are found in granitic rocks derived from the melting of sedimentary material.
[10]Only in very rapidly cooled, dry rocks are peak equilibration temperatures preserved. For example, basalts from Hawaii and lunar basalts both preserve temperatures in excess of 1,000°C.

and will be most pronounced (from a stable isotope perspective) in regions where meteoric water is extremely light.[11] For this reason, the majority of hydrothermal interaction studies have been made at high latitudes or altitudes. Modern examples of interaction between intrusions and meteoric water are associated with geothermal systems worldwide. Simply picture the spectacular thermal features at Yellowstone National Park to gain a sense of the magnitude of this effect, wherein shallow-level plutons supply the heat for all of the magnificent hot springs and geysers seen at the surface today.

Intrusive igneous rocks that are hydrothermally altered by meteoric water share a number of characteristics, illustrated by an example from the Isle of Skye (Forester and Taylor, 1977). The $\delta^{18}O$ values of the altered rock are less than those of unaltered rocks, dropping, in the Skye example, to below $-5‰$. There is a characteristic concentric depletion of $\delta^{18}O$ values towards the center of the complex. Low $\delta^{18}O$ values are seen in both the intrusive and the country rocks, demonstrating that the hydrothermal exchange occurred after emplacement. More alteration occurs in samples of higher permeability. δD values of amphiboles and chlorites are uniformly low, ranging from -132 to $-119‰$. Both quartz ($\delta^{18}O = -2.7$ to $+7.6‰$) and plagioclase ($\delta^{18}O = -6.7$ to $+6.0‰$) have $\delta^{18}O$ values well below those of normal granites, but the degree of hydrothermal alteration and isotopic lowering is far greater for feldspars. $\Delta^{18}O$(quartz $-$ feldspar) values are much larger than in normal granites, indicating that feldspars continued to be altered well into the postsolidus stage. Low $\delta^{18}O$ values of exchange-resistant zircon (Gilliam and Valley, 1997) confirm earlier work by Forester and Taylor that the pre-emplacement $\delta^{18}O$ values of some of the later intrusions were lower than those of normal granites.

A spectacular example of a fossil hydrothermal system is found centered around a series of shallow-level granitic plutons in the Idaho batholith (Criss and Taylor, 1983). The $\delta^{18}O$ values of altered plagioclase are as low as $-8.2‰$, compared with the initial values of $9.3‰$. The dimensions of the alteration cover over 15,000 km. The $\delta^{18}O$ values can be contoured, with a general bull's-eye pattern centered on each intrusion. In at least one case, there is a 60-km-diameter high-permeability ring fracture zone associated with caldera collapse where the lowest $\delta^{18}O$ values are found (Fig. 11.11).

The preceding results can be used to produce a general picture of shallow-level hydrothermal alteration of plutonic intrusive rocks. First, however, it should be recognized that meteoric waters do not alter the $\delta^{18}O$ value of a pluton until it has crystallized. This is because the partially molten pluton is at lithostatic pressure, while water in the country rock is most likely at hydrostatic pressure. Furthermore, the viscosity of a siliceous magma is high, and diffusion of water into the pluton is a slow and inefficient process. Finally, the amount of water necessary to change the isotopic composition of the magma significantly is on the order of 15%, far larger than the water solubility of a silicate liquid.

Once the magma crystallizes, fracture networks can develop, leading to intense hydrothermal convection systems. In the case of the Isle of Skye system, there appear to be two mechanisms that could explain the low $\delta^{18}O$ values: an initially low $\delta^{18}O$ magma and postemplacement hydrothermal exchange. The progressively lower *initial* $\delta^{18}O$ values of each magma suite are explained by successive intrusions having incorporated wall-rock material of previously emplaced magmas whose $\delta^{18}O$ values had already been lowered by hydrothermal interaction. Postsolidus, a giant hydrothermal circulation system was set up, resulting in a continued lowering of the $\delta^{18}O$ values of feldspar. The convection system would be focused

---

[11]In terms of ore deposits associated with pluton emplacement, magmatic water, sedimentary brines, and seawater are all far more important than meteoric water.

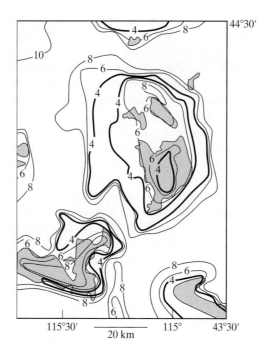

FIGURE 11.11: $\delta^{18}O$ contours surrounding Eocene plutons (gray shaded areas) in the Sawtooth Ring Zone (SRZ) of the Idaho Batholith, United States. The SRZ is a high-permeability zone formed when the caldera collapsed and is characterized by low $\delta^{18}O$ values (<4‰ shown by thick contour lines—values as low as −8‰ are measured) due to the massive influx of meteoric water. (Simplified from Criss and Taylor, 1983.)

towards the center of the pluton, which retained its heat longest, resulting in higher fluid/rock ratios in the central zones and the observed concentrically lower $\delta^{18}O$ values towards the center. This characteristic pattern, with lower $\delta^{18}O$ values of feldspars towards the center, has been seen in a large number of igneous complexes.

11.3.3 High-$\delta^{18}O$ Igneous Rocks

There are three ways in which an igneous rock can attain high $\delta^{18}O$ values: (1) through inheritance from the original magma (e.g., anatexis of original high-$\delta^{18}O$ material), (2) by high-temperature hydrothermal exchange with high-$\delta^{18}O$ country rocks, and (3) via low-temperature exchange with a meteoric fluid. As we saw in Chapter 5 with regard to the oceanic crust, the fractionations between minerals and water become so large at low temperatures that water–rock reactions *raise* the $\delta^{18}O$ value of the rock. Various methods have been used to ascertain which mechanism is responsible for causing high $\delta^{18}O$ values and are briefly explained in the next section.

The two distinct processes of anatexis (melting of country rock) and high-temperature exchange with country rock can usually be distinguished by considering spatial variability. The isotopic composition of original homogeneous high-$\delta^{18}O$ magmas may vary between intrusive events, but should be constant for each distinct intrusion. Even if the magma body was heterogeneous, it should not correlate with distance to the country rock. In contrast, if the high-$\delta^{18}O$ character of an intrusion results from exchange with the country rocks following emplacement, then we should expect to see a spatial relationship between $\delta^{18}O$ values and distance from the contact. Combining oxygen, hydrogen, and strontium isotope data (as well as old-fashioned mineralogy!) may also be used to distinguish these two mechanisms.

Typical S-type (sedimentary) granites, with high Al contents and muscovite, are generated from the melting of pelitic material or have incorporated material that has been through the sedimentary cycle. Sediments generally have high $\delta^{18}O$, δD, and $^{87}Sr/^{86}Sr$ values, all features that are seen in S-type granites. $\delta^{18}O$ values of S-type granites are over 10‰, compared with 7–9‰ for I-type (igneous) granites (e.g., O'Neil and Chappell, 1977). The $^{87}Sr/^{86}Sr$ ratios of I-type granites are generally 0.704 to 0.706, while S-type granites have $^{87}Sr/^{86}Sr$ ratios over 0.708 and up to 0.720. Groundmass minerals show the predicted isotopic enrichment trends, but in some cases, such as the Tuscan Magmatic Province in Italy, large sanidine crystals (1–5 cm in length) have $\delta^{18}O$ values that are 0.5 to 1.5‰ too high for the groundmass, indicating mixing of two distinct magmas (Taylor and Turi, 1976). These data cannot be explained in terms of either low-temperature alteration or exchange with country rocks.

Several clear-cut examples of high-temperature exchange post-emplacement have been found. Southern California batholiths intruding high-grade pelitic rocks have high $\delta^{18}O$ values only at the margins (Turi and Taylor, 1971). Because quartz, as well as feldspar and biotite, have high $\delta^{18}O$ values, exchange with the country rock probably occurred at high temperatures. If the alteration had occurred at low temperatures, then the far-more-resistant quartz would have retained its original isotopic composition. An important condition of high-temperature exchange with country rock is that the country rock must be hot prior to intrusion. Otherwise, the heat necessary to melt the rock would be more than the intrusion could supply.

Examples of high-$\delta^{18}O$ granites caused by low-temperature hydrothermal alteration are found worldwide. There are a number of mineralogical features that are characteristic of low-temperature alteration. In terms of stable isotopes, the strongest evidence comes from the large "isotopic reversal" between quartz and feldspar: Low-temperature hydrothermal exchange occurs with feldspar, but not quartz. In the Butler Hill pluton in Missouri, the $\Delta^{18}O_{(quartz-feldspar)}$ values can be contoured throughout the pluton (Wenner and Taylor, 1976). They are near equilibrium values in the northeast and decrease to negative values in the southwest. Meteoric waters responsible for the low-temperature exchange are calculated to have had $\delta^{18}O = -6$ to 0‰ and $\delta D = -25$ to 0‰. These values are too high for modern meteoric water, but are equivalent to saline formation waters that existed in the region. Extensive Na and Fe metasomatism in the area supports a brine origin for the fluids.

11.4 CALCULATING FLUID/ROCK RATIOS

It is clear that fluid–rock interaction has occurred in many intrusive bodies. The low $\delta^{18}O$ values that are commonly seen attest to the fact that large amounts of fluid have passed through the rock. As a first estimate, the amounts of fluid can be determined using the simple mass-balance relationship (Taylor, 1978)

$$X_{rock} \cdot \delta^i_{rock} + (1 - X_{rock})\delta^i_{water} = X_{rock} \cdot \delta^f_{rock} + (1 - X_{rock})\delta^f_{water}, \quad (11.1)$$

where X_{rock} is the molar fraction of the element in the rock, $(1 - X_{rock})$ is the molar fraction of the element in water, i is the initial value, and f is the final value. The water/rock ratio can be calculated by rearranging this equation to the form

$$\frac{W}{R} = \frac{\delta^f_{rock} - \delta^i_{rock}}{\delta^i_{water} - (\delta^f_{rock} - \Delta_{r-w})}, \quad (11.2)$$

where W/R is the water/rock ratio (in mole fraction) and Δ_{r-w} is the equilibrium fractionation between the final rock and water (i.e., $\Delta_{r-w} = \delta^f_{rock} - \delta^f_{water}$). Although this equation seems to allow us to quantify the water/rock ratio, keep in mind that several of the variables in it cannot be measured. First, the initial $\delta^{18}O_{rock}$ value is estimated either on the basis of similar rock types, or from measured values of a sample far removed from the effects of alteration. Likewise, the initial $\delta^{18}O_{water}$ value is estimated either from the paleogeographic reconstruction of meteoric water values or for a range of assumed values. The Δ_{r-w} value is estimated from an assumed temperature of reaction, from an assumed equilibrium fractionation between water and rock, and by making one last assumption that the water and rock were in isotopic equilibrium.[12] It is worthwhile to mention these ambiguities not so much to discourage researchers from making such calculations, but to impress upon them the limitations and uncertainties involved. Fluid/rock ratio estimates can be highly informative, but only as a rough indication of what actually occurred in the rock.

Equation (11.2) applies only to the physical condition in which all water enters a rock as a single event, equilibrates with it, and is then expelled. This is the "closed system," or *single*-pass case, of water–rock interaction based on mass balance. In the more dynamic "open-system," or *multi*pass, case, an infinitesimally small amount of water infiltrates the rock, equilibrates with it, and is then expelled. This model is more realistic in situations in which fluids are streaming through the rock. Mathematically, the open-system case is obtained by differentiating and integrating equation (11.2) to give

$$\frac{W}{R} = \ln\left(\frac{\delta^i_{water} + \Delta_{r-w} - \delta^i_{rock}}{\delta^i_{water} - (\delta^f_{rock} - \Delta_{r-w})}\right), \tag{11.3}$$

or

$$\frac{W}{R} = \ln\left[\frac{W}{R}_{closed\ system} + 1\right]. \tag{11.4}$$

A more detailed examination of the caveats relating to these equations and additional one-dimensional mixing models are given in Chapter 12.

Taylor (1978) considered the effects of water–rock interaction on the combined isotopic systems of oxygen and hydrogen. Because the hydrogen concentration of a rock is only a fraction of the oxygen concentration, interaction with meteoric water affects the isotopic composition of hydrogen much more strongly than that of oxygen. That is to say, for a fixed amount of water infiltration, the actual molar *W/R* ratios for hydrogen will be much larger than for oxygen.[13] The combined effects of hydrothermal alteration on hydrogen and oxygen isotopic compositions is shown in Fig. 11.12 for infiltration by a light meteoric water. Even for very low *W/R* ratios, the δD values of the hydrous minerals change rapidly, while oxygen isotope ratios remains essentially unchanged. Only when *W/R* ratios become quite high do we see the $\delta^{18}O$ values of the rock change, by which time hydrogen in the hydrous phases has completely equilibrated with the infiltrating fluid. This behavior is analogous to what we see during carbonate diagenesis, in which carbon changes more slowly than oxygen (Figure 6.8). The result is a dogleg-shaped profile of hydrogen–oxygen isotope ratios for a complete sequence

---

[12]The details are quite complex. Different minerals interact with hydrothermal fluids at different rates. Some minerals, such as feldspar, equilibrate quickly, while others—notably, zircon (King et al., 1997)—may never reequilibrate with the fluid.

[13]This, of course, assumes that all the interstitial water is equilibrated with the hydrous minerals.

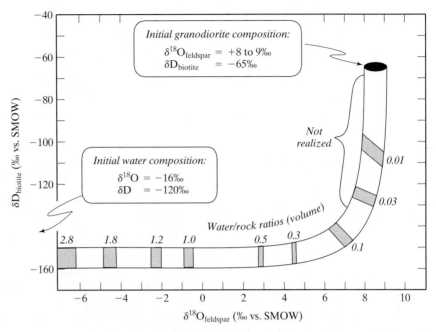

FIGURE 11.12: Combined oxygen–hydrogen isotope composition of a granodiorite infiltrated by meteoric water with an initial $\delta^{18}O = -16‰$ and $\delta D = -120‰$. The δD values of the biotites are affected by even small amounts of water, whereas far larger W/R ratios are required to change the $\delta^{18}O$ value of the rock. Very low W/R ratios are not realized, because the water will not flow when present in such small quantities. The trajectories are calculated for $\Delta^{18}O_{feldspar-H_2O} = +2‰$ and $\Delta D_{biotite-H_2O} = -30$ to $-40‰$. (After Taylor, 1978.)

of W/R ratios. Measured $\delta D/\delta^{18}O$ values of intrusive complexes have been shown to follow the predicted trend (e.g., Criss and Taylor, 1986).

11.5 OTHER PROCESSES: DEGASSING, ASSIMILATION, AND FRACTIONAL CRYSTALLIZATION

11.5.1 Magmatic Volatiles

Volatiles in magmas are the primary cause of explosive eruptions. Volatiles also represent the major flux of some elements from the mantle—notably, carbon. Calculating the speciation of volatile phases and measuring their abundance in magmas is difficult, because solubilities change with pressure and temperature, phase changes occur in response to redox state and degassing or late contamination will alter abundances and stable isotope ratios of magmatic volatiles. The major volatile species measured in volcanic gases are H_2O, CO_2, SO_2, H_2S, H_2, and HCl (Taylor, 1986). Gases measured in basaltic glass vesicles consist primarily of CO_2. Gases dissolved in glasses can be extracted by heating, but there is a risk of auto-oxidation–reduction reactions and the formation of gas phases that may not actually be present in the original sample.

Volatile degassing has essentially no effect on the $\delta^{18}O$ value of a magma, because the proportion of oxygen leaving the system is too small to change the value of the silicate reservoir. In contrast, the carbon, hydrogen, and sulfur isotope ratios of both of the evolved gas and of the residual magma, can change dramatically during degassing, because the proportion of the element entering the gas phase can be significant. This effect is seen, for example, in the $\delta^{13}C$ values of fluid inclusions in MORB. The $\delta^{13}C$ values range from -9 to $-4‰$ (Pineau and Javory, 1983), which can be explained in terms of fractionation between a mafic melt and CO_2 gas under closed-system conditions.

Devolatilization effects can be especially large for hydrogen isotopes of residual melts and hydrous phases. Hydrogen isotope ratios in basaltic glasses range from the expected primary basalt value of $\sim -80‰$ up to $-30‰$ (compilation of Taylor, 1986). These data are consistent with a kinetic model of degassing, in which the lighter isotopologues of water diffuse out of the melt faster than the heavier ones. In granitic plutons, there is a well-recognized correlation of δD values with water (Taylor, 1986). As an example of this effect, Nabelek et al. (1983) measured the $\delta^{18}O$ and δD values of samples from the composite granitic Notch Peak stock in Utah. Oxygen isotope ratios are more or less constant, but hydrogen isotope ratios vary by more than 50‰. The hydrogen isotope variations are well correlated with total water content and are concentrically zoned within each intrusion. The data can be explained entirely by a Rayleigh fractionation process. Loss of H_2O vapor will lead to crystallization of hydrous minerals with ever-*decreasing* δD values.[14] By the time 90% of the H_2O is lost from the system, the δD values of biotites will decrease by 40 to 50‰.

Taylor et al. (1983) measured the water content and δD values of water-rich obsidian ejecta associated with rhyolite volcanism. There is a regular lowering of δD values associated with degassing (wt % water), as is seen in Figure 11.13. The data are best explained by open-system Rayleigh degassing. A major conclusion of Taylor et al. is that low δD values do not require an influx of isotopically light meteoric water. Loss of hydrogen—as opposed to water—can lead to similar behavior, but because H_2 is depleted in deuterium

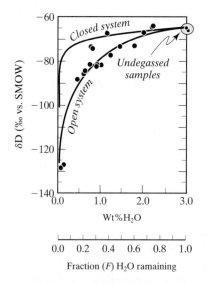

FIGURE 11.13: δD values vs. water content of fresh obsidian from Little Glass Mountain, Medicine Lake Highland, California, United States. The decrease in δD values from the undegassed samples follows an open-system Rayleigh fractionation trajectory, suggesting that the low values are primary features of open-system degassing, not of interaction with meteoric water. (After Taylor et al., 1983.)

---

[14]Remember that the δD value of water is *higher* than that of the coexisting melt, in contrast to $\delta^{18}O$.

compared with minerals or melt, the δD values of partially dehydrogenated biotites may have extremely high δD values (Feeley and Sharp, 1996). Overall, it is clear that the minor volatile elements in magmas—H, C, N, and S—which strongly partition into the fluid–vapor phase, can undergo large fractionations under conditions of volatile loss.

11.5.2 Assimilation–Fractional Crystallization (AFC) Processes

Taylor (1980) and DePaolo (1981) derived models for evaluating changes in isotope ratios of magmas due to assimilation of country rock and the effects of fractional crystallization. In particular, the authors showed that simple binary mixing models of $^{18}O/^{16}O$ and $^{87}Sr/^{86}Sr$ ratios do not adequately describe trends associated with assimilation. In the discussion that follows, the mechanisms of AFC are considered first in terms of simple fractional crystallization of a cooling magma and then by considering assimilation of country rock.

Crystallization and removal (by settling) of minerals in a magma chamber can be modeled with closed-system Rayleigh fractionation, where

$$\delta_{melt} - \delta^o_{melt} = (1000 + \delta^o_{melt})(F^{(\alpha-1)} - 1). \tag{11.5}$$

Here, F refers to the fraction of element remaining in the melt, δ^o_{melt} is the initial melt composition, δ_{melt} refers to the isotopic composition of the melt at F, and α is the fractionation between melt and crystal. For oxygen isotopes, values of α are on the order of 1.001 ($\Delta^{18}O_{melt-solid} = 1‰$), and even 95% crystallization will only nominally affect the $\delta^{18}O$ value of the remaining melt. Similarly, the δD value of the melt will change by less than 20‰ even if 80% of the hydrogen is removed to the melt phase (assuming a melt–mineral fractionation of no more than 10‰). In other words, the fractionation of H and O isotopes during crystallization cannot be large. Fractional crystallization will have no effect on Sr isotope ratios, because there is no measurable fractionation of the heavy isotopes between phases.

Assimilation of wall-rock material can have much larger effects if the initial isotopic composition of the wall-rock is very different from the magma, and a large proportion of wall-rock is incorporated. How much wall rock is assimilated is limited by the available heat, which is generated from the latent heat of crystallization of the magma. The heat can then be made available to melt the country rock. The initial background temperature of a country rock determines how much of it will melt; obviously, less heat is needed to melt a rock that is already hot.

Characteristic mixing for heavy radiogenic isotopes differs from that for stable isotopes in several respects. First, there is no isotopic fractionation between different phases, and second, the *concentrations* of the elements and isotopic ratios in the assimilant (wall rock), magma, and crystals can be very different. Therefore, even small amounts of mixing can drastically affect radiogenic isotope ratios. Thus, assimilation and fractional crystallization will affect oxygen and strontium isotope ratios in very different ways. The equations governing AFC processes are complex and can be found in DePaolo (1981) and Taylor (1986). The utility of AFC models is that combined isotopic trends in complex magmatic suites can be interpreted in terms of either simple mixing between end members or more complicated scenarios in which the assimilation of wall rock is required to explain the data. As an example, Dallai et al. (2003) posed the question whether a suite of Cambrian–Ordovician gabbros, diorites, granodiorites, and granites forming the Mt. Abbott composite intrusions (in Northern Victoria Land, Antarctica) could be generated from simple fractional crystallization and mixing or whether the assimilation of country rock was necessary. The combined oxygen and strontium

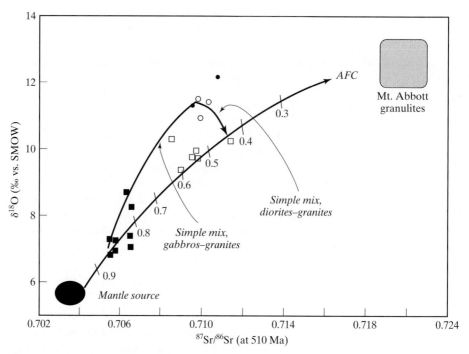

FIGURE 11.14: $\delta^{18}O-{}^{87}Sr/{}^{86}Sr$ plot of gabbros (filled squares), diorites (open squares), granodiorites (filled circles), and granites (open circles) from the Granite Harbour Intrusives, Antarctica. The diorite data cannot be the result of mixing between granite and gabbro. Instead, they require an assimilation component of the local Mt. Abbott granulites. (After Dallai et al., 2003.)

isotope data clearly require assimilation of the locally abundant Mt. Abbott granulites (Fig. 11.14). Simple mixing of the gabbros and granites will generate a hyperbolic mixing line that does not include the diorite data.

PROBLEMS SET

1. Construct a diagram similar to Fig. 11.12, assuming the same conditions as the original diagram, but infiltration of seawater instead of meteoric water. (See Field and Fifarek, 1985, for other examples.)
2. For olivine with a composition of $Mg_{1.5}Fe_{0.5}SiO_4$, how many micromoles of O_2 are extracted when 1 mg is fluorinated (e.g., with 100% yield)?
3. If all carbon in the Earth's crust is derived from the mantle without fractionation, and the $\delta^{13}C$ value of the mantle is given as -5.7‰, then what are the proportions of carbonate to reduced carbon if the $\delta^{13}C$ value of each is 0‰ and -25‰, respectively?
4. Explain why mantle eclogites have such a wide range of $\delta^{18}O$ values, while peridotites have a very narrow range?
5. Why do diamonds show such a large range of $\delta^{13}C$ values, while carbonatites have a very narrow range?

6. a. The $\delta^{13}C$ values of methane and CO_2 from fluid inclusions in olivine from a mantle xenolite are -9.25 and $-5.60‰$, respectively. Find the temperature of equilibration, given that the equation of fractionation is

$$1000 \ln \alpha = \frac{-0.62 \times 10^9}{T^3} + \frac{6.616 \times 10^6}{T^2} + \frac{6.04 \times 10^3}{T} - 3.08.$$

   b. Why didn't the methane and CO_2 reequilibrate during cooling?

7. In a scenario in which a pluton is immediately emplaced at a shallow level and meteoric water circulates through the pluton at 450°C, (a) what is the $\delta^{18}O$ values of the first fluid to fully equilibrate with the rock and (b) how would you expect the isotopic composition to change over time? Use the quartz–water fractionation equation

$$1000 \ln \alpha_{qz-H_2O} = \frac{3.13 \times 10^6}{T^2} - 2.94 \quad \text{(Matsuhisa et al., 1979)},$$

a $\delta^{18}O$ value of 7.5‰, and $-8‰$ for the quartz in granite and meteoric water, respectively.

ANSWERS

1. Figure 11.12 is based on equation (11.3),

$$W/R = \ln\left(\frac{\delta^i_{water} + \Delta_{r-w} - \delta^i_{rock}}{\delta^i_{water} - (\delta^f_{rock} - \Delta_{r-w})}\right),$$

which can be rearranged to give

$$\delta_{f,r} = \frac{\delta_{i,r} - (\delta_{i,w} + \Delta_{r,w}) + (\delta_{i,w} + \Delta_{r,w})e^{W/R}}{e^{W/R}}. \tag{11.P1}$$

This equation can be solved for either hydrogen or oxygen. The only additional variable to consider is that the W/R ratio for a given quantity of water exchanging with the rock is not the same for oxygen and hydrogen. While there are approximately equal molar fractions of oxygen in water and rock by volume, there may be at least 20 times the molar fraction of H in water compared with a relatively anhydrous rock. Therefore, the W/R ratio of hydrogen will be 20 times that of oxygen for the same passage of water. Equation (11.P1) reduces to

$$\delta_{f,r} = \frac{6.5 + (2)e^{W/R}}{e^{W/R}}$$

for oxygen and

$$\delta_{f,r} = \frac{-40 - (25)e^{W/R}}{e^{W/R}}$$

for hydrogen, given initial variables $\Delta^{18}O_{feldspar-water} = 2$, $\Delta D_{biotite-water} = -25$, $\delta^{18}O_{i,r} = 8.5$, $\delta^{18}O_{i,w} = 0$, $\delta D_{i,r} = -65$, and $\delta D_{i,w} = 0$. If the volumetric W/R abundance of H is 1/30 that of O, then, for each *W/R* value used in equation (11.P1) for oxygen, the value for hydrogen must be 30 times as great. The final plot is given by

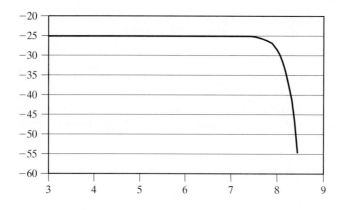

FIGURE 11.P1: Solution to problem 1.

2. The molecular weights of Mg = 24.31, Fe = 55.85, Si = 28.09, and O = 16.0. The molecular weight of $Mg_{1.5}Fe_{0.5}SiO_4$ is 156.48. For every mole of olivine, we get 2 moles of O_2. Therefore, 156.48 mg of olivine gives 2 moles of O_2. This translates to $\frac{2}{156.48} 1000 = 12.78$ μm O_2/mg olivine.

5. The total amount of carbon (hosted in diamonds) in a mantle rock is at trace levels. Therefore, any addition, mixing, devolatilization, or incorporation of exotic material can change the $\delta^{13}C$ value of the reservoir from which the diamond ultimately forms. Carbon is a major phase in carbonatites, which can be quite massive. Therefore, slight degassing, mixing with an exotic reservoir, or the addition of small amounts of CO_2, methane, etc., will have very little effect on the $\delta^{13}C$ value of carbonatites.

6. Part (a) T = 1,235°C. Part (b) The high temperatures are reasonable, and the reason that could be preserved during cooling is that (i) cooling was rapid and (ii) exchange between methane and CO_2 is very slow. The latter is true because, in order for exchange to occur, a molecule of CO_2 and CH_4 would need to collide, and during the collision all bonds would need to be broken and re-form around the other carbon. The activation energy for such a reaction is too high without a catalyst. Methane–CO_2 carbon isotope exchange will not occur in the gas phase.

7. The fractionation between quartz and water at 450°C is 3.05‰. For a rock of 7.5‰, the initial water should have a $\delta^{18}O$ value of 4.45‰. With time, however, the rock will see ever higher amounts of water, so the *W/R* ratio will increase. Eventually (if the system remains hot for a long enough period of time), the rock will be buffered by the water, so that the geothermal waters will retain their initial $\delta^{18}O$ values. Overall, the waters should decrease from an initial composition of 4.45‰ to −8‰.

REFERENCES

Baldridge, S. W., Sharp, Z. D., and Reid, K. D. (1996). Quartz-bearing basalts: Oxygen isotopic evidence for crustal contamination of continental mafic rocks. *Geochimica et Cosmochimica Acta* **60**, 4765–4772.

Beaumont, V., and Robert, F. (1999). Nitrogen isotope ratios of kerogens in Precambrian cherts: A record of the evolution of atmosphere chemistry? *Precambrian Research* **96**, 63–82.

Boettcher, A. L., and O'Neil, J. R. (1980). Stable isotope, chemical and petrographic studies of high-pressure amphiboles and micas: Evidence for metasomatism in the mantle source regions of alkali basalt and kimberlites. *American Journal of Science* **280-A**, 594–621.

Bottinga, Y. (1968). Calculation of fractionation factors for carbon and oxygen isotopic exchange in the system calcite–carbon dioxide–water. *Journal of Physical Chemistry* **72**, 800–808.

Bottinga, Y. (1969). Carbon isotope fractionation between graphite, diamond and carbon dioxide. *Earth and Planetary Science Letters* **5**, 301–307.

Caporuscio, F. A. (1990). Oxygen isotope systematics of eclogite mineral phases from South Africa. *Lithos* **25**, 203–210.

Cartigny, P., Harris, J. W., and Javoy, M. (1998). Eclogitic diamond formation at Jwaneng: No room for a recycled component. *Science* **280**, 1421–1424.

Chacko, T., Mayeda, T. K., Clayton, R. N., and Goldsmith, J. R. (1991). Oxygen and carbon isotope fractionations between CO_2 and calcite. *Geochimica et Cosmochimica Acta* **55**, 2867–2882.

Chaussidon, M., Albarede, F., and Sheppard, S. M. F. (1987). Sulphur isotope heterogeneity in the mantle from ion microprobe measurements of sulphide inclusions in diamonds. *Nature* **330**, 242–244.

Clayton, R. N., Onuma, N., and Mayeda, T. K. (1976). A classification of meteorites based on oxygen isotopes. *Earth and Planetary Science Letters* **30**, 10–18.

Criss, R. E., and Taylor, H. P., Jr. (1983). An $^{18}O/^{16}O$ and D/H study of Tertiary hydrothermal systems in the southern half of the Idaho Batholith. *Geological Society of America Bulletin* **94**, 640–663.

Criss, R. E., and Taylor, H. P., Jr. (1986). Meteoric-hydrothermal systems. In *Stable Isotopes in High Temperature Geological Processes,* Vol. 16 (ed. J. W. Valley, H. P. J. Taylor, and J. R. O'Neil), pp. 373–424. Chelsea, MI: Mineralogical Society of America.

Dallai, L., Ghezzo, C., and Sharp, Z. D. (2003). Oxygen isotope evidence for crustal assimilation and magma mixing in the Granite Harbour Intrusives, Northern Victoria Land, Antarctica. *Lithos* **67**, 135–151.

Deines, P. (1980a). The carbon isotopic composition of diamonds: Relationship to diamond shape, color, occurrence and vapor composition. *Geochemica et Cosmochimica Acta* **44**, 943–961.

Deines, P. (1980b). The isotopic composition of reduced organic carbon. In *Handbook of Environmental Isotope Geochemistry; Volume 1, The Terrestrial Environment* (ed. P. Fritz and J. C. Fontes), pp. 329–406. Amsterdam: Elsevier.

Deines, P. (2002). The carbon isotope geochemistry of mantle xenoliths. *Earth-Science Reviews* **58**, 247–278.

Deines, P., and Gold, D. P. (1973). The isotopic composition of carbonatite and kimberlite carbonates and their bearing on the isotopic composition of deep-seated carbon. *Geochimica et Cosmochimica Acta.* **37**, 1709–1733.

Deines, P., and Wickman, F. E. (1975). A contribution to the stable carbon isotope geochemistry of iron meteorites. *Geochimica et Cosmochimica Acta.* **39**, 547–557.

Deloule, E., Albarede, F., and Sheppard, S. M. F. (1991). Hydrogen isotope heterogeneities in the mantle from ion probe analysis of amphiboles from ultramafic rocks. *Earth and Planetary Science Letters* **105**, 543–553.

DePaolo, D. J. (1981). Trace element and isotopic effects of combined wallrock assimilation and fractional crystallization. *Earth and Planetary Science Letters* **53**, 189–202.

DesMarais, D. (1985). Carbon exchange between the mantle and the crust, and its effect upon the atmosphere: Today compared to Archean time. In *The Carbon Cycle and Atmospheric CO_2: Natural Variations Archean to Present,* Vol. 32, pp. 602–611.

Dobson P. F., and O'Neil, J. R. (1987). Stable isotope compositions and water contents of boninite series volcanic rocks from Chichi-Jima, Bonin Islands, Japan. *Earth and Planetary Science Letters* **82**, 75–86.

Eiler, J. M. (2001). Oxygen isotope variations of basaltic lavas and upper mantle rocks. In *Stable Isotope Geochemistry,* Vol. 43 (ed. J. W. Valley and D. R. Cole), pp. 319–364. Washington, DC: Mineralogical Society of America.

Eiler, J. M., Farley, K. A., Valley, J. W., Stolper, E. M., Haurl, E. H., and Craig, H. (1995). Oxygen isotope evidence against bulk recycled sediment in the mantle sources of Pitcairn Island lavas. *Nature* **377**, 138–141.

Exley, R. A., Mattey, D. P., Clague, D. A., and Pillinger, C. T. (1986). Carbon isotope systematics of a mantle "hotspot": A comparison of Loihi Seamount and MORB glasses. *Earth and Planetary Science Letters* **78**, 189–199.

Farquhar, J., Wing, B. A., McKeegan, K. D., Harris, J. W., Cartigny, P., and Thiemens, M. H. (2002). Mass-independent sulfur of inclusions in diamond and sulfur recycling on early Earth. *Science* **298**, 2369–2372.

Feeley, T. C., and Sharp, Z. D. (1996). Chemical and hydrogen isotope evidence for in situ dehydrogenation of biotite in silicic magma chambers. *Geology* **24**, 1021–1024.

Feldstein, S. N., Lange, R. A., Vennemann, T., and O'Neil, J. R. (1996). Ferric–ferrous ratios, H_2O contents and D/H ratios of phlogopite and biotite from lavas of different tectonic regimes. *Contributions to Mineralogy and Petrology* **126**, 51–66.

Field, C. W., and Fifarek, R. H. (1985). Light stable-isotopes systematics in the epithermal environment. *Reviews in Economic Geology* **2**, 99–128.

Fischer, T. P., Hilton, D. R., Zimmer, M. M., Shaw, A. M., Sharp, Z. D., and Walker, J. A. (2002). Contrasting nitrogen isotope behavior along the Central America margin: Implications for the nitrogen balance of the Earth. *Science* **297**, 1154–1157.

Forester, R. W., and Taylor, H. P., Jr. (1977). $^{18}O/^{16}O$, D/H and $^{13}C/^{12}C$ studies of the Tertiary igneous complex of Skye, Scotland. *American Journal of Science* **277**, 136–177.

Garlick, G. D. (1966). Oxygen isotope fractionation in igneous rocks. *Earth and Planetary Science Letters* **1**, 361–368.

Giggenbach, W. F. (1992). Isotopic shifts in waters from geothermal and volcanic systems along convergent plate boundaries and their origin. *Earth and Planetary Science Letters* **113**, 495–510.

Gilliam, C. E., and Valley, J. W. (1997). Low $\delta^{18}O$ magma, Isle of Skye, Scotland: Evidence from zircons. *Geochimica et Cosmochimica Acta* **61**, 4975–4981.

Gregory, R. T., and Criss, R. E. (1986). Isotopic exchange in open and closed systems. In *Stable Isotopes in High Temperature Geological Processes,* Vol. 16 (ed. J. W. Valley, H. P. Taylor Jr., and J. R. O'Neil), pp. 91–128. Chelsea, MI: Mineralogical Society of America.

Grunder, A. L. (1987). Low $\delta^{18}O$ silicic volcanic rocks at the Calabozos Caldera Complex, Southern Andes; evidence for upper-crustal contamination. *Contributions to Mineralogy and Petrology* **95**, 71–81.

Harmon, R. S., and Hoefs, J. (1995). Oxygen isotope heterogeneity of the mantle deduced from global ^{18}O systematics of basalts from different geotectonic settings. *Contributions to Mineralogy and Petrology* **120**, 95–114.

Hart, S. R., Hauri, E. H., Oschmann, L. A., and Whitehead, J. A. (1992). Mantle plumes and entrainment; isotopic evidence. *Science* **256**, 517–520.

Hoefs, J. (1973). Ein Beitrag zur Isotopengeochemie des Kohlenstoffs in magmatischen Gesteinen. *Contributions to Mineralogy and Petrology* **41**, 277–300.

Hulston, J. R., and Thode, H. G. (1965). Variations in the S^{33}, S^{34}, and S^{36} contents of meteorites and their relation to chemical and nuclear effects. *Journal of Geophysical Research* **70**, 3475–3484.

Ito, E., White, W. M., and Goepel, C. (1987). The O, Sr, Nd and Pb isotope geochemistry of MORB. *Chemical Geology* **62**, 157–176.

Javoy, M. (1980). $^{18}O/^{16}O$ and D/H ratios in high temperature peridotites. *Colloques Internationaux du C.N.R.S.* **272**, 279–287.

Javoy, M. (1997). The major volatile elements of the Earth: Their origin, behavior, and fate. *Geophysical Research Letters* **24**, 177–180.

Kerridge, J. F. (1985). Carbon, hydrogen and nitrogen in carbonaceous chondrites; abundances and isotopic compositions in bulk samples. *Geochimica et Cosmochimica Acta* **49**, 1707–1714.

King, E. M., Barrie, C. T., and Valley, J. W. (1997). Hydrothermal alteration of oxygen isotope ratios in quartz phenocrysts, Kidd Creek mine, Ontario: Magmatic values are preserved in zircon. *Geology* **25**, 1079–1082.

Kung, C. C., and Clayton, R. N. (1978). Nitrogen abundances and isotopic compositions in stony meteorites. *Earth and Planetary Science Letters* **38**, 421–435.

Kuroda, Y., Suzuoki, T., and Matsuo, S. (1977). Hydrogen isotope composition of deep-seated water. *Contributions to Mineralogy and Petrology* **60**, 311–315.

Kyser, T. K. (1986). Stable isotope variations in the mantle. In *Stable Isotopes in High Temperature Geological Processes,* Vol. 16 (ed. J. W. Valley, H. P. Taylor, and J. R. O'Neil), pp. 141–164. Chelsea, MI: Mineralogical Society of America.

Kyser, T. K., and O'Neil, J. R. (1984). Hydrogen isotope systematics of submarine basalts. *Geochimica et Cosmochimica Acta* **48**, 2123–2133.

Kyser, T. K., O'Neil, J. R., and Carmichael, I. S. E. (1982). Genetic relations among basic lavas and ultramafic nodules; evidence from oxygen isotope compositions. *Contributions to Mineralogy and Petrology* **81**, 88–102.

MacGregor, I. D., and Manton, W. I. (1986). Roberts Victor eclogites; ancient oceanic crust. *Journal of Geophysical Research, B* **91**, 14063–14079.

Marty, B., and Dauphas, N. (2003). The nitrogen record of crust–mantle interaction and mantle convection from Archean to Present. *Earth and Planetary Science Letters* **206**, 397–410.

Marty, B., and Humbert, F. (1997). Nitrogen and argon isotopes in oceanic basalts. *Earth and Planetary Science Letters* **152**, 101–112.

Matsuhisa, Y., Goldsmith, J. R., and Clayton, R. N. (1979). Oxygen isotopic fractionation in the system quartz–albite–anorthite–water. *Geochimica et Cosmochimica Acta.* **43**, 1131–1140.

Mattey, D., Lowry, D., and Macpherson, C. (1994). Oxygen isotope composition of mantle peridotite. *Earth and Planetary Science Letters* **128**, 231–241.

Mohapatra, R. K., and Murty, S. V. S. (2004). Nitrogen isotopic composition of the MORB mantle: A reevaluation. *Geochemistry, Geophysics, Geosystems* **5**, doi:10.1029/2003GC000612.

Nabelek, P. I., O'Neil, J. R., and Papike, J. J. (1983). Vapor phase exsolution as a controlling factor in hydrogen isotope variation in granitic rocks; the Notch Peak granitic stock, Utah. *Earth and Planetary Science Letters* **66**, 137–150.

O'Neil, J. R., and Chappell, B. W. (1977). Oxygen and hydrogen isotope relations in the Berridale batholith. *Journal of the Geological Society of London* **133**, 559–571.

Ongley, J. S., Basu, A. R., and Kyser, T. K. (1987). Oxygen isotopes in coexisting garnets, clinopyroxenes and phlogopites of Roberts Victor eclogites; implications for petrogenesis and mantle metasomatism. *Earth and Planetary Science Letters* **83**, 80–84.

Pineau, F., and Javory, M. (1983). Carbon isotopes and concentrations in mid-oceanic ridge basalts. *Earth and Planetary Science Letters* **62**, 239–257.

Perkins, G. B., Sharp, Z. D., and Selverstone, J. (2006). Oxygen isotope evidence for subduction and rift-related mantel metasomatism beneath the Colorado Plateau–Rio Grande rift transition. *Contributions to Mineralogy and Petrology* (in press).

Poreda, R. (1985). Helium-3 and deuterium in back-arc basalts; Lau Basin and the Mariana Trough. *Earth and Planetary Science Letters* **73**, 244–254.

Reuter, J. H., Epstein, S., and Taylor, H. P., Jr. (1965). O^{18}/O^{16} ratios of some chondritic meteorites and terrestrial ultramafic rocks. *Geochimica et Cosmochimica Acta* **29**, 481–488.

Richet, P., Bottinga, Y., and Javoy, M. (1977). A review of hydrogen, carbon, nitrogen, oxygen, sulphur, and chlorine stable isotope fractionation among gaseous molecules. *Annual Review of Earth and Planetary Science* **5**, 65–110.

Sakai, H., DesMarais, D. J., Ueda, A., and Moore, J. G. (1984). Concentrations and isotope ratios of carbon, nitrogen and sulfur in ocean-floor basalts. *Geochemica et Cosmochimica Acta* **48**, 2433–2441.

Sano, Y. N., Takahata, N., Nishio, Y., and Fischer, T. P. (2001). Volcanic flux of nitrogen from the earth. *Chemical Geology* **171**, 263–271.

Scheele, N., and Hoefs, J. (1992). Carbon isotope fractionation between calcite, graphite and CO_2: An experimental study. *Contributions to Mineralogy and Petrology* **112**, 35–45.

Sheppard, S. M. F., and Dawson, J. B. (1975). Hydrogen, carbon and oxygen isotope studies of megacryst and matrix minerals from Lesothan and South African kimberlites. *Physics and Chemistry of the Earth* **9**, 747–763.

Sheppard, S. M. F., and Epstein, S. (1970). D/H and $^{18}O/^{16}O$ ratios of minerals of possible mantle or lower crustal origin. *Earth and Planetary Science Letters* **9**, 232–239.

Shervais, J. W., Taylor, L. A., Lugmair, G., Clayton, R. N., Mayeda, T. K., and Korotev, R. (1988). Archean oceanic crust and the evolution of sub-continental mantle: Eclogites from southern Africa. *Geological Society of America Bulletin* **100**, 411–423.

Snyder, G. A., Taylor, L. A., Jerde, E. A., Clayton, R. N., Mayeda, T. K., Deines, P., Rossman, G. R., and Sobolev, N. V. (1995). Archean mantle heterogeneity and the origin of diamondiferous eclogites, Siberia; evidence from stable isotopes and hydroxyl in garnet. *American Mineralogist* **80**, 799–809.

Swart, P. K., Pillinger, C. T., Milledge, H. J., and Seal, M. (1983). Carbon isotopic variation within individual diamonds. *Nature* **303**, 793–795.

Taylor, B. E. (1986). Magmatic volatiles: Isotopic variation of C, H, and S. In *Stable Isotopes in High Temperature Geological Processes,* Vol. 16 (ed. J. W. Valley, H. P. J. Taylor, and J. R. O'Neil), pp. 185–225. Chelsea, MI: Mineralogical Society of America.

Taylor, B. E., Eichelberger, J. C., and Westrich, H. R. (1983). Hydrogen isotopic evidence of rhyolitic magma degassing during shallow intrusion and eruption. *Nature* **306**, 541–545.

Taylor, H. P., Jr. (1974). The application of oxygen and hydrogen isotope studies to problems of hydrothermal alteration and ore deposition. *Economic Geology* **69**, 843–883.

Taylor, H. P., Jr. (1978). Oxygen and hydrogen isotope studies of plutonic granitic rocks. *Earth and Planetary Science Letters* **38**, 177–210.

Taylor, H. P., Jr. (1980). The effects of assimilation of country rocks by magmas on $^{18}O/^{16}O$ and $^{87}Sr/^{86}Sr$ systematics in igneous rocks. *Earth and Planetary Science Letters* **47**, 243–254.

Taylor, H. P., Jr., and Epstein, S. (1963). O^{18}/O^{16} ratios in rocks and coexisting minerals of the Skaergaard intrusion, East Greenland. *Journal of Petrology* **4**, 51–74.

Taylor, H. P., Jr., Frechen, J., and Degens, E. T. (1967). Oxygen and carbon isotope studies of carbonatites from the Laacher See district, west Germany and the Alnö district, Sweden. *Geochimica et Cosmochimica Acta.* **31**, 407–430.

Taylor, H. P., Jr., and Sheppard, S. M. F. (1986). Igneous rocks: I. Processes of isotopic fractionation and isotope systematics. In *Stable Isotopes in High Temperature Geological Processes,* Vol. 16 (ed. J. W. Valley, H. P. J. Taylor, and J. R. O'Neil), pp. 227–271. Chelsea, MI: Mineralogical Society of America.

Taylor, H. P., Jr., and Turi B. (1976). High-^{18}O igneous rocks from the Tuscan Magmatic Province, Italy. *Contributions to Mineralogy and Petrology* **55**, 33–54.

Turi, B., and Taylor, H. P. J. (1971). An oxygen and hydrogen isotope study of a granodiorite pluton from the Southern California Batholith. *Geochimica et Cosmochimica Acta* **35**, 383–406.

van Keken, P. E., Hauri, E. H., and Ballentine, C. J. (2002). Mantle mixing: The generation, preservation, and destruction of chemical heterogeneity. *Annual Review of Earth and Planetary Science* **30**, 493–525.

Viljoen, K. S., Smith, C. B., and Sharp, Z. D. (1996). Stable and radiogenic isotope study of eclogite xenoliths from the Orapa kimberlite, Botswana. *Chemical Geology* **131**, 235–255.

Wenner, D. B., and Taylor, H. P., Jr. (1976). Oxygen and hydrogen isotope studies of a Precambrian granite–rhyolite terrane, St. Francois Mtns., S.E. Missouri. *Geological Society of America Bulletin* **87**, 1587–1598.

Wilson, M. R., Kyser, T. K., and Fagan R. (1996). Sulfur isotope systematics and platinum group element behavior in REE-enriched metasomatic fluids; a study of mantle xenoliths from Dish Hill, California, USA. *Geochimica et Cosmochimica Acta* **60**, 1933–1942.

Woodhead, J. D., Greenwood, P., Harmon, R. S., and Stoffers, P. (1993). Oxygen isotope evidence for recycled crust in the source of EM-type ocean island basalts. *Nature* **362**, 809–813.

Xia, Q.-K., Deloule, E., Wu, Y.-B., Chen, D.-G., and Cheng, H. (2002). Anomalously high δD values in the mantle. *Geophysical Research Letters* **29**, 4-1–4-4.

Zindler, A., and Hart, S. (1986). Chemical Geodynamics. *Annual Review of Earth and Planetary Sciences* **14**, 493–571.

CHAPTER 12

METAMORPHIC GEOLOGY

12.1 INTRODUCTION

Early advances in the stable isotope geochemistry of rocks were certainly not limited to carbonates. In remarkable pioneering studies, Baertschi and Silverman (Baertschi, 1950; Baertschi and Silverman, 1951; Silverman, 1951) were able to analyze the oxygen isotope composition of silicates in metamorphic and igneous rocks, recognizing that significant differences existed between different rock types. But after these early publications, further work on high-temperature rocks languished for the better part of a decade, during which time major breakthroughs were being made in hydrogen, carbon, oxygen, sulfur, and nitrogen isotope studies of carbonates, meteoric water, organic matter, etc. The stagnation was partly related to analytical difficulties having to do with fluorination and partly to the idea that the range of oxygen isotope values in rocks equilibrated at high temperatures was simply too small to be of interest.

Then, in 1958, Clayton and Epstein measured the oxygen isotope values of quartz, calcite, and magnetite from various metamorphic rocks and found that the fractionations varied systematically with temperature of formation. Follow-up studies made it apparent that predictable oxygen isotope fractionations exist between minerals which vary with temperature (Clayton and Epstein, 1961; James and Clayton, 1962; Taylor and Epstein, 1962a; Taylor and Epstein, 1962b; Taylor et al., 1963; O'Neil and Clayton, 1964). The field of stable isotope geochemistry in metamorphic studies was on its way.

Metamorphic rocks form by recrystallization of an existing protolith in response to changes in pressure, temperature, and fluid composition. During metamorphism, processes such as deformation, volatilization, solid–solid mineral reactions and metasomatism occur at time scales that can be in excess of 100 My. The thermodynamic driving force of these chemical processes is far larger than that of isotopic substitution, so stable isotope geochemistry passively follows the more energetic mineral reactions, and not the reverse. The application of stable isotope geochemistry to metamorphic rocks requires special considerations, but also offers unique opportunities for retrieving information about metamorphic phenomena. The problems of rock protoliths, fluid sources and fluid/rock ratios, mineral reaction progress, extent of isotopic exchange between contrasting lithologies, metamorphic temperatures and cooling rates have all been addressed with the methods of stable isotope geochemistry.

12.2 STABLE ISOTOPES AS GEOCHEMICAL TRACERS

12.2.1 Closed System: Protolith Identification and Alteration

In the simplest case, a rock behaves as a closed system during metamorphism, and the bulk rock isotopic value is nearly unchanged from that of the protolith. For example, quartz sand generally has $\delta^{18}O$ values ranging from 13 to 18‰, whereas cherts have $\delta^{18}O$ values well over 30‰. The protolith of a highly recrystallized quartzite could therefore be determined from the measured $\delta^{18}O$ value (assuming that the system remained closed to fluid infiltration). As an example, Vennemann et al. (1992) measured the $\delta^{18}O$ values of quartz pebbles from the gold- and uranium-bearing conglomerates from the Witwatersrand district of South Africa. The meta-chert pebbles have a $\delta^{18}O$ value of 9 to 11.5‰, far lower than expected from a marine chert. Accordingly, the authors proposed that, instead, auriferous iron formations and exhalatives were the source of the ore-bearing pebbles.

Attempts have been made to distinguish orthogneisses from paragneisses on the basis of their $\delta^{18}O_{whole\ rock}$ values. Paragneisses are derived from a mixture of carbonates and pelitic sediments and should have higher $\delta^{18}O$ values than metamorphosed granitoids. Schwarcz and Clayton (1965) were able to clearly distinguish ortho- and para-amphibolites in one well-constrained example from Ontario, but in other localities the isotopic compositions were ambiguous. They attributed the similar $\delta^{18}O$ values to reequilibration with external buffering fluids during metamorphism. Later studies met with only mixed success. In a reconnaissance study of 10,000 km² of Archean Canadian Shield, the $\delta^{18}O$ values of ortho- and paragneisses were found to be indistinguishable (Shieh and Schwarcz, 1977). The authors concluded that the rocks had exchanged with a metamorphic fluid or, more likely, that the sedimentary protolith to the paragneisses consisted essentially of detrital igneous material which had the same isotopic composition as its precursor.

Marbles cover a broad range of $\delta^{18}O$ values, but cluster around 20–25‰ (SMOW), typical of diagenetically altered limestones. Granulite-facies marbles may have slightly lower $\delta^{18}O$ values (Baertschi, 1951), and siliceous marbles and calc–silicates are the lowest of all, due to their higher permeability than pure marbles and the contribution of a low-$\delta^{18}O$ clastic component (Fig 12.1). The very low $\delta^{13}C$ values of siliceous carbonates can be explained in terms of higher permeability and interaction with organic matter. Carbonatites are of igneous origin and have $\delta^{18}O$ values in the range of those of typical magmatic rocks (6 to 10‰) and $\delta^{13}C$ values thought to represent the mantle (-8 to -4‰) (Taylor et al., 1967). The origin of marbles—either carbonatitic or biogenic—can be constrained with the use of stable isotope geochemistry.

The origin of sulfur in metamorphosed ore deposits can also be addressed with sulfur isotope geochemistry. Whereas thermochemical reduction of seawater sulfate will result in $\delta^{34}S$ values that are far greater than 0‰, the reaction of metalliferous fluids with magmatic sulfide (e.g., sulfide delivered from hydrothermal vents at Mid Ocean Ridges—so called seafloor hydrothermal fluids) have values averaging $+1.5$‰. High-spatial-resolution analyses

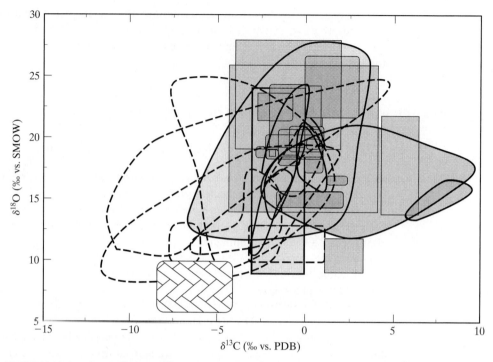

FIGURE 12.1: Compilation of literature on marbles. There is a clustering of $\delta^{18}O$ and $\delta^{13}C$ values around 20‰ (SMOW) and 0‰ (PDB), respectively. Siliceous marbles and calc–silicates (shown as dashed lines) have lower $\delta^{18}O$ and, especially, $\delta^{13}C$ values. Also shown is the igneous carbonatite box (hatched area). (Each shaded area represents the range of values from an individual publication).

of individual sulfides from a seafloor hydrothermal spire, Axial Seamount, Juan de Fuca Ridge, in the eastern Pacific, cover a 5‰ interval, requiring mixing of hydrothermal sulfide with seawater sulfate (Crowe and Valley, 1992).

The source(s) of sulfur in metamorphosed ore deposits can be deduced from careful stable isotope analyses. In a sulfur isotope study of the Canadian Sullivan Pb–Zn–Ag sedimentary exhalative (Sedex) deposit, which was metamorphosed to upper-greenschist conditions, the author concluded that thermogenic sulfate reduction, bacterial sulfate reduction, and remobilization of preexisting sulfide grains all contributed to the sulfur in the deposit (Taylor, 2004).

12.2.2 Open Systems: Volatilization and Fluid Infiltration Processes

In an *open system*, bulk rock isotopic composition can change as a result of fluid or melt entering and leaving the system. Open-system behavior is common, occurring over the full range of geological conditions. Examples include carbonate diagenesis, in which the infiltration of meteoric waters and the expulsion of pore waters occur during compaction and recrystallization; the incorporation of wall rock by ascending magmas; and hydrothermal activity following the emplacement of a shallow-level pluton. There is no limit on how much fluid may pass through a rock, but there are limits on how much internally generated fluids can be lost to the surroundings. In the absence of a fluid or melt *entering* the rock, the bulk chemistry of the rock limits how much material can leave the system and, consequently, how much the isotopic composition of a rock can change. Examples of fluid loss by dewatering, dehydration, and decarbonation are given in the remainder of this section.

Above subgreenschist facies, loss of fluids takes place by volatilization or decarbonation reactions. The isotopic shifts caused by volatilization reactions depend on the bulk composition of the rock and the *P–T* path followed during metamorphism. Fluids released during volatilization reactions may escape either through a fracture network or, in cases of high permeability, pervasively. Two end-member models can be constructed for volatilization–decarbonation reactions, with natural examples almost certainly lying between the two extremes (Valley, 1986). In the first model, termed *batch* volatilization, fluid is evolved and remains in the system for the entirety of the volatilization episode. Equilibrium between fluid and rock is maintained until the reaction is complete, at which time the fluid escapes. While episodic fluid loss almost certainly occurs, the retention of *all* fluid until decarbonation or volatilization reactions go to completion can be realized only in cases where permeability is low and the total fluid production is minimal. The second end-member case is *Rayleigh* volatilization, in which fluid leaves the system as it is formed; this model is analogous to the Rayleigh fractionation models for precipitation (Chapter 4). By continuous removal of infinitely small "packets" of fluid that are in equilibrium with the rock, the isotopic composition of the rock and the bulk composition of the residual rock–fluid system change continuously.

The batch volatilization model relies on calculations of equilibrium fractionation between all species on the basis of simple mass-balance constraints. (See Box 12.1.) When fluid leaves the system, the change in the isotopic composition of the rock can be calculated by

$$\delta_{\text{initial rock}} = \delta_{\text{final rock}}(x) + \delta_{\text{fluid}}(1 - x), \tag{12.1}$$

where x is the mole fraction of the element (e.g., oxygen) remaining in the rock during dehydration or volatilization and δ_{fluid} is the composition of newly formed fluid in equilibrium with the final rock. The isotopic composition of the final rock is related to that of the fluid by the equation

$$\delta_{\text{final rock}} - \delta_{\text{fluid}} \approx 1000 \ln \alpha_{r-f}, \tag{12.2}$$

where α_{r-f} is the average fractionation between the final rock and fluid at the temperature of interest. Equation (12.1) can be expressed as

$$\delta_{\text{fluid}} = \delta_i - (x)1000 \ln \alpha_{r-f}. \tag{12.3}$$

If the $\delta^{18}O$ values of both the initial and final rock can be determined, and $1000 \ln \alpha_{\text{final rock-fluid}}$ is estimated, then both the amount and the $\delta^{18}O$ value of fluid lost from the system can be calculated. Generally, α is referred to the stoichiometrically weighted average fractionation between the fluid and rock (Valley, 1986), although this is not a precise solution of equation (12.3). The fractionation factor changes in relation to the changing proportions of the solid components and with temperature (Rumble, 1982). Nevertheless, the approximation of a constant α is generally sufficient to model fluid loss.

> **BOX 12.1 Mass (material) balance relationships; the lever rule**
>
> The concept of material balance is fundamental in all geochemical applications, but is not well understood by many students. The isotopic compositions of smaller reservoirs change more than those of larger reservoirs in a system involving the interaction between two or more reservoirs of the same element. Removal of a vapor from the ocean will not measurably change the composition of the ocean, even though the light isotopes are preferentially incorporated into the vapor phase. The ocean must get heavier because light material is lost. But because there is so much more ocean than vapor, the increase is infinitely small. If we take enough vapor out of the ocean (i.e., via the formation of ice caps), we *will* begin to see a change in the $\delta^{18}O$ (or δD) value of the ocean. In this case, the abundance of the lesser component—the ice caps, which make up over 2% of the hydrosphere—begins to approach that of the major component—the ocean (97.2%). In the same way, adding a small amount of hydrothermal fluid to a rock will not change the $\delta^{18}O$ value of the rock appreciably. The fluid–rock ratio is simply too low. Only when a great deal of fluid interacts with the rock do we begin to see a change in the $\delta^{18}O$ (rock) value.
>
> The δ value of a binary mixture a and b is given by
>
> $$\delta(\text{mixture}) = \frac{N_a\delta_a + N_b\delta_b}{(N_a + N_b)} = X_a\delta_a + X_b\delta_b,$$
>
> where N is the abundance of the element (e.g., carbon, oxygen) and X is the mole fraction. The general equation for n components is
>
> $$\delta(\text{system}) = \sum_{i=1}^{n} X_i\delta_i.$$
>
> The molar proportion of oxygen in metamorphic rock is >50%. Hydrogen, by contrast, is present only in minor hydrous phases. The infiltration of an aqueous fluid will cause a larger change in the δD values than in the $\delta^{18}O$ values because of the mass-balance rule. Only when the fluid–rock ratio gets very high will the $\delta^{18}O$ values of the rock be affected.
>
> Finally, there is a common misunderstanding that the fluid phase in a metamorphic rock is an important oxygen reservoir. That is not at all the case: The amount of fluid normally present along grain boundaries and fluid inclusions is too small to be of much consequence *as a fluid reservoir*. Therefore, a fluid can change the $\delta^{18}O$ value of a mineral appreciably only if there is a large flux of fluid passing through the rock.
>
> Keeping a simple mental image of the mass-balance rule is very useful in interpreting isotopic results. "Is it reasonable on mass-balance considerations?" is a question that we should always keep in the back of our minds.

The Rayleigh model, or *open-system* volatilization model, will be approached if permeability is high and the P–T conditions result in a strong buoyancy of the expelled fluids. Lord Rayleigh derived the solution of the distillation equation (see Chapter 4) and applied it to the separation of different gases by diffusion through porous media. A modification of Rayleigh's equation, applicable to changes in the isotopic composition of a phase undergoing vaporization (or volatilization), is

$$\delta_f - \delta_i = (1000 + \delta_i)(F^{(\alpha-1)} - 1), \tag{12.4}$$

where δ_f and δ_i are, respectively, the final and initial δ values of the rock, α is the fractionation between rock and fluid, and F is the variable defined by the fraction of the element remaining; F ranges from 1 (unreacted) to 0 (all of the element is lost to the vapor or fluid phase). That is, F equals 1 prior to any removal of the element of interest and approaches 0 as all of the element is removed from the system. Equation (12.4), often called the *Rayleigh equation*, is a perfect solution for isothermal reactions where there is one reactant and one product, such as the decarbonation reaction (12.5) in terms of carbon, or the transformation of water to water vapor at constant temperature (Chapter 4). The equation becomes far more complicated where there are multiple reactants and products or changing temperatures, because the fractionation factor α changes with temperature and as a function of the relative proportions of the products and reactants (Young, 1993). Nonetheless, an average α value adequately approximates most geological situations.

The end-member batch and Rayleigh fractionation model paths are illustrated in Fig. 12.2. Several fundamental consequences are apparent from the figure. First, for moderate degrees of volatilization, the difference between batch and Rayleigh behavior is nearly indistinguishable. Only when the reactions are taken to near completion do the Rayleigh and batch curves diverge significantly. Second, the magnitude by which volatilization reactions affect the isotopic composition of a system depends on the reactions that occur, the mineral mode, the fractionation α, and the P–T conditions of the system. (See Valley, 1986.) However, there are fundamental limits as to how much oxygen can be removed from the system by decarbonation or dehydration, and thus how far along the Rayleigh fractionation curve a reaction can proceed. For a simple decarbonation reaction such as

$$\underset{\text{calcite}}{CaCO_3} + \underset{\text{quartz}}{SiO_2} \longrightarrow \underset{\text{wollastonite}}{CaSiO_3} + CO_2, \tag{12.5}$$

a maximum of 40% of the oxygen present in the reactants is lost to the fluid phase, so F cannot decrease to below 0.6, termed the *calc–silicate limit*.[1] This limit is, in fact, rarely reached. It requires that the reactants are in perfect stoichiometric proportions and that the reaction goes to completion. It is clear from Fig. 12.2 that decarbonation reactions alone cannot change the bulk rock $\delta^{18}O$ value of a rock by more than several per mil.

The maximum F value for oxygen loss from common dehydration reactions is far less than from decarbonation reactions. As an example, consider the following two volatilization reactions involving muscovite and talc:

$$\underset{\text{muscovite}}{KAl_3Si_3O_{10}(OH)_2} + \underset{\text{quartz}}{SiO_2} = \underset{\text{microcline}}{KAlSi_3O_8} + \underset{\text{kyanite}}{Al_2SiO_5} + H_2O \tag{12.6}$$

and

$$\underset{\text{talc}}{Mg_3Si_4O_{10}(OH)_2} = 3\,\underset{\text{enstatite}}{MgSiO_3} + \underset{\text{quartz}}{SiO_2} + H_2O. \tag{12.7}$$

F values cannot go below 0.93 and 0.92, respectively. An extreme example that is rarely encountered is the dehydration of brucite or portlandite, in which F can approach 50%. The maximum shift in the $\delta^{18}O_{WR}$ value due to dehydration will generally be less than 1‰. The bottom line is this: **Volatilization reactions in metamorphic rocks cannot change the**

---

[1] A maximum of 66% O_2 loss is possible for the reaction $CaCO_3 \rightarrow CaO + CO_2$, but this reaction is extremely rare.

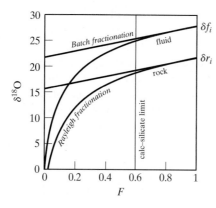

FIGURE 12.2: Changes in $\delta^{18}O$ value of rock (δr) and evolving CO_2 fluid (δf) for equation (12.4). with $\alpha = 1.006$. For high values of F, the batch and Rayleigh fractionation models have the same effect. Only for low F values (reactions proceeding to near completion) do the batch and Rayleigh curves diverge. For oxygen, the maximum degree of reaction for decarbonation reactions is given by the calc–silicate limit: F values cannot be less than 0.6.

$\delta^{18}O_{WR}$ **values of a lithology by more than several per mil under any conditions.** Larger shifts in $\delta^{18}O_{WR}$ values in geologic studies require the infiltration of a fluid or melt.

The effect of volatilization reactions on the isotopic composition of minor elements, such as hydrogen and carbon, can be much larger. In the case of these two elements, volatilization reactions can proceed until F approaches zero *for that element*, as is the case with reaction (12.5) for carbon. Under such circumstances, there are no limits on how much the δ value of the element is changed.

In typical pelites, a series of dehydration reactions occurs with increasing temperatures until, by the time granulite-facies conditions are reached, nearly all hydrogen is driven from the rock. Hoernes and Hoffer (1979) found that the δD values of biotites in a prograde metamorphic sequence from Namibia decrease from $-50‰$ at rocks metamorphosed at $\sim 480°C$ to $-80‰$ for rocks metamorphosed at $650°C$. The decreasing δD values can be explained by the removal of water with high δD values during volatilization. In theory, as F values approach zero, delta values will decrease asymptotically towards $-1,000‰$ under Rayleigh fractionation conditions. Obviously, such extreme lowering is never seen in metamorphic systems, but examples of Rayleigh volatilization trends for carbon isotope ratios are commonly observed in carbonate rocks, as discussed in the next section.

Nabelek et al. (1984) measured the $\delta^{18}O$ and $\delta^{13}C$ values of argillaceous sediments intruded by a granitic stock. The $\delta^{13}C$ values decreased by almost $12‰$ as the intrusion was approached, concomitant with a decrease in the percentage of carbonate. All data fit inside an envelope defined by the end-member batch and Rayleigh decarbonation models (Fig. 12.3a), suggesting that the carbon isotope changes can be explained entirely by decarbonation reactions. In contrast, the low $\delta^{18}O$ values cannot be explained by decarbonation alone. Decarbonation would lower the $\delta^{18}O$ value of the sediments by a maximum of only $2‰$, far less than the $12‰$ actually measured (Fig. 12.3b). The $\delta^{18}O$ values falling below $18‰$ require open-system behavior, explained by a low-$\delta^{18}O$ fluid from the crystallizing granite infiltrating the permeable sediments.

The study of the Elkhorn contact aureole serves as a second example of complex volatilization and fluid infiltration behavior (Bowman et al., 1985). Changes in $\delta^{13}C$ and $\delta^{18}O$ values during decarbonation are calculated for both the batch and Rayleigh equations (curves D–D′ and D–R, respectively, in Fig. 12.4). Neither model reproduces the measured $\delta^{13}C$ and $\delta^{18}O$ values of the metamorphosed carbonates, shown as filled black circles. Bowman et al. proposed a two-stage model, incorporating partial decarbonation along a Rayleigh path

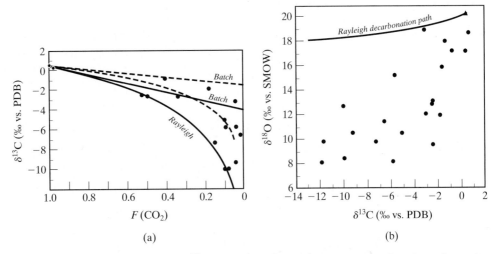

FIGURE 12.3: (a) Changes in $\delta^{13}C$ values of carbonate layers as a function of reaction progress for argillaceous sediments surrounding the Notch Peak Stock, Utah. Dashed lines are for $\alpha_{CO_2\text{-}cc} = 1.0025$; solid lines represent $\alpha_{CO_2\text{-}cc} = 1.0045$. All data fall between the end-member batch and Rayleigh fractionation models, but not for any one value of α. (b) In contrast to the carbon data, $\delta^{18}O$ values near the stock are too low to be explained by Rayleigh fractionation during decarbonation reactions (curve). Instead, light fluids emanating from the stock must have infiltrated the permeable argillaceous sediments. (After Nabelek et al., 1984.)

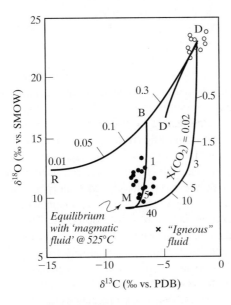

FIGURE 12.4: $\delta^{13}C$ vs. $\delta^{18}O$ values for the Elkhorn contact aureole, Montana, United States (Bowman et al., 1985). The unmetamorphosed limestone plots at D. Batch and Rayleigh volatilization curves are shown by trajectories D–D′ and D–R, respectively. The trajectory for mixing with a magmatic fluid is given by D–M. The metamorphosed calcites can only be fit with a two-stage model—first Rayleigh volatilization (D–B), followed by fluid infiltration (B–M).

(curve D–B) followed by mixing with a magmatic fluid (curve B–M). The data in Fig. 12.3b plot in a similar field, although with more scatter, in support of the two-stage hypothesis.

Valley (1986) plotted trends for 16 studies of contact metamorphism, along with the predicted trends for pure volatilization. None of the coupled $\delta^{13}C$–$\delta^{18}O$ trends can be explained purely in terms of volatilization. In all cases, infiltration of an external fluid must be invoked.

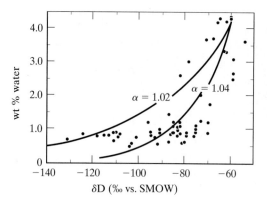

FIGURE 12.5: Water content vs. δD value for the prograde metamorphic sequence of pelitic rocks, Duluth Complex, Minnesota, United States. Lower δD values are found in higher grade samples with lower water contents, consistent with a Rayleigh fractionation associated with loss of water. (*Note*: Different types of rock are not differentiated.) Rayleigh fractionation curves are calculated for $\alpha_{water-rock}$ = 1.02 and 1.04. (Modified from Ripley, 1992.)

A relationship between $\delta D_{mineral}$ and total wt % water consistent with Rayleigh volatilization has been observed in crystallizing magmas and in contact metamorphic rocks (Nabelek et al., 1983; Ripley et al., 1992). In a study of pelitic rocks in the contact aureole of the Duluth Complex, Minnesota, Ripley et al. (1992) found hydrogen isotope variations of over 70‰ which were well correlated with total water content (Fig. 12.5). The authors calculated the theoretical δD vs. water content trends for different mixed CH_4–H_2O fluids at a variety of temperatures and concluded that the samples underwent Rayleigh fractionation coupled with an influx of fluid containing 0 to 15 mole % methane.

12.3 FLUID SOURCES AND FLUID–ROCK INTERACTION

12.3.1 Oxygen and Hydrogen

Through a combination of oxygen, hydrogen, and carbon isotope analyses, it is often possible to constrain, or even unambiguously define, the origin of an infiltrating fluid. To place numerical constraints on the *amount* of an infiltrating fluid, two conditions must be met: First, the composition of the infiltrating fluid must be different from that of the unaltered rock; second, the isotopic composition of the infiltrating fluid and unaltered rock must be known.

A general compartmentalization of the hydrogen–oxygen isotope composition of fluids from different sources is given in Fig. 12.6. The isotopic compositions of meteoric waters, ocean water, and brines have been measured directly. Metamorphic and igneous fluids are generally inferred. Direct measurements of metamorphic and igneous waters can be made on fumaroles and fluid inclusions, but this is rare. More often, the isotopic composition of a metamorphic fluid is calculated from the measured isotopic composition of solid phases and an assumed equilibrium fluid–rock fractionation (Sheppard, 1986).

Meteoric and ocean waters have distinct $\delta^{18}O$ values: Ocean waters have unambiguously high δD values, whereas meteoric waters may have extremely low δD and $\delta^{18}O$ values, depending upon latitude and elevation. Meteoric fluid–rock interactions occur prior to metamorphism (Fourcade and Javoy, 1973; Valley and O'Neil, 1982; Sturchio and Muehlenbachs, 1985; Yui et al., 1995), during peak metamorphism (Fricke et al., 1992), as a retrograde feature (Frey et al., 1976; Mora and Valley, 1991; Jenkin et al., 1992; Jenkin, Craw, and Fallick, 1994), and in contact metamorphic rocks (Bowman et al., 1985; Jamtveit et al., 1992). It is reasonable to state with some certainty that a metamorphic rock interacted with meteoric waters at some time in its past if the whole rock $\delta^{18}O$ or δD values are less than 3 to 4‰ and −100‰, respectively.

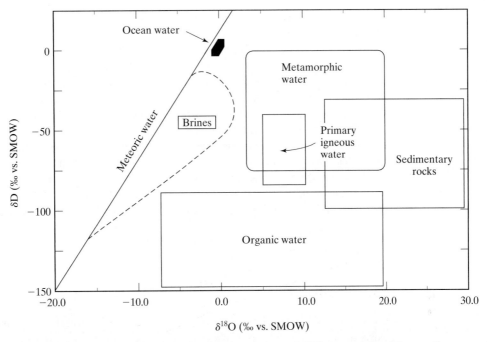

FIGURE 12.6: "Generic" isotopic fields for waters of different origins. The small range of ocean water is explained by changes in volume of the ice pack over time. Meteoric waters become lighter at higher latitudes and altitudes. Primary igneous waters are derived from the mantle, with some degree of modification due to vapor loss. Organic water is related to the oxidation of organic material, but is a minor constituent in metamorphic rocks. Metamorphic waters are buffered by metamorphic rocks. Because of the extreme variety of metamorphic rock types, the metamorphic box is large. The sedimentary box is *not* sedimentary fluids, but rather the $\delta^{18}O–\delta D$ values of the rock itself. During sedimentation in the marine environment, the $\delta^{18}O–\delta D$ values are near zero (seawater). They increase during metamorphism as the fractionations between rock and fluid decrease. (Diagram after Sheppard, 1986.)

Probably the most striking example of meteoric water alteration in a metamorphic rock comes from the work of Yui et al. (1995) on the ultrahigh-pressure coesite–diamond-bearing rocks of the Dabie and Sulu terranes of eastern China. These rocks were buried to depths far in excess of 100 km. The $\delta^{18}O$ values of all minerals are in equilibrium for the expected metamorphic temperatures. What is so remarkable are the $\delta^{18}O$ values themselves: They range from -10.4 to $-9.0‰$ and -10.2 to $-7.3‰$ in eclogites and quartzites, respectively. These are certainly some of the lowest $\delta^{18}O$ values ever found in metamorphic rocks. The authors suggest that the rocks underwent *high-latitude or high-altitude hydrothermal alteration with meteoric waters prior to subduction and metamorphism*. This hypothesis requires that the system remained closed to external fluids throughout the extended journey from the surface to depths and back to the surface again.

Waters of oceanic origin have been recognized, the most definitive line of evidence being the uniquely high δD values. Wickham and Taylor (1985) found isotopic evidence for large-scale seawater infiltration to depths of 6–12 km. Homogeneous $\delta^{18}O$ values and very high δD values of the muscovite $(-25‰)$ point to an oceanic source.

The distinct isotopic signature of ocean fluids can be traced throughout a rock's metamorphic or tectonic history. Oceanic hydrothermal alteration produces hydrous phases with high δD values. Subsequent dehydration of the hydrous phases during later subduction-related metamorphism will evolve a fluid with δD and $\delta^{18}O$ values characteristic of seawater. The high δD values (~−30) and low $\delta^{18}O$ values (8.3‰ for quartz) from some Alpine eclogites are characteristic of rocks that have undergone seafloor metamorphism (Sharp et al., 1993). Dehydration of seafloor metamorphosed rocks during subduction presumably caused the release of fluids of an ocean affinity, which then interacted with the surrounding rocks. Similar characteristic signatures of volatilization and infiltration from a subducting slab are seen at shallower levels (Bebout and Barton, 1989), where δD values as high as −25‰ are measured.

The compositional fields for waters other than those of oceanic or meteoric origin overlap and are therefore ambiguous. The primary magmatic water box represents mantle aqueous fluids that have been modified by degassing during their ascent. (See Chapter 11.) This box overlaps with the far larger metamorphic water box, the range reflecting the varied compositions of the protoliths and their temperatures of metamorphism. The metamorphic water box could be subdivided further on the basis of different rock types and water sources. For example, the high δD end of the metamorphic water box is defined by hydrothermally altered ocean basalts, while the high $\delta^{18}O$ end of the metamorphic water box is for waters that have equilibrated with high $\delta^{18}O$ metasediments.

12.3.2 Carbon

Carbon isotope systematics differ from those of oxygen in several ways. First, carbon generally exists in only two solid forms, either as carbonate minerals or as reduced graphite (or organic matter at lower grades and diamonds at ultrahigh pressures). In carbonate- and graphite-free rocks, carbon is a minor or even trace element. While the molar abundance of oxygen is relatively constant for all but the most unusual rock types, carbon concentrations range from zero up to 12 wt % (in a pure marble) and approach 100% in graphite or coal deposits. Oxygen is the major component of most aqueous metamorphic fluids, while carbon is generally a trace component of such fluids. Finally, carbon can exist in a reduced or an oxidized form in the fluid phase. The carbon isotope fractionation between CO_2 and CH_4 is huge (Fig. 12.7), so the oxidation state of the fluid will strongly affect the $\delta^{13}C$ value of a solid during fluid–rock interaction.

Carbon isotope measurements are normally made on carbonates or graphite.[2] The diffusion rate of carbon in graphite is too slow for any exchange to occur at "metamorphic"

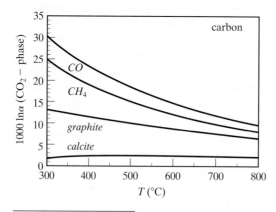

FIGURE 12.7: Carbon isotope fractionation between CO_2 and other phases. (See Fig. 11.4 for references.)

---

[2] $\delta^{13}C$ determinations have been made on other noncarbonate phases, such as cordierite (Vry et al., 1988), scapolite (Hoefs et al., 1981; Moecher et al., 1994), and fluid inclusions (Hoefs and Touret, 1975; Rye et al., 1976; Kreulen, 1980).

temperatures; carbon exchange between graphite and carbonate occurs only during the recrystallization of graphite. Calcite–graphite thermometry has been used extensively, as will be illustrated with an example in Section 12.6.3.

Decarbonation reactions can change the $\delta^{13}C$ value of a carbonate, but only if a significant fraction of the carbonate is reacted away. The insensitivity of the carbon isotope composition to decarbonation is due to the small fractionation between CO_2 and carbonate (Fig. 12.7). A maximum (positive) fractionation of 2.6‰ occurs at ~450°C, and fractionation decreases with increasing temperature. 50% decarbonation will lower the $\delta^{13}C$ value of a carbonate by less than 2‰.

Another mechanism that changes the $\delta^{13}C$ value of carbonate is exchange with an exotic CO_2- or CH_4-rich fluid (Deines and Gold, 1969; Kreulen, 1980; Pineau et al., 1981). The behavior can be quite diagnostic of fluid infiltration. The generally accepted $\delta^{13}C$ value for mantle carbon is approximately −6‰. Pineau et al. (1981) measured carbonates from deep granulites from Bamble, Norway, and found $\delta^{13}C$ values of −8.2 ± 1‰. They interpreted these low values as evidence of extensive infiltration of a mantle CO_2-rich fluid. Simple decarbonation of a typical carbonate (see Fig. 11.1) would not lead to such low $\delta^{13}C$ values.

A $\delta^{13}C$ value close to −6‰ is by no means proof of mantle origin. Any one of a number of processes, such as exchange with CO_2 derived by the oxidation of organic matter, will lower the $\delta^{13}C$ value of the carbonate (Frezzotti et al., 2000). Finally, mysteries remain as to carbon sources when $\delta^{13}C$ values have unexplained origins. Ghent and O'Neil (1985) found Precambrian marbles with $\delta^{13}C$ values of 8.9‰, which they tentatively explained as a primary, ancient sedimentary value. (See Fig. 7.15.) Deines (1968) found the record heavy sample: Carbonate inclusions in a mica peridotite that have $\delta^{13}C$ values as high as +25‰! These high values may be related to volatilization and loss of a CH_4-rich fluid leaving behind a ^{13}C-enriched residue, but the explanation is not totally satisfactory.

12.3.3 Sulfur

Sulfur isotope systematics are similar to those of carbon for the following reasons: (1) In most rocks, sulfur is a trace component occurring in both oxidized (sulfate) and reduced (sulfide) form. The concentration of sulfur in most metamorphic fluids is low, and sulfur-bearing species in the fluid can exist in a number of oxidation states.

At first glance, isolated sulfides appear to be far less "reactive" than silicates or carbonates, although the cause of this is somewhat deceiving. In a study of regionally metamorphosed graphitic sulfidic schists from south-central Maine, Oliver et al. (1992) found that the $\delta^{34}S$ values of coexisting pyrite and pyrrhotite did not approach equilibrium during metamorphism, even in rocks that had reached 500°C. Instead, the grains preserved their premetamorphic $\delta^{34}S$ values, averaging about −27‰ and decreasing only slightly, if at all, with increasing metamorphic grade. The authors had expected equilibrium to be attained on the scale of tens to hundreds of meters, on the basis of published fluid fluxes for the region and the fact that diffusion and exchange of sulfur between coexisting sulfides occurs rapidly in sulfide deposits, even at low temperatures. What Oliver et al. realized is that the absence of equilibrium on even the hand-sample scale is caused by the low sulfur concentrations in the fluid, analogous to what had been proposed by Monster et al. (1979). Although fluid flow was extensive and calculated *aqueous* fluid fluxes were high, the absence of sulfur species in the fluid resulted in a *sulfide* fluid flux that was vanishingly low. Indeed, sulfur concentrations in

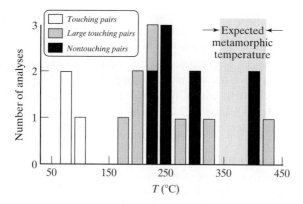

FIGURE 12.8: Temperature estimates from $\Delta^{34}S$ values of coexisting pyrrhotite–chalcopyrite pairs from the greenschist facies Rua Cove deposit, Alaska, United States. While isolated grains may preserve fractionations corresponding to peak temperatures, samples in direct contact have reequilibrated during cooling. (After Crowe, 1994.)

the fluid were simply so low that the fluids did not even act as a conduit for sulfur transport between isolated sulfide grains!

The exchangeability of sulfur during metamorphism was further refined in an elegant in situ laser study of metamorphosed volcanogenic massive sulfide deposits (Crowe, 1994). Both touching and isolated chalcopyrite and pyrrhotite grains were analyzed from a suite of greenschist- to amphibolite-grade rocks from Alaska. Sulfide grains that were completely enclosed in quartz preserved their original hydrothermal $\delta^{34}S$ values. Touching sulfide pairs were reset to less-than-peak metamorphic temperatures during retrogression (Fig. 12.8). The degree of resetting was more pronounced in small touching pairs than in large ones. Crowe concluded that the blocking temperature for sulfur isotope exchange is low and that primary hydrothermal sulfur isotope compositions are preserved only in isolated grains that could not "communicate" with each other during retrogression.

12.4 SCALES OF EQUILIBRATION DURING METAMORPHISM

Remarkably varied types of fluid movement have been identified in metamorphic terranes with the aid of stable isotope geochemistry. There are examples of fluid exchange at the kilometer scale, channeled fluid flow restricted to the centimeter scale, meteoric or seawater fluids penetrating down to depths in excess of 10 kilometers, and channeled fluids in regional terranes in tightly clustered "hot spots" or controlled by large-scale structural features or lithological boundaries. It is clear from the many different studies of metamorphic rocks that no simple rules apply to fluid flow behavior. Each terrane (and lithology) needs to considered on a case-by-case basis.

12.4.1 Regional-Scale Exchange

Taylor et al. (1963) were the first to propose large-scale isotopic homogenization attending metamorphism. They measured the $\delta^{18}O$ values of mineral separates from three different bulk assemblages in kyanite zone pelitic schists in Central Vermont, United States. Samples were collected over distances of several hundred meters from two metasedimentary units. The samples had a minimum of retrograde alteration and were all in textural equilibrium at the thin-section scale. The measured isotopic fractionations between coexisting mineral pairs in a single lithology were the same in three of the four samples, indicating that they had equilibrated under the same temperature conditions. More surprising was the fact that the $\delta^{18}O$ values of minerals common to all of the samples were nearly identical. This could be due to initial sedimentary isotopic compositions and modal abundances that coincidentally led to identical isotopic compositions at peak metamorphic conditions, or, as the authors actually proposed, a fluid phase had promoted complete oxygen isotope homogenization between the different units. They suggested that there was a pervasive metamorphic fluid that led to isotopic homogenization at the scale of more than 100 meters. One way of testing this hypothesis would be to run mineral separates from rock types that were quite different, such as metapelites and coexisting calc–pelites. In this way, one could more reliably state that the initial $\delta^{18}O$ value of the two lithologies were different, yet the similar $\delta^{18}O$ values of hornblendes, epidotes, garnets, etc., were a strong indication of fluid homogenization.

Shieh and Schwarcz (1977) found essentially identical bulk $\delta^{18}O$ values of different ortho- and para-gneisses over a 10,000-km^2 area. There is no mechanism that currently explains fluid circulation and homogenization on such a large scale; rather, these data support the idea of a "fortuitously" homogeneous protolith. Bowman and Ghent (1986) measured the oxygen and hydrogen isotope composition of metapelites from staurolite to sillimanite zones in British Columbia and concluded that the general homogeneity of the $\delta^{18}O_{quartz}$ values (13.3 ± 0.3 1σ, $n = 9$, 3 outliers) was caused by a homogeneous protolith. Demonstrating the presence of a pervasive fluid is difficult and is best carried out when adjacent lithologies have clearly different protolith $\delta^{18}O$ values, such as an amphibolite–metapelite association.

Isotope homogenization is probably more common for hydrogen than for oxygen. First, the buffering capacity of a rock for hydrogen is not very strong, given the low concentration of hydrogen in rocks. Second, the exchange rate of hydrogen between minerals and fluids is much more rapid than that of oxygen. In the classic example of this phenomenon, in order to explain the homogeneity of hydrogen isotope ratios and the heterogeneity of oxygen isotope ratios, Frey et al. (1976) state that rocks of the metamorphic Monta Rosa granite in the central Swiss Alps have "stewed in their own juices."[3]

12.4.2 Localized Exchange

Fluid interaction with what are commonly deemed "impermeable rock types"—particularly marbles and cherts—is extremely limited. Marbles often maintain their original $\delta^{18}O$ and $\delta^{13}C$ values, even when surrounded by lithologies that have completely exchanged with an external fluid. Marbles are most susceptible to isotopic exchange in samples where the permeability was enhanced by metamorphic decarbonation reactions (Rumble and Spear, 1983), which could produce a volatile phase that increases the fluid pressure. At the same time, the change in volume of the solid phases for the reaction, ΔV_{solids}, is negative, reflecting increasing porosity. The degree of fluid infiltration is often proportional to the magnitude of calc–silicate reactions that have occurred. It should be pointed out, however, that ductile marbles probably cannot accommodate increased fluid pressures without flowing, and there is a very serious issue regarding wetting angles in order for this process to persist for any significant period of time. (See Holness and Graham, 1991, for a further discussion.)

Rye et al. (1976) made a comprehensive stable isotope study of interbedded schist and marble units in a prograde metamorphic sequence at Naxos, Greece. The $\delta^{18}O$ values of

---

[3]One of the great "lines" from a renowned stable isotope geochemist!

the quartz and muscovite in the schists correlate well with metamorphic-grade rock. The $\delta^{18}O$ values of quartz decrease from a typical sedimentary value of 19‰ in the least metamorphosed rocks to 9.4‰ in the migmatized[4] samples. The marbles, on the other hand, have relatively constant $\delta^{18}O$ values throughout the entire sequence. Rye et al. concluded that there was significant fluid infiltration into the schists from "deep-seated" sources, while the marbles were impermeable to the evolving fluids. The only significant change in the $\delta^{18}O$ values of the marbles occurs during recrystallization and decarbonation reactions at intermediate grades (~500°C). At these grades, the permeability of the marbles had temporarily increased.

Fluid flow is often controlled by large-scale structural features, such as thrust ramps, thermal domes related to igneous intrusives, and antiforms, as well as by kilometer-scale vein systems. Shear zones often act as important conduits for enormous amounts of fluid movement. Different $\delta^{18}O$ values between mylonites or ultramylonites and host rock clearly indicate that fluids passed through the shear zone. Both downward penetration of fluids—a meteoric source (McCaig et al., 1990; Morrison and Anderson, 1998)—and upward flow from deeper levels (Kerrich, 1986; Fourcade et al., 1989; Burkhard and Kerrich, 1990; Crespo-Blanc et al., 1995) have been identified.

12.5 QUANTIFYING FLUID–ROCK RATIOS AND FLUID FLUXES

12.5.1 Simple Mixing Models: Zero-Dimensional Water–Rock Interaction Models

Taylor (1979) proposed a model for estimating fluid/rock ratios by assuming instantaneous and constant isotopic equilibration between a fluid and rock. His model was developed for fluids infiltrating a granite (Chapter 11), but has been applied to metamorphic rocks as well. The basic *mass-balance* equation for mixing a rock and fluid is

$$W\delta^{i}_{H_2O} + R\delta^{i}_{rock} = W\delta^{f}_{H_2O} + R\delta^{f}_{rock}, \qquad (12.8)$$

where i is the initial value, f is the final value, W is the atom fraction of the element in water that has participated in the interaction, and R is the atom fraction of the element in rock. It is necessary to *assume* a value for $\delta^{i}_{H_2O}$, such as local meteoric water values. This value can almost never be measured, because the initial fluid is no longer present in the rock, other than as fluid inclusions that most likely have $\delta^{18}O$ values quite different from those responsible for fluid–rock interaction. Similarly, the $\delta^{18}O$ value of the unaltered rock must be estimated. For shear zones, the nonsheared equivalent can be taken as the reasonable δ^{i}_{rock} value. In hydrothermal studies of epithermal granites, a normal granite is used for the δ^{i}_{rock} variable. The δ^{f}_{rock} value is measured. The only variable not known is $\delta^{f}_{H_2O}$, which must be calculated for an *assumed* temperature on the basis of known fractionation factors between rock and fluid ($\Delta_{rock-fluid}$).

Equation (12.8) can be rearranged to give the *closed-system* water–rock relationship

---

[4]Migmatites are rocks that have gotten sufficiently hot to partially melt.

$$\frac{W}{R} = \frac{\delta^f_{rock} - \delta^i_{rock}}{\delta^i_{H_2O} - (\delta^f_{rock} - \Delta_{rock-fluid})}. \qquad (12.9)$$

where $\Delta_{rock-fluid}$ is the equilibrium fractionation between the rock and the fluid at the temperature of interest. Equation (12.9) is valid for the special case where fluid enters a system, equilibrates with the rock, and is expelled in a single event. It is equivalent to the *batch fractionation* described earlier for volatilization or dehydration reactions. Ohmoto (1986) compares the Taylor one-pass model to "an autoclave which contains water and rock, and is maintained at a constant temperature." The other end-member scenario—the multipass model—is equivalent to Rayleigh fractionation. It models the process of an infinitely small quantity of fluid coming into contact with a rock, equilibrating with it, and then being expelled. This model can be derived by writing equation (12.8) in differential form and integrating it as follows:

$$\frac{W}{R} = \ln\left[\frac{\delta^i_{H_2O} + \Delta - \delta^i_{rock}}{\delta^i_{H_2O} - (\delta^f_{rock} - \Delta)}\right] = \ln\left[\left(\frac{W}{R}\right)_{closed\ sytem} + 1\right]. \qquad (12.10)$$

This equation is discussed in more detail by Nabelek (1991). Care should be taken in using all these water–rock ratio models. The simple *W/R* ratio equations presented thus far are too often misapplied to metamorphic systems. In deep-seated metamorphic rocks, estimates of the $\delta^i_{H_2O}$ and δ^i_{rock} values, as well as of the temperature of interaction, may not be well constrained, and the researcher should be aware of the potentially large uncertainties that may be associated with calculated fluid/rock ratios.

In all cases, the *W/R* ratios calculated with equations (12.9) and (12.10) are considered *minimum* values. Fluids can pass through a rock and either not interact with it or interact with it only partially. In terms of the effect on the rock, it is as if the fluid had never passed through the rock. Such a scenario can be imagined where quartz precipitates along the walls of a newly forming vein. Once a veneer of quartz has precipitated, more fluid could still flow through the vein, yet never "communicate" with the host rock, because it is isolated by the wall of newly formed quartz.

12.5.2 One-Dimensional (Directional) Water–Rock Interaction Models

The previously described fluid–rock interactions can be considered as zero-dimensional models. They consist of three stages: (1) mixing a certain amount of rock and water, (2) allowing the components to react and equilibrate isotopically, and, finally, (3) removing the water. They offer a simple mass-balance treatment and do not consider the direction of flow. This works well in some cases, but does not adequately describe continuous fluid–rock interaction for fluids travelling unidirectionally. Instead, it is necessary to consider a one-dimensional system, in which fluid comes from a source with a distinct isotopic composition and reacts with the rock along a specific *flow path*. When we consider fluid–rock interaction in this way, we see how the isotopic compositions of the rock and the fluid change with time *and space*.

The concept behind a one-dimensional flow model is based in chromatographic theory and can easily be described by simple "laboratory" analogs. First, we'll consider the already discussed zero-dimensional model. In our hypothetical example, we will consider only mixing between freshwater and salt water. It is easy to visual this system, yet it is no different from mixing two reservoirs with different isotopic compositions. Our system consists of a beaker of freshwater and infiltrating seawater with a salinity of 3.5%. If we add a little bit of seawater to the beaker, the salinity increases slightly. The more we add, the higher the

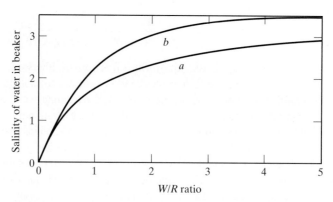

FIGURE 12.9: Simple mixing of freshwater and seawater. The salinity of our beaker will only reach the salinity of the infiltrating fluid (3.5%) for infinite *W/R* ratios. Curve *a* "one pass"; curve *b* "multipass."

salinity rises. Using equation (12.9), we have δ_{rock}^i as the beaker's initial salt-free water, and $\delta_{\text{H}_2\text{O}}^i$ as seawater with a salinity of 3.5%. The salinity of the mixture is δ_{rock}^f, and $\Delta = 0$. The beaker filling up with ever-greater amounts of seawater represents the one-pass model. If the inflow of seawater is balanced by an outflow from the beaker, so that the level of water never changes, then the multipath model is realized. A graph using equations (12.9) and (12.10) to depict the *W/R* ratio is given in Fig. 12.9 for one pass (curve *a*) and multipass (curve b) models. With the multipass model, salinity at first increases rapidly with increasing *W/R* ratios, but finally approaches the 3.5% salinity of the incoming fluid asymptotically. We need a *W/R* ratio of ∞ to reach the same salinity as the infiltrating fluid.

Now, instead of a beaker, consider a long, thin capillary tube filled with fresh water. Entering from one end is our seawater, with a salinity of 3.5%. As the seawater enters the tube, freshwater is expelled from the other end. If there is no dispersion or diffusion at the seawater–freshwater interface, then a sharp boundary exists at that interface. The interface migrates forward with time, with the rate of forward progress related to the *flux* (in units of volume/area), or rate of flow. This construction is illustrated in Fig. 12.10. At $t = 0$, the salinity of the entire tube is 0 and seawater is just beginning to enter from the left-hand side. As the flow continues, the seawater–freshwater front moves to the right, so that at time $t = x$ there is a salinity front

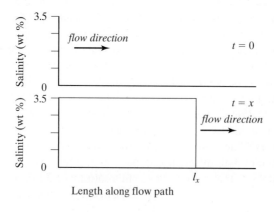

FIGURE 12.10: One-dimension flow model of seawater infiltrating a column of freshwater. At time $t = 0$, the entire length of the column will have freshwater. As the seawater flows into the column, a sharp front will migrate forward, at a rate proportional to the fluid flux. Although the salinity is not the same along the entire column, any plane perpendicular to the length of the column receives the same amount of fluid (e.g., had the same *W/R* ratio).

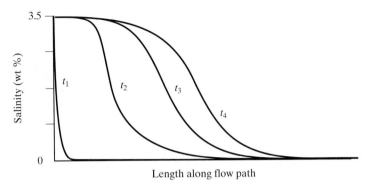

FIGURE 12.11: One-dimensional infiltration model, illustrating the "smoothing" effects of dispersion, diffusion, and finite reaction kinetics. t_x refers to the time after the beginning of mixing. As time proceeds, the boundary becomes more diffuse.

at position l_x. All of the fluid at $l < l_x$ has a salinity of 3.5%, and all fluid at $l > l_x$ has a salinity of 0%. If we were to apply the simple mixing equations, either (12.9) or (12.10), to this system, we would calculate a W/R ratio of 0 at all points $l > l_x$ and W/R ratios of ∞ at all points $l < l_x$. In fact, all points along the flow path receive the same fluid flux!

In nature, the sharp front depicted in Fig. 12.10 becomes diffuse by a number of processes, including diffusion, dispersion, sluggish exchange kinetics (between rock and fluid), and changing fractionation factors with temperature. As a result, the infiltration front becomes less sharp (Fig. 12.11) as flow proceeds.

Spooner et al. (1977) first modeled stable isotope data with flow-path calculations. They divided the flow path up into a finite number of *en echelon* boxes, and considered each as a zero-dimensional system. A finite quantity of fluid enters the first box, where it equilibrates with the rock volume, the $\delta^{18}O$ values of the rock and fluid being determined by the fractionation factor $\Delta^{18}O_{rock-fluid}$ and the W/R ratio. The amount of fluid entering the rock is determined by the porosity (ϕ) of the rock. When a small amount of fluid enters the first box, the $\delta^{18}O$ value of the rock changes imperceptibly, while the tiny packet of fluid equilibrates "instantly" with the rock. The simple mass-balance equation (12.9) applies. As the newly equilibrated fluid enters the second box, it has an isotopic composition that is *almost*—but not exactly—in equilibrium with the original rock, so the $\delta^{18}O$ value of both the rock and the fluid box change very little. As each new increment of fluid enters the system, the integrated W/R ratio of the first box increases, and the $\delta^{18}O$ (rock) value slowly shifts towards equilibrium with the initial fluid. If we have an infinite number of infinitely small boxes, the migrating front will be sharp. Increasing the size of each box causes the diffusion front to smooth out.

A number of authors have developed mathematical solutions to the problem of one-dimensional fluid flow (Lichtner, 1985; Baumgartner and Rumble, 1988; Blattner and Lassey, 1989). These solutions are in the form of partial differential equations that must be solved numerically. The models can be extremely complex, and only the most intrepid geochemists–mathematicians dare brave the mathematics! Nonetheless, some very nice applications have been made, particularly to fluid–rock interaction in shallow-level contact environments (Nabelek and Labotka, 1993; Bowman et al., 1994; Barnett and Bowman, 1995; Cook et al., 1997) and between contrasting lithologies (Ganor et al., 1989; Bickle and Baker, 1990; Vyhnal and Chamberlain, 1996).

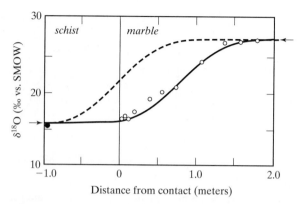

FIGURE 12.12: Oxygen isotope profile of calcite across a schist–marble contact at Naxos, Greece. Data are shown by circles; the solid line is derived from a combined infiltration–diffusion model in which the fluid is moving from the schist into the marble. The arrows show the assumed initial compositions of the schist and marble, respectively. Note that if there were only pure diffusional exchange, the sigmoidal profile would be symmetrical about the contact (as shown by the dashed line). The displacement into the marble requires a component of flow from schist to marble. (From Bickle and Baker, 1990.)

As an example, Bickle and Baker (1990) measured a detailed isotopic profile across a schist–marble contact in a Tertiary metamorphosed sequence from Naxos, Greece. The sigmoidal-shaped diffusion pattern is clearly displaced into the marble (Fig. 12.12). The authors concluded that the pattern requires there to have been a net transport (advection) from the schist into the marble. The data can be explained either by a very modest degree of fluid *advection* toward the marble from the schist, combined with diffusion, or by a much larger degree of fluid flow *parallel* to bedding with only a small component perpendicular to bedding. The two cannot be distinguished. Only the contact-perpendicular component can be computed.

In a similar study, Cartwright and Valley (1990) measured a sigmoidal-shaped isotopic profile that was indeed centered on a granite–marble contact in a regionally metamorphosed setting, in contrast to Bickle and Baker's displaced profile. They inferred this pattern to be related to diffusional exchange via an intergranular pore fluid, without any advective flow perpendicular to the contact. The symmetry about the contact indicates that there was no preferential direction of flow. (See Box 12.2.)

Models using stable isotope profiles have been proposed to quantify fluid flow directions and fluxes in a contact metamorphic event. The situation has an additional complexity in comparison to regional metamorphism, in that significant temperature changes occur along the flow path (Dipple and Ferry, 1992), and consequently, so does isotopic fractionation between the fluid and rock. Two end-member flow models can be envisioned: fluid flow out of the intrusion—the so-called down-temperature flow path; or fluid flow towards, and eventually into, the crystallized intrusion—the up-temperature flow path. Both models have been rigorously defended by various researchers. In fact, it seems likely that both flow regimes occur at a single contact aureole at different times. Fluids flows *out* of a pluton as it crystallizes and expels water. Once the pluton has crystallized, however, the temperature differential between the intrusive and the country rock creates a buoyant driving force for fluids to flow *toward* the pluton and, eventually, upwards. The complexities of fluid–rock interaction in a contact metamorphic setting are discussed in Nabelek (1991).

> **BOX 12.2 Comparison of isotopic data across lithological contacts**
>
> It is common to consider isotopic variations across a lithological contact and use differences to interpret mixing between the two lithologies. Care must be taken in terms of what is being compared, however! Many studies exist in which comparisons are made incorrectly. Consider the juxtaposition of a quartzite and an iron formation. The $\delta^{18}O_{\text{bulk rock}}$ values of the two lithologies are completely different and will show a marked decrease across the boundary. This decrease could mistakenly be interpreted as isotopic disequilibrium. In reality, the iron formation has a lower $\delta^{18}O_{\text{bulk rock}}$ simply because it has a high percentage of magnetite. More meaningful information regarding cross-lithology equilibrium is obtained when either similar lithologies or like minerals are compared across lithologies. If the $\delta^{18}O$ values of quartz in the iron formation are the same as those in the quartzite, then isotopic equilibrium exists between the two lithologies. It doesn't matter what the $\delta^{18}O$ value of the bulk rock is. If fluid exchange has resulted in isotopic equilibrium between the two lithologies, then the $\delta^{18}O$ values of each mineral will be the same in each rock type, even though the bulk rock values for each lithology could be completely different.
>
> In a similar way, comparisons of the bulk rock $\delta^{18}O$ values of mafic and more felsic igneous rocks are flawed, yet frequently made. A higher proportion of low-$\delta^{18}O$ minerals, such as olivine and pyroxene, in the more mafic magma will lead to a low $\delta^{18}O$ value, even when mafic and felsic commingled magmas are actually in isotopic equilibrium.

12.6 THERMOMETRY

12.6.1 Introduction

Stable isotope thermometry began in the late 1940s, with applications primarily to low-temperature carbonates. Baertschi and Silverman (Baertschi, 1951; Baertschi and Silverman, 1951) first measured the $\delta^{18}O$ values of whole-rock silicates, but it wasn't until some years later that Clayton and Epstein demonstrated that the fractionations between coexisting minerals in high-temperature rocks were large enough to be valuable for thermometry (Clayton and Epstein, 1958; 1961). Garlick and Epstein (1967) made a detailed study of prograde metamorphic sequences and found that mineral-pair fractionations decreased with increasing metamorphic grade (Fig. 12.13). There was a systematic enrichment of $\delta^{18}O$ values in the following order: magnetite, ilmenite, chlorite, biotite, garnet, muscovite, quartz. Garlick and Epstein were able to quantify temperatures of metamorphism on the basis of quartz–magnetite fractionations, which had been calibrated experimentally. Myriad thermometric studies have since been made of a wide range of metamorphic rock types, ranging from subgreenschist facies to rocks heated to well over 1,000°C (Sharp et al., 1992; Sharp, 1998).

Stable isotope thermometry applied to metamorphic rocks is complicated by the slow cooling that occurs during retrogression. Taylor et al. (1963) and Garlick and Epstein (1967) recognized the problem of retrograde reequilibration. Taylor et al. asked whether chemical and mineralogical equilibrium is a prerequisite for oxygen isotope equilibrium. They reasoned that if recrystallization processes are strong enough to break Si–O bonds, they will surely result in cation reorganization as well. Taylor et al. wrote "If the O^{18}/O^{16} ratios of these minerals are each 'frozen in' at a different, but constant, characteristic temperature, then similar Δ-values might be obtained in each rock." This was the first mention of retrograde exchange in the sense

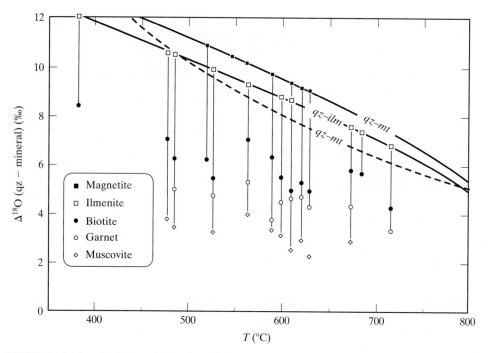

FIGURE 12.13: **Garlick and Epstein's (1967) measured fractionations of minerals relative to quartz for several metamorphic terranes.** The authors noted that there is a general decrease in fractionation as temperatures increase. The data are plotted relative to temperature, as determined by the quartz–magnetite fractionation available at the time (top curve). A more recent calibration is shown as the dashed line. In general, fractionations for all minerals decrease with increasing grade and there is an isotopic enrichment of one mineral relative to another. Note that the smooth variation in magnetite is a result of the construction of the curve, where the X-axis is defined by the $\Delta^{18}O_{qz-mt}$ value.

of the concept of "closure temperatures." Although complications associated with retrograde exchange exist, careful treatment of oxygen isotope data has led to numerous successful thermometric studies in metamorphic rocks. In fact, with judicious sampling, more information than just peak metamorphic temperatures can be "teased out" of a rock that has undergone partial retrograde exchange, as is explained in Section 12.7.

Two questions must be addressed in order to extract metamorphic temperature estimates by means of stable isotope theromometry: (1) Was isotopic equilibrium attained during peak metamorphism, and (2) have the $\delta^{18}O$ values corresponding to peak temperatures been preserved during slow retrograde cooling? In low-grade metamorphic rocks, the first condition is rarely met: Inheritance of premetamorphic values is all too common. In very high grade rocks, problems associated with retrograde exchange are a far greater problem, related to the second consideration. There are a number of processes that can disturb the isotopic equilibrium of an assemblage. Retrograde diffusion, deformation and recrystallization, and the introduction of a disequilibrium fluid will all enhance reequilibration during cooling. If the diffusion rate of oxygen for a pair of minerals is rapid, then some degree of resetting must occur during slow cooling from a high temperature. The isotope community is divided over the accuracy of high-temperature estimates obtained by stable isotope fractionation. Some groups have had great success in retrieving peak metamorphic temperatures; others maintain that peak-temperature equilibrium is almost never retained. Recent work indicates that certain minerals are extremely "refractory," in the sense that they do not reequilibrate during cooling (Sharp, 1995; Valley, 2001). It is safe to say that careful selection of lithologies and minerals is crucial in any effort to extract peak-temperature data from high-T metamorphic rocks (e.g., Farquhar et al., 1993).

12.6.2 Oxygen Isotope Thermometry in Metamorphic Rocks: Testing for Equilibrium

The fractionation between coexisting minerals decreases with increasing temperature, as is expected from thermodynamic relations. The isotopic fractionation between two solid phases can be expressed by the equation

$$1000 \ln \alpha_{x-y} = \frac{a_{x-y} \times 10^6}{T^2} + b_{x-y}, \tag{12.11}$$

where a is the temperature coefficient of fractionation for phases x and y; b is a constant, usually equal to zero for anhydrous minerals; and T is in kelvins. Remember that the term $1{,}000 \ln \alpha_{x-y}$ is very close to $\delta_x - \delta_y$ for small values of $1{,}000 \ln \alpha$. (See equation 2.17.) In theory, any two phases can be used to determine the temperature of metamorphism. In order to make such measurements, minerals must be physically separable and the mineral-pair thermometer must be calibrated.[5] Several strategies have been developed to extract accurate temperature estimates from metamorphic rocks:

1. *Measure multiple minerals in a single sample.* If all mineral pairs give the same temperature, then there is more confidence in the validity of the results. Equation (12.11) can be rearranged in the simple linear $(y = mx)$ format

$$1000 \ln \alpha - b = a \times \left(\frac{10^6}{T^2}\right), \tag{12.12}$$

where $y = 1{,}000 \ln \alpha - b$, $m = 10^6/T^2$, and $x = a$. If all minerals in an assemblage are in isotopic equilibrium at T (in kelvins), then a plot of $1{,}000 \ln \alpha - b$ vs. a must define a straight line with a slope $= 10^6/T^2$ (Javoy et al., 1970). Consider the mineral suite quartz, plagioclase, biotite, muscovite, and magnetite (Fig. 12.14). In the plot of $1{,}000 \ln \alpha_{\text{qz-mineral}} - b$ relative to the reference mineral quartz vs. the temperature coefficient a relative to quartz, all minerals plot on a straight line passing through the origin with a slope of $10^6/T^2$, except for magnetite. Thus, all minerals except magnetite appear to be in isotopic equilibrium.

This type of representation has been used in numerous studies. It provides a statistical method for averaging temperatures in a multimineral system (assuming that they are all in equilibrium) and clearly indicates any potential disequilibrium by one or more mineral.

2. *Analyze "refractory" minerals.* Oxygen self-diffusion, or intracrystalline diffusion, is the rate at which oxygen diffuses through a crystal lattice. The oxygen self-diffusion rates of

---

[5] A half century of research has gone into determining the temperature coefficients a and b (equation (12.11)) for mineral pairs. Even so, there are still a number of phases for which the calibration is poorly known. An extensive list of fractionation factors is given in Chacko et al. (2001) and at http://www.ggl.ulaval.ca/cgi-bin/isotope/generisotope.cgi.

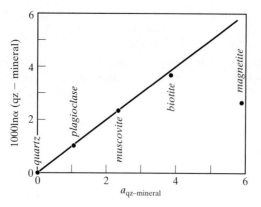

FIGURE 12.14: **Typical isotherm plot for coexisting minerals.** All mineral data are plotted relative to quartz (although the choice of reference mineral is arbitrary and will not change the result). In the example illustrated, all minerals except magnetite define an isotherm, and magnetite is clearly out of equilibrium with the other phases. The slope of the best-fit line is equal to $10^6/T^2$ (with T in kelvins).

minerals are highly variable. The diffusion rates of biotite, leucite, and feldspar[6] are so rapid that isotopic exchange with other phases will occur well below peak metamorphic temperatures during slow cooling. Other minerals, such as garnet, olivine, zircon, and kyanite, have very slow self-diffusion rates. For these minerals, the $\delta^{18}O$ value of the mineral should not change during slow cooling, and peak temperature estimates can be extracted under certain conditions. Bimineralic assemblages containing at least one refractory phase are most likely to preserve isotope fractionations that correspond to peak metamorphic temperatures. Refractory minerals, defined as those which are least likely to exchange once crystallized, have their own set of problems, however. There are clear examples of refractory minerals forming during the prograde path of a metamorphic event and not reequilibrating during peak metamorphic conditions. Larson and Sharp (2003, 2005) demonstrated that garnet and the aluminum silicates andalusite, sillimanite, and kyanite may form at different times and that they preserve $\delta^{18}O$ values corresponding to the temperature of their growth such that they do not reflect peak metamorphic temperatures. So, while refractory minerals may circumvent the problem of retrograde diffusional exchange, they carry their own set of potential problems. It is thus necessary to consider phase equilibria and mineral reaction sequences in interpreting stable isotope data of refractory phases.

3. *Avoid "disturbed" samples.* Late fluid infiltration and deformation will act to reset the $\delta^{18}O$ values of minerals. Fluids may preferentially interact with one mineral relative to another. It is not uncommon to find $\delta^{18}O$ values of feldspars that are higher than those of quartz, indicating that hydrothermal alteration has raised the $\delta^{18}O$ value of the feldspar, but not the quartz. On the other hand, by analyzing mineral phases that are sensitive to hydrothermal exchange (e.g., micas, feldspar, carbonates), a late fluid infiltration event can often be detected and the $\delta^{18}O$ value of the infiltrating fluid determined. Deformation and recrystallization also enhance reequilibration rates. Oxygen isotope disequilibrium is commonly observed in shear zones, where kinetically controlled reequilibration overwhelms equilibrium thermodynamic controls (e.g., Crespo-Blanc et al., 1995).

---

[6]The diffusion rate of oxygen in feldspar is strongly dependent on the presence or absence of an aqueous fluid—"wet" vs. "dry" diffusion, respectively. (See Elphick, et al., 1988; Sharp, 1991, for more information.)

12.6.3 Applications of Stable Isotope Thermometry

The most successful applications of stable isotope thermometry are generally made in rocks of amphibolite and lower granulite facies. At greenschist and sub-greenschist facies, mineral separation becomes difficult due to the fine grain size, and temperatures may not have been high enough for all mineral phases to attain isotopic equilibrium.[7] Rocks of amphibolite facies and higher generally have coarse-grained minerals, which can easily be separated. Temperatures are sufficiently high during amphibolite-facies metamorphism that there is a strong likelihood that isotopic equilibrium was reached during metamorphism, but not so high that retrograde diffusional effects should be a serious factor. In very high temperature rocks, retrograde resetting is common, although there are examples of successful applications to very hot rocks (Fourcade and Javoy, 1973; Sharp et al., 1992; Farquhar et al., 1993).

Dunn and Valley (1992) used calcite–graphite thermometry in an elegant application of thermometry using refractory minerals. Graphite stands out as one of the minerals most resistant to retrograde resetting. Once prograde recrystallization ceases, there is no change in the $\delta^{13}C$ value of graphite. Diffusion rates of carbon in graphite are so slow that experimental estimates of diffusion rates or fractionation factors have been made only with great difficulty. Dunn and Valley compared temperature estimates from calcite–graphite fractionations ($\Delta^{13}C_{calcite-graphite}$) with those from calcite–dolomite solvus thermometry in a contact aureole that had received a regional amphibolite-grade overprint (Fig. 12.15). The calcite–dolomite solvus thermometer gives uniform temperatures that correspond to the regional amphibolite-grade overprinting event. The calcite–graphite isotope thermometer also records the amphibolite-facies event far from the gabbroic intrusion. But as the intrusion is approached, the calcite–graphite temperatures increase systematically, in accordance with the predicted thermal halo associated with the premetamorphic intrusion. In this instance, the calcite–graphite thermometer has preserved early high temperatures in spite of a later, protracted amphibolite-facies overprint.

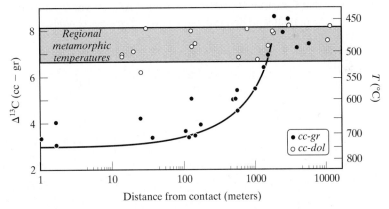

FIGURE 12.15: Profile of calcite–graphite carbon isotope thermometry and calcite–dolomite solvus thermometry away from a gabbro–marble contact (Tudor Gabbro, Ontario, Canada). The carbon isotope thermometer preserves contact temperatures (curve), while the calcite–dolomite solvus thermometer has been completely reset during later regional metamorphism (shaded area). (After Dunn and Valley, 1992.)

---

[7]In low-grade rocks, a strategy of analyzing *less* refractory phases may be the most successful approach. For example, quartz–calcite pairs can provide information about temperatures less than 250°C (Kirschner et al., 1995).

12.7 RETROGRADE EXCHANGE: "GEOSPEEDOMETRY"

Deformation, exsolution, and recrystallization during late fluid infiltration processes can completely reset the isotopic composition of a mineral phase. In the absence of these processes, however, the degree of retrograde exchange is controlled by intracrystalline diffusion, which can be treated quantitatively if diffusion rate data are available. Although retrograde diffusion-based resetting hinders our ability to obtain the peak temperatures of many metamorphic systems, it can be used to extract cooling-rate information, the so-called stable isotope geospeedometer.

The basic idea of determining cooling rates was first presented by Giletti (1986) and is illustrated in Fig. 12.16. A bimineralic assemblage a–b is heated to some high temperature $T(p)$ (the peak temperature). Diffusion rates are rapid enough at $T(p)$ that the two minerals quickly approach isotopic equilibrium by diffusional exchange.[8] The equilibrium fractionation at $T(p)$ is $\Delta(p)$. Now, if the rock is cooled rapidly to room temperature, the peak-temperature equilibrium fractionation $\Delta(p)$ is preserved, because the diffusion rate at room temperature is immeasurably slow. Now consider what happens if the rock is cooled very slowly. Starting at $T(p)$, the temperature is lowered by a small increment ΔT, and the new temperature $(T(p) - \Delta T)$ is maintained for an extended period of time. The diffusion rate of oxygen in the two minerals is rapid enough that they will eventually reequilibrate at the new, lower temperature. Now the rock is cooled again by a small increment ΔT and remains at the new temperature until isotopic equilibrium is again attained. It is clear that by cooling a rock slowly, the $\Delta(a-b)$ value will track along the equilibrium curve, shown by the arrow in Fig. 12.16. At some point, however, the diffusion rate becomes so slow that no further exchange occurs. This narrow temperature interval over which the assemblage goes from one in equilibrium to one in which the diffusion rate effectively stops is defined as the **closure temperature** $T(c)$ (Dodson, 1973). Further cooling does not change the $\delta^{18}O$ values of either phase. The fractionation that is measured in the laboratory is $\Delta(c)$, given by the closure temperature of the system. It is unrelated to the peak temperature $T(p)$, but *is* related to the cooling rate.

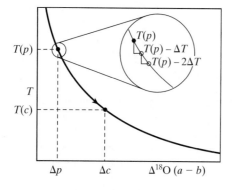

FIGURE 12.16: Schematic plot of $\Delta^{18}O$ value for phases a and b vs. temperature. The initial, peak temperature is $T(p)$. If the sample is cooled rapidly to room temperature, it will preserve fractionations Δp, corresponding to equilibrium at $T(p)$. If the sample is cooled slowly, it will continue to reequilibrate during cooling, following the equilibrium fractionation-vs.-temperature path (e.g., $T(p)$ to $T(p) - \Delta T$, etc.). The sample will continue to reequilibrate until the closure temperature $T(c)$ is reached, at which point intracrystalline diffusion effectively ceases.

---

[8]Diffusional exchange in this case is limited by the **intracrystalline** diffusion rate, or self-diffusion rate. This is the diffusion rate of oxygen within a crystal. The **intercrystalline**, or grain-boundary, diffusion rate, which allows oxygen to migrate between phases, is considered to be orders of magnitude more rapid than the intracrystalline diffusion rate. Limited experimental studies support this assumption (e.g., Farver and Yund, 1991a).

In both cooling scenarios, the mineralogy, peak temperature, and final temperature (room temperature) were the same. The different measured fractionations were related only to differences in cooling rates. If we could quantify this effect, then we would have a *"geospeedometer,"* a tool to calculate how fast a rock cooled.

Again, it should be stressed that oxygen diffusion does not stop at a single temperature. Diffusion rates slow and then effectively stop with decreasing temperature. But because diffusion rate is related exponentially to temperature, the *temperature interval* over which diffusion behavior changes from being rapid to insignificant is small, and a single closure temperature can be defined for purposes of calculations. In a simple bimineralic assemblage, the *closure temperature T(c)* is determined by the mineral with the *slower* diffusion rate. During cooling, the $T(c)$ of one mineral is reached. Diffusion then stops for that mineral, and its isotopic composition does not change during further cooling. The diffusion rate of oxygen in the second phase may still be rapid, but because there is nothing for it to exchange with, its isotopic composition is also fixed.

The closure temperature of a mineral is a function of (1) the cooling rate dT/dt; (2) the grain size r (for radius); (3) the grain geometry (spherical, cylindrical, or platelike); (4) the diffusion rate; and (5) the modal abundance of the minerals. Dodson (1973) derived the following equation relating the cooling rate to the closure temperature for a phase in an infinite reservoir:

$$T_c = \frac{Q/R}{\ln\left[\dfrac{-ART_c^2(D_o/a^2)}{Q(dT/dt)}\right]}. \tag{12.13}$$

In this equation, T_c is the closure temperature, dT/dt is the cooling rate, a is the grain radius, R is the gas constant, A is a diffusional anisotropy parameter (8.7 for a sphere, 55 for a cylinder), Q is the activation energy for diffusion, and D_o is the preexponential factor. Q and D_o are related to the diffusion rate by the Arrhenius relation

$$D = D_o \times \exp\left(\frac{-Q}{RT}\right) \tag{12.14}$$

and are normally determined experimentally. An extensive list of diffusion data is compiled in Cole and Chakraborty (2001).

Figure 12.17 illustrates the evolution of $\delta^{18}O$ values of a bimineralic assemblage during slow cooling. The top figure shows the progression of $\delta^{18}O$ values for a bimineralic quartz–magnetite mixture of equal oxygen proportions cooling from an initial temperature of 750°C. The solid lines represent the $\delta^{18}O$ values for a cooling rate of 10°C/M.Y., and the dashed line is for the same system cooled at 100°C/M.Y. The closure temperature for the rapidly and slowly cooled rocks are 608°C and 545°C, respectively. Above 608°C, the cooling curves are identical. The only difference is the temperature at which diffusion effectively stops. The lower figure shows the same conditions, except that now there is 90-atom % O in quartz and 10-atom % O in magnetite. In this case, the $\delta^{18}O$ value of magnetite changes considerably, while the $\delta^{18}O$ value of quartz is almost unchanged, a consequence of mass-balance considerations. In both cases, the closure temperatures are the same.

Things become more complicated (and more interesting) when a rock has multiple phases. Cooling rate curves for a four-phase assemblage are plotted in Fig. 12.18, which uses

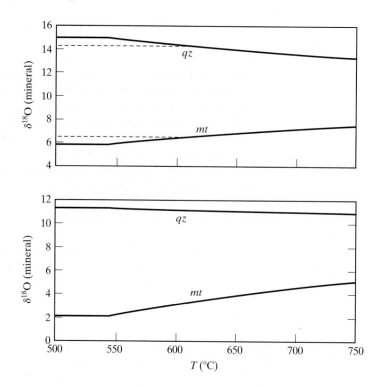

FIGURE 12.17: Variations in $\delta^{18}O$ values of coexisting quartz and magnetite (initial $\delta^{18}O_{bulk\ rock}$ = 10.34‰, grain radius 2.5 mm) during slow cooling. Top figure shows the progression of $\delta^{18}O$ values for equal atomic proportions of O in both phases. Solid line = cooling rate of 10°C/M.Y.; dashed line is for cooling rate of 100°C/M.Y. Bottom figure is for cooling rate of 10°C/M.Y., with 90% modal O_2 in quartz and only 10% in magnetite. Closure temperatures are the same, but the change in the $\delta^{18}O$ values of each phase depend critically on modal abundance. (Diffusion data from Table 12.1.)

modal proportions and grain-size and diffusion rate data from Table 12.1. All phases maintain isotopic equilibrium until the closure temperature of hornblende is reached (Tc_{hnbd}). The $\delta^{18}O$ value of hornblende does not change with further cooling. All other phases continue to exchange until the closure temperature of quartz is reached (Tc_{qz}). The composition of quartz is then fixed, and magnetite and plagioclase continue to exchange until the closure temperature of magnetite is reached, after which all phases retain their $\delta^{18}O$ values down to room temperature.

TABLE 12.1: Parameters for diffusion calculations for Figs 12.17 and 12.18. Coefficients a and b are from equation (12.11) relative to quartz, D_o, Q, radius r, and anisotropy parameter a from equation (12.13).

Mineral	D_o (cm²/sec)	Q (kJ/mol)	Grain radius (mm)	Anisotropy parameter	a	b	Fraction O_2
Quartz	190	284.0	2.5	8.7	0	0	0.25
Plagioclase	1.4 E-7	109.6	0.5	55	−1.39	0	0.25
Hornblende	1.0 E-7	171.5	0.5	8.7	−3.15	0.3	0.25
Magnetite	2.7 E-5	202.5	0.5	55	−6.11	0	0.25

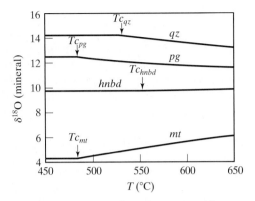

FIGURE 12.18: Model calculation for the four-phase assemblage consisting of quartz (*qz*), plagioclase (*pg*), hornblende (*hnbd*), and magnetite (*mt*) cooled at 20°C/M.Y. Initial temperatures are 650°C; the closure temperature of hornblende is reached at 551°C, after which no further exchange occurs. The closure temperatures of quartz and magnetite are reached at 528°C and 484°C, respectively. Although the closure temperature of plagioclase is 221°C, its isotopic composition is fixed by the closure temperature of magnetite. The plagioclase is still able to exchange with other phases, but there is nothing left for it to exchange with. (Data used for calculations are from Table 12.1.)

The significance of Fig. 12.18 is that the measured $\delta^{18}O$ values bear no relationship to peak metamorphic temperatures. Instead they represent "apparent temperatures" related to the effects of retrograde cooling. The cooling rate of a rock can be estimated by finding the best fit between the measured and calculated $\delta^{18}O$ values of each phase by varying the cooling rate estimates.

The simple equation of Dodson includes certain assumptions that are not strictly correct. For example, it applies only to minerals hosted by an infinite reservoir, such as a single mineral in a fluid, or to argon diffusion, whereby the element diffuses into the surroundings, but not back into the crystal from the host. More rigorous solutions for diffusion considering modal abundance and a finite reservoir have been made (Eiler et al., 1992; Jenkin, Farrow, Fallick, and Higgins, 1994). Although the calculations using these more exact solutions are complex, simple computer models have been developed, making their application relatively straightforward (Eiler et al., 1995). For many common mineral assemblages, the solutions attained with equation (12.13) and the more complex computer programs generally give the same answer.

The diffusion rate of oxygen in minerals appears to be strongly affected by the presence or absence of an aqueous fluid. Both experimentally and empirically, there are clear indications of an aqueous fluid enhancing diffusion rates (Elphick et al., 1988; Farver and Yund, 1991b; Sharp et al., 1991; Sharp and Moecher, 1994). Some authors have suggested that H_2O fugacity rather than the actual presence of a fluid phase may control diffusion rates (Kohn, 1999), although such an interpretation requires that diffusion of oxygen in most metamorphic terranes follow the "wet" diffusion data, which is certainly not the case.

12.8 STATE OF THE ART

The application of stable isotope geochemistry to metamorphic studies is in an exciting state of transition. New methods of analysis are being combined with new strategies for interpretation, and new fractionation and diffusion data are constantly being generated. The laser extraction technique for oxygen isotope analysis of silicates and sulfur analysis of sulfides has drastically reduced sample sizes (Crowe et al., 1990; Kelley and Fallick, 1990; Sharp, 1990). For the first time, isotopic mineral zonation at a submillimeter scale can be measured and interpreted in a geologically meaningful context (Chamberlain and Conrad, 1991; Conrad and Chamberlain, 1992; Kirschner et al., 1993; Kohn et al., 1993; Young and Rumble, 1993). The recent application of ultraviolet lasers and gas chromatography techniques has increased the spatial resolution of analyses down to 50 microns (Young et al., 1998). Ion microprobe techniques have also been refined to the point where they can be applied to stable isotope studies (Valley and Graham, 1998).

Our methods of interpretation have also improved. Models have been developed to calculate the effects of net-transfer reactions on the isotopic composition and zoning of minerals that form during porphyroblast growth in a prograde event (Kohn et al., 1993; Young, 1993). The closure temperature concept of Giletti has been refined to account for a number of variables, including modal abundance and variable cooling rate. Strategies that combine numerical modeling techniques, as well as specific sampling strategies, have been developed to extract peak-temperature information by means of stable isotope thermometry. Finally, complex equations have been developed to estimate fluid fluxes, directions, and temperatures of fluid flow. Although analyzing metamorphic rocks can be quite complicated, our understanding of how to approach these rocks is now better than ever.

PROBLEM SET

1. Measured $\delta^{18}O$ values of mineral separates from a single sample are listed in Table 12.P2. Using the coefficients of fractionation in Table 12.P1, answer the following questions.
   a. Which phase is out of equilibrium with the others?
   b. Excluding that phase, what is the temperature of equilibration?

TABLE 12.P1: Some commonly used coefficients of oxygen isotope fractionation relative to quartz. In form of equation 12.11 (after Javoy, 1977).

Mineral	a	b
Plagioclase	$0.97 + 1.04\beta$	0
Olivine	2.75	0
Muscovite	2.2	−0.6
Biotite	3.69	−0.6
Amphibole	3.15	−0.3
Ilmenite	5.29	0
Magnetite	5.57	0
Garnet	2.88	0
Chlorite	5.44	−1.63

β is the fraction of anorthite in plagioclase.

TABLE 12.P2: Isotopic data for problem 12.1.

Mineral	$\delta^{18}O$ value
Quartz	10
Plag (An 30)	8
Garnet	5
Muscovite	4.5
Biotite	4.4
Magnetite	1.8

TABLE 12.P3: Data for problem 12.2.

Mineral	$\delta^{18}O$ value	Grain radius	Q (kJ/mole)	D_o (cm^2/sec)
Quartz	10	1.2 mm	284	190
Magnetite	2	0.5	203	2.7×10^{-5}

    c. What is the temperature of equilibration between quartz and the phase that is out of equilibrium with the other phases? Fractionation factors are given in Table 12.P1.

2. a. What is the cooling rate of a bimineralic quartz–magnetite rock, given the information in Table 12.P3?

    (*Hint*: Assume that the closure temperature is defined by the fractionation between the two phases.)

    b. The preceding calculations are made using the diffusion coefficients assuming the presence of an aqueous fluid. Under anhydrous conditions, the preexponential factor is two orders of magnitude smaller. What is the cooling rate, assuming anhydrous (dry) cooling?

3. The $\delta^{18}O_{\text{whole rock}}$ values of adjacent gabbro and mica schist are 8‰ and 12‰, respectively. Could the two lithologies be in isotopic equilibrium? How could you test this hypothesis?

4. A thin layer of quartzite is sandwiched between layers of metasediments. You measure the $\delta^{18}O$ values of quartz in the quartzite and metasediment (pelite) in order to address the question whether the quartzite protolith was a chert or sandstone. The $\delta^{18}O$ value of both lithologies is 13.4‰. What can you conclude?

5. Cavosie et al. (2002) measured the $\delta^{18}O$ values of quartz, andalusite, and sillimanite from a metamorphic vein. If all phases were in equilibrium at the temperature of their formation (both isotopically and in terms of phase equilibria), what is the *temperature* and *pressure* of vein formation. Use

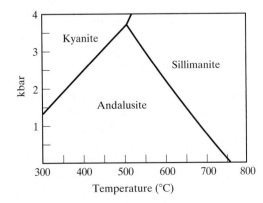

FIGURE 12.P1: Phase equilibrium diagram for the 3 aluminum silicates, kyanite, andalusite, and sillimanite.

the relationship $1000 \ln \alpha_{qz-AS} = \dfrac{2.25 \times 10^6}{T^2}$, where AS stands for aluminum silicate (either andalusite or kyanite), and Fig. 12.P1.

Measured $\delta^{18}O$ values are $\delta^{18}O_{qz} = 13.39\%_o$, $\delta^{18}O_{and} = 10.68\%_o$ and $\delta^{18}O_{sil} = 10.67\%_o$.

6. One volume of calcitic limestone (average $\delta^{18}O = 23.72\%_o$) underwent complete oxygen isotope exchange with three volumes of meteoric water ($\delta^{18}O = -8.40\%_o$) at 370°C and 1 kbar to produce an isotopically homogenous marble. What is the $\delta^{18}O$ value of this marble? The following data will be needed:

Densities in g/cm³ at 1 kb: calcite = 2.71   water = 0.72
Molecular weights: calcite = 100   water = 18
$10^3 \ln \alpha(CaCO_3 - H_2O) = 2.78(10^6 T^{-2}) - 2.89$

ANSWERS

1. a. Muscovite is out of equilibrium.
   b. Excluding muscovite, temperatures are 544°C.
   c. Quartz–muscovite fractionation corresponds to a temperature of 330°C.

2. The closure temperature is 561°C. Solving equation 12.13 for the conditions given in Table 12.P3 gives a cooling rate of 120°C/Ma for quartz and almost the same for magnetite.

3. Although the whole-rock values are different, the two lithologies could be in isotopic equilibrium. Because the mica schist has a higher silica content, it will tend to have more minerals that concentrate ^{18}O. This does not mean that the two rocks are not in equilibrium, however. One way of testing this possibility is to analyze minerals that are common to both lithologies. If they have the same $\delta^{18}O$ values, then the lithologies could be in equilibrium.

4. Cherts are chemical sediments, and their $\delta^{18}O$ values are generally over $30\%_o$. But because the two lithologies are in equilibrium ($\delta^{18}O_{qz}$ is the same for both units), you *cannot* determine whether the quartzite was a sandstone (whose normal $\delta^{18}O$ values are around $13\%_o$) or a chert that has reequilibrated with the volumetrically dominant surrounding pelites. Thus, your measurements are inconclusive.

5. Temperatures indicate that all three minerals are in equilibrium at 638°C. Projecting this hypothesis onto the diagram, with the additional constraint that the temperature and pressure must lie on the andalusite–sillimanite univariant curve, corresponds to a very low pressure of about 1.6 kbar.

REFERENCES

Baertschi, P. (1950). Isotopic composition of the oxygen in silicate rocks. *Nature* **166**, 112–113.

Baertschi, P. (1951). Relative abundances of oxygen and carbon isotopes in carbonate rocks. *Nature* **168**, 288–289.

Baertschi, P., and Silverman, S. R. (1951). The determination of relative abundances of the oxygen isotopes in silicate rocks. *Geochimica et Cosmochimica Acta* **1**, 4–6.

Barnett, D. E., and Bowman, J. R. (1995). Coupled mass transport and kinetically limited isotope exchange: Applications and exchange mechanisms. *Geology* **23**, 225–228.

Baumgartner, L. P., and Rumble, D. III (1988). Transport of stable isotopes; 1, Development of a kinetic continuum theory for stable isotope transport. *Contributions to Mineralogy and Petrology* **98**, 417–430.

Bebout, G. E., and Barton, M. D. (1989). Fluid flow and metasomatism in a subduction zone hydrothermal system; Catalina schist terrane, California. *Geology* **17**, 976–980.

Bickle M. J., and Baker, J. (1990). Advective–diffusive transport of isotopic fronts; an example from Naxos, Greece. *Earth and Planetary Science Letters* **97**, 78–93.

Blattner, P., and Lassey, K. R. (1989). Stable-isotope exchange fronts, Damköhler numbers, and fluid to rock ratios. *Chemical Geology* **78**, 381–392.

Bowman, J. R., and Ghent, E. D. (1986). Oxygen and hydrogen isotope study of minerals from metapelitic rocks, staurolite to sillimanite zones, Mica Creek, British Columbia. *Journal of Metamorphic Geology* **4**, 131–141.

Bowman, J. R., O'Neil, J. R., and Essene, E. J. (1985). Contact skarn formation at Elkhorn, Montana. II: Origin and evolution of C–O–H skarn fluids. *American Journal of Science* **285**, 621–660.

Bowman, J. R., Willett, S. D., and Cook, S. J. (1994). Oxygen isotopic transport and exchange during fluid flow: One-dimensional models and applications. *American Journal of Science* **294**, 1–55.

Burkhard, M., and Kerrich R. (1990). Fluid–rock interactions during thrusting of the Glarus nappe—evidence from geochemical and stable isotope data. *Schweizerische Mineralogische und Petrographische Mitteilungen* **70**, 77–82.

Cartwright, I., and Valley, J. W. (1990). Fluid–rock interaction in the North-west Adirondack Mountains, New York State. In *High Temperature Metamorphism And Crustal Anatexis*, Vol. 2 (ed. J. R. Ashworth and M. Brown), pp. 180–197. London and Boston: Unwin Hyman.

Cavosie, A., Sharp, Z. D., and Selverstone, J. (2002). Co-existing aluminum silicates in quartz veins: A quantitative approach for determining andalusite–sillimanite equilibrium in natural samples using oxygen isotopes. *American Mineralogist* **87**, 417–423.

Chacko, T., Cole, D. R., and Horita, J. (2001). Equilibrium oxygen, hydrogen and carbon isotope fractionation factors applicable to geologic systems. In *Stable Isotope Geochemistry*, Vol. 43 (ed. J. W. Valley and D. R. Cole), pp. 1–81. Washington, DC: Mineralogical Society of America.

Chamberlain, C. P., and Conrad, M. E. (1991). Oxygen isotope zoning in garnet. *Science* **254**, 403–406.

Clayton, R. N., and Epstein, S. (1958). The relationship between O^{18}/O^{16} ratios in coexisting quartz, carbonate, and iron oxides from various geological deposits. *Journal of Geology* **66**, 352–373.

Clayton, R. N., and Epstein, S. (1961). The use of oxygen isotopes in high-temperature geological thermometry. *Journal of Geology* **69**, 447–452.

Cole, D. R., and Chakraborty, S. (2001). Rates and mechanisms of isotopic exchange. In *Stable Isotope Geochemistry*, Vol. 43 (ed. J. W. Valley and D. R. Cole), pp. 83–223. Washington, DC: Mineralogical Society of America.

Conrad, M. E., and Chamberlain, C. P. (1992). Laser-based, in situ measurements of fine-scale variations in the $\delta^{18}O$ values of hydrothermal quartz. *Geology* **20**, 812–816.

Cook, S. J., Bowman, J. R., and Forster, C. B. (1997). Contact metamorphism surrounding the Alta stock: Finite element model simulation of heat- and $^{18}O/^{16}O$ mass-transport during prograde metamorphism. *American Journal of Science* **297**, 1–55.

Crespo-Blanc, A., Masson, H., Sharp, Z. D., Cosca, M., and Hunziker, J. (1995). A stable and $^{40}Ar/^{39}Ar$ isotope study of a major thrust in the Helvetic nappes. *Geological Society of America Bulletin* **107**, 1129–1144.

Crowe, D. E. (1994). Preservation of original hydrothermal $\delta^{34}S$ values in greenschist and amphibolite volcanogenic massive sulfide deposits. *Geology* **22**, 873–876.

Crowe, D. E., and Valley, J. W. (1992). Laser microprobe study of sulfur isotope variation in a sea-floor hydrothermal spire, Axial Seamount, Juan de Fuca Ridge, eastern Pacific. *Chemical Geology* **101**, 63–70.

Crowe, D. E., Valley, J. W., and Baker, K. L. (1990). Micro-analysis of sulfur-isotope ratios and zonation by laser microprobe. *Geochimica et Cosmochimica Acta* **54**, 2075–2092.

Deines, P. (1968). The carbon and oxygen isotopic composition of carbonates from a mica peridotite dike near Dixonville, Pennsylvania. *Geochimica et Cosmochimica Acta* **32**, 613–625.

Deines, P., and Gold, D. P. (1969). The change in carbon and oxygen isotopic composition during contact metamorphism of Trenton limestone by the Mount Royal pluton. *Geochimica et Cosmochimica Acta* **33**, 421–424.

Dipple, G. M., and Ferry, J. M. (1992). Fluid flow and stable isotopic alteration in rocks at elevated temperatures with applications to metamorphism. *Geochimica et Cosmochimica Acta* **56**, 3539–3550.

Dodson, M. H. (1973). Closure temperature in cooling geochronological and petrological systems. *Contributions to Mineralogy and Petrology* **40**, 259–274.

Dunn, S. R., and Valley, J. W. (1992). Calcite–graphite isotope thermometry; a test for polymetamorphsim in marble, Tudor gabbro aureole, Ontario, Canada. *Journal of Metamorphic Geology* **10**, 487–501.

Eiler, J. M., Baumgartner, L. P., and Valley, J. W. (1992). Intercrystalline stable isotope diffusion; a fast grain boundary model. *Contributions to Mineralogy and Petrology* **112**, 543–557.

Eiler, J. M., Baumgartner, L. P., and Valley, J. W. (1995). Fast Grain Boundary: A Fortran-77 program for calculating the effects of retrograde interdiffusion of stable isotopes. *Computers and Geosciences* **20**, 1415–1434.

Elphick, S. C., Graham, C. M., and Dennis, P. F. (1988). An ion microprobe study of anhydrous oxygen diffusion in anorthite; a comparison with hydrothermal data and some geological implications. *Contributions to Mineralogy and Petrology* **100**, 490–495.

Farquhar, J., Chacko, T., and Frost, B. R. (1993). Strategies for high-temperature oxygen isotope thermometry; a worked example from the Laramie anorthosite complex, Wyoming, USA. *Earth and Planetary Science Letters* **117**, 407–422.

Farver, J. B., and Yund, R. A. (1991a). Measurement of oxygen grain boundary diffusion in natural, fine-grained, quartz aggregates. *Geochimica et Cosmochimica Acta* **55**, 1597–1607.

Farver, J. R., and Yund, R. A. (1991b). Oxygen diffusion in quartz: Dependence on temperature and water fugacity. *Chemical Geology* **90**, 55–70.

Fourcade, S., and Javoy, M. (1973). Rapports $^{18}O/^{16}O$ dans les roches du vieux socle catazonal d'In Ouzzal (Sahara algérien). *Contributions to Mineralogy and Petrology* **42**, 235–244.

Fourcade, S., Marquer, D., and Javoy, M. (1989). $^{18}O/^{16}O$ variations and fluid circulation in a deep shear zone: The case of the Alpine ultramylonites from the Aar massif (Central Alps, Switzerland). *Chemical Geology* **77**, 119–131.

Frey, M., Hunziker, J. C., O'Neil, J. R., and Schwander, H. W. (1976). Equilibrium–disequilibrium relations in the Monte Rosa Granite, Western Alps: Petrological, Rb–Sr and stable isotope data. *Contributions to Mineralogy and Petrology* **55**, 147–179.

Frezzotti, M. L., Dallai, L., and Sharp, Z. D. (2000) Fluid-inclusion and stable-isotope evidence for fluid infiltration and veining during metamorphism in marbles and metapelites. *European Journal of Mineralogy* **12**, 231–246.

Fricke, H. C., Wickham, S. M., and O'Neil, J. R. (1992). Oxygen and hydrogen isotope evidence for meteoric water infiltration during mylonitization and uplift in the Ruby Mountains–East Humboldt Range core complex, Nevada. *Contributions to Mineralogy and Petrology* **111**, 203–221.

Ganor, J., Matthews, A., and Paldor, N. (1989). Constraints on effective diffusivity during oxygen isotope exchange at a marble–schist contact, Sifnos (Cyclades), Greece. *Earth and Planetary Science Letters* **94**, 208–216.

Garlick, G. D., and Epstein, S. (1967). Oxygen isotope ratios in coexisting minerals of regionally metamorphosed rocks. *Geochimica et Cosmochimica Acta* **31**, 181–214.

Ghent, E. D., and O'Neil, J. R. (1985). Late Precambrian marbles of unusual carbon-isotope composition, southeastern British Columbia. *Canadian Journal of Earth Sciences* **22**, 324–329.

Giletti, B. J. (1986). Diffusion effects on oxygen isotope temperatures of slowly cooled igneous and metamorphic rocks. *Earth and Planetary Science Letters* **77**, 218–228.

Hoefs, J., Coolen, J. J. M. M. M., and Touret, J. (1981). The sulfur and carbon isotope composition of scapolite-rich granulites from southern Tanzania. *Contributions to Mineralogy and Petrology* **78**, 332–336.

Hoefs, J., and Touret, J. (1975). Fluid inclusion and carbon isotope study from Bamble granulites (South Norway); a preliminary investigation. *Contributions to Mineralogy and Petrology* **52**, 165–174.

Hoernes, S., and Hoffer, E. (1979). Equilibrium relations of prograde metamorphic mineral assemblages. A stable isotope study of rocks of the Damara Orogen, Namibia. *Contributions to Mineralogy and Petrology* **68**, 377–389.

Holness, M. B., and Graham, C. M. (1991). Equilibrium dihedral angles in the system H_2O–CO_2–NaCl–calcite, and implications for fluid flow during metamorphism. *Contributions to Mineralogy and Petrology* **108**, 368–383.

James, H. L., and Clayton, R. N. (1962). Oxygen isotope fractionation in metamorphosed iron formations of the Lake Superior region and in other iron-rich rocks. In *Petrologic Studies, Buddington Volume* (ed. A. E. J. Engel, H. L. James, and B. F. Leonard), pp. 217–233. New York: Geological Society of America.

Jamtveit, B., Bucher, N. K., and Stijfhoorn-Derk, E. (1992). Contact metamorphism of layered shale–carbonate sequences in the Oslo Rift; I, Buffering, infiltration, and the mechanisms of mass transport. *Journal of Petrology* **33**, 377–422.

Javoy, M. (1977). Stable isotopes and geothermometry. *Journal of the Geological Society of London* **133**, 609–636.

Javoy, M., Fourcade, S., and Allegre, C. J. (1970). Graphical method for examination of $^{18}O/^{16}O$ fractionations in silicate rocks. *Earth and Planetary Science Letters* **10**, 12–16.

Jenkin, G. R. T., Craw, D., and Fallick, A. E. (1994). Stable isotopic and fluid inclusion evidence for meteoric fluid penetration into an active mountain belt; Alpine Schist, New Zealand. *Journal of Metamorphic Geology* **12**, 429–444.

Jenkin, G. R. T., Fallick, A. E., and Leake, B. E. (1992). A stable isotope study of retrograde alteration in SW Connemara, Ireland. *Contributions to Mineralogy and Petrology* **110**, 269–288.

Jenkin, G. R. T., Farrow, C. M., Fallick, A. E., and Higgins, D. (1994). Oxygen isotope exchange and closure temperatures in cooling rocks. *Journal of Metamorphic Geology* **12**, 221–235.

Kelley, S. P., and Fallick, A. E. (1990). High precision spatially resolved analysis of $\delta^{34}S$ in sulphides using a laser extraction technique. *Geochimica et Cosmochimica Acta* **54**, 883–888.

Kerrich, R. (1986). Fluid transport in lineaments. *Philosophical Transactions of the Royal Society of London, Series A* **317**, 219–251.

Kirschner, D. L., Sharp, Z. D., and Masson, H. (1995). Oxygen isotope thermometry of quartz–calcite veins: Unraveling the thermal-tectonic history of the subgreenschist facies Morcles nappe (Swiss Alps). *Geological Society of America Bulletin* **107**, 1145–1156.

Kirschner, D. L., Sharp, Z. D., and Teyssier, C. (1993). Vein growth mechanisms and fluid sources revealed by oxygen isotope laser microprobe. *Geology* **21**, 85–88.

Kohn, M. J. (1999). Why most "dry" rocks should cool "wet." *American Mineralogist* **84**, 570–580.

Kohn, M. J., Valley, J. W., Elsenheimer, D., and Spicuzza, M. J. (1993). O isotope zoning in garnet and staurolite: Evidence for closed-system mineral growth during regional metamorphsim. *American Mineralogist* **78**, 988–1001.

Kreulen, R. (1980). CO_2-rich fluids during regional metamorphism on Naxon (Greece): Carbon isotopes and fluid inclusions. *American Journal of Science* **280**, 745–771.

Larson, T., and Sharp, Z. D. (2003). Stable isotope constraints on the Al_2SiO_5 "triple-point" rocks from the Proterozoic Priest pluton contact aureole, New Mexico. *Journal of Metamorphic Geology* **21**, 785–798.

Larson, T. E., and Sharp, Z. D. (2005). Interpreting prograde-growth histories of Al_2SiO_5 triple-point rocks using oxygen-isotope thermometry: an example from the Truchas Mountains, USA. *Journal of Metamorphic Geology* **23**, 847–863.

Lichtner, P. C. (1985). Continuum model for simultaneous chemical reactions and mass transport in hydrothermal systems. *Geochimica et Cosmochimica Acta* **49**, 779–800.

McCaig, A. M., Wickham, S. M., and Taylor, H. P., Jr. (1990). Deep fluid circulation in Alpine shear zones, Pyrenees, France: Field and oxygen isotope studies. *Contributions to Mineralogy and Petrology* **106**, 41–60.

Moecher, D. P., Valley, J. W., and Essene, E. J. (1994). Extraction and carbon isotope analysis of CO_2 from scapolite in deep crustal granulites and xenoliths. *Geochimica et Cosmochimica Acta* **58**, 959–967.

Monster, J., Appel, P. W. U., Thode, H. G., Schidlowski, M., Carmichael, C. M., and Bridgwater, D. (1979). Sulfur isotope studies in early Archaean sediments from Isua, West Greenland; implications for the antiquity of bacterial sulfate reduction. *Geochimica et Cosmochimica Acta.* **43**, 405–413.

Mora, C. I., and Valley, J. W. (1991). Prograde and retrograde fluid–rock interaction in calc–silicates northwest of the Idaho Batholith; stable isotopic evidence. *Contributions to Mineralogy and Petrology* **108**, 162–174.

Morrison, J., and Anderson, J. L. (1998). Footwall refrigeration along a detachment fault; implications for the thermal evolution of core complexes. *Science* **279**, 63–66.

Nabelek, P. I. (1991). Stable isotope monitors. In *Contact Metamorphism*, Vol. 26 (ed. D. M. Kerrick), pp. 395–435. Chelsea, MI: Mineralogical Society of America.

Nabelek, P. I., and Labotka, T. C. (1993). Implications of geochemical fronts in the Notch Peak contact-metamorphic aureole, Utah, USA. *Earth and Planetary Science Letters* **119**, 539–559.

Nabelek, P. I., Labotka, T. C., O'Neil, J. R., and Papike, J. J. (1984). Contrasting fluid/rock interaction between the Notch Peak granitic intrusion and argillites and limestones in western Utah; evidence from stable isotopes and phase assemblages. *Contributions to Mineralogy and Petrology* **86**, 25–34.

Nabelek, P. I., O'Neil, J. R., and Papike, J. J. (1983). Vapor phase exsolution as a controlling factor in hydrogen isotope variation in granitic rocks; the Notch Peak granitic stock, Utah. *Earth and Planetary Science Letters* **66**, 137–150.

Ohmoto, H. (1986). Stable isotope geochemistry of ore deposits. In *Stable Isotopes in High Temperature Geological Processes*, Vol. 16 (ed. J. W. Valley, H. P. Taylor, Jr., and J. R. O'Neil), pp. 491–559. Chelsea, MI: Mineralogical Society of America.

Oliver, N. H. S., Hoering, T. C., Johnson, T. W., Rumble, D. III, and Shanks, W. C. III (1992). Sulfur isotopic disequilibrium and fluid–rock interaction during metamorphism of sulfidic black shales from the Waterville–Augusta area, Maine, USA. *Geochimica et Cosmochimica Acta* **56**, 4257–4265.

O'Neil, J. R., and Clayton, R. N. (1964). Oxygen isotope geothermometry. In *Isotopic and Cosmic Chemistry*, pp. 157–168. Amsterdam: North Holland Publishing Co.

Pineau, F., Javoy, M., Behar, F., and Touret, J. (1981). La géochimie isotopique du facies granulite du Bamble (Norvège) et l'origine des fluides carbones dans la croûte profonde. *Bulletin de Mineralogie* **104**, 630–641.

Ripley, E. M., Butler, B. K., and Taib, N. I. (1992). Effects of devolatilization on the hydrogen isotopic composition of pelitic rocks in the contact aureole of the Duluth Complex, northeastern Minnesota, U.S.A. *Chemical Geology* **102**, 185–197.

Rumble, D. III (1982). Stable isotope fractionation during metamorphic devolatilization reactions. In *Characterization of Metamorphism through Mineral Equilibria*, Vol. 10 (ed. J. M. Ferry), pp. 327–353. Chelsea, MI: Mineralogical Society of America.

Rumble, D. III, and Spear, F. S. (1983). Oxygen-isotope equilibration and permeability enhancement during regional metamorphism. *Journal of the Geological Society of London* **140**, 619–628.

Rye, R. O., Schuiling, R. D., Rye, D. M., and Jansen, J. B. H. (1976). Carbon, hydrogen, and oxygen isotope studies of the regional metamorphic complex at Naxos, Greece. *Geochimica et Cosmochimica Acta.* **40**, 1031–1049.

Schwarcz, H. P., and Clayton, R. N. (1965). Oxygen isotopic studies of amphibolites. *Canadian Journal of Earth Science* **2**, 72–84.

Sharp, Z. D. (1990). A laser-based microanalytical method for the *in situ* determination of oxygen isotope ratios of silicates and oxides. *Geochimica et Cosmochimica Acta* **54**, 1353–1357.

Sharp, Z. D. (1991). Determination of oxygen diffusion rates in magnetite from natural isotopic variations. *Geology* **19**, 653–656.

Sharp, Z. D. (1995). Oxygen isotope geochemistry of the Al_2SiO_5 polymorphs. *American Journal of Science* **295**, 1058–1076.

Sharp, Z. D. (1998). Application of stable isotope geochemistry to low grade metamorphic rocks. In *Low-Grade Metamorphism* (ed. M. Frey and D. Robinson), pp. 227–260. London: Blackwell.

Sharp, Z. D., Essene, E. J., and Hunziker, J. C. (1993). Stable isotope geochemistry and phase equilibria of coesite-bearing whiteschists, Dora Maira Massif, Western Alps. *Contributions to Mineralogy and Petrology* **114**, 1–12.

Sharp, Z. D., Essene, E. J., and Smyth, J. R. (1992). Ultra-high temperatures from oxygen isotope thermometry of a coesite–sanidine grospydite. *Contributions to Mineralogy and Petrology* **112**, 358–370.

Sharp, Z. D., Giletti, B. J., and Yoder, H. S., Jr. (1991). Oxygen diffusion rates in quartz exchanged with CO_2. *Earth and Planetary Science Letters* **107**, 339–348.

Sharp, Z. D., and Moecher, D. P. (1994). O-isotope variations in a porphyroclastic meta-anorthosite; diffusion effects and false isotherms. *American Mineralogist* **79**, 951–959.

Sheppard, S. M. F. (1986). Characterization and isotopic variations in natural waters. In *Stable Isotopes in High Temperature Geological Processes*, Vol. 16 (ed. J. W. Valley, H. P. Taylor, Jr. and J. R. O'Neil), pp. 165–183. Chelsea, MI: Mineralogical Society of America.

Shieh, Y., and Schwarcz, H. P. (1977). An estimate of the oxygen isotope composition of a large segment of the Canadian Shield in northwestern Ontario. *Canadian Journal of Earth Science* **14**, 927–931.

Silverman, S. R. (1951). The isotope geology of oxygen. *Geochimica et Cosmochimica Acta* **2**, 26–42.

Spooner, E. T. C., Beckinsale, R. D., England, P. C., and Senior, A. (1977). Hydration, oxygen 18 enrichment and oxidation during ocean floor hydrothermal metamorphism of ophiolitic metabasic rocks from E. Liguria, Italy. *Geochimica et Cosmochimica Acta* **41**, 857–871.

Sturchio, N. C., and Muehlenbachs, K. (1985). Origin of low-^{18}O metamorphic rocks from a late Proterozoic shear zone in the Eastern Desert of Egypt. *Contributions to Mineralogy and Petrology* **91**, 188–195.

Taylor, B. E. (2004). Biogenic and thermogenic sulfate reduction in the Sullivan Pb–Zn–Ag deposit, British Columbia (Canada): Evidence from micro-isotopic analysis of carbonate and sulfide in bedded ores. *Chemical Geology* **204**, 215–236.

Taylor, H. P., Jr. (1979). Oxygen and hydrogen isotope relationships in hydrothermal mineral deposits. In *Geochemistry of Hydrothermal Ore Deposits* (ed. H. L. Barnes), pp. 236–277. New York: J. Wiley and Sons.

Taylor, H. P., Jr., Albee, A. L., and Epstein, S. (1963). O^{18}/O^{16} ratios of coexisting minerals in three assemblages of kyanite-zone pelitic schist. *Journal of Geology* **71**, 513–522.

Taylor, H. P., Jr., and Epstein, S. (1962a). Relationship between O^{18}/O^{16} ratios in coexisting minerals of igneous and metamorphic rocks, Part 1. Principles and experimental results. *Geological Society of America Bulletin* **73**, 461–480.

Taylor, H. P., Jr., and Epstein, S. (1962b). Relationship between ratios in coexisting minerals of igneous and metamorphic rocks, Part 2. Application to petrologic problems. *Geological Society of America Bulletin* **73**, 675–694.

Taylor, H. P., Jr., Frechen, J., and Degens, E. T. (1967). Oxygen and carbon isotope studies of carbonatites from the Laacher See district, west Germany and the Alnö district, Sweden. *Geochimica et Cosmochimica Acta.* **31**, 407–430.

Valley, J. W. (1986). Stable isotope geochemistry of metamorphic rocks. In *Stable Isotopes in High Temperature Geological Processes*, Vol. 16 (ed. J. W. Valley, H. P. Taylor, Jr., and J. R. O'Neil), pp. 445–490. Chelsea, MI: Mineralogical Society of America.

Valley, J. W. (2001). Stable isotope thermometry at high temperatures. In *Stable Isotope Geochemistry*, Vol. 43 (ed. J. W. Valley and D. R. Cole), pp. 365–413. Washington, DC: Mineralogical Society of America.

Valley, J. W., and Graham, C. M. (1998). Ion microprobe analysis of oxygen, carbon, and hydrogen isotope ratios. In *Applications of Microananalytical Techniques to Understanding Mineralizing Processes*, Reviews in Economic Geology 7 (ed. M. A. McKibben, W. C. Shanks III, and W. I. Ridlye), pp. 73–98. Chelsea, MI: Bookcrafters, Inc.

Valley J. W., and O'Neil, J. R. (1982). Oxygen isotope evidence for shallow emplacement of Adirondack anorthosite. *Nature* **300**, 497–500.

Vennemann, T. W., Kesler, S. E., and O'Neil, J. R. (1992). Stable isotope compositions of quartz pebbles and their fluid inclusions as tracers of sediment provenance; implications for gold- and uranium-bearing quartz pebble conglomerates. *Geology* **20**, 837–840.

Vry, J. K., Brown, P. E., Valley, J. W., and Morrison, J. (1988). Constraints on granulite genesis from carbon isotopic compositions of cordierite and graphite. *Nature* **332**, 66–68.

Vyhnal, C. R., and Chamberlain, C. P. (1996). Preservation of early isotopic signatures during prograde metamorphism, eastern Vermont. *American Journal of Science* **296**, 394–419.

Wickham, S. M., and Taylor, H. P., Jr. (1985). Stable isotopic evidence for large-scale seawater infiltration in a regional metamorphic terrane; the Trois Seigneurs Massif, Pyrenees, France. *Contributions to Mineralogy and Petrology* **91**, 122–137.

Young, E. D. (1993). On the $^{18}O/^{16}O$ record of reaction progress in open and closed metamorphic systems. *Earth and Planetary Science Letters* **117**, 147–167.

Young, E. D., Coutts, D. W., and Kapitan, D. (1998). UV laser ablation and irm–GCMS microanalysis of $^{18}O/^{16}O$ and $^{17}O/^{16}O$ with application to a calcium–aluminium-rich inclusion from the Allende Meteorite. *Geochimica et Cosmochimica Acta* **62**, 3161–3168.

Young, E. D., and Rumble, D. III. (1993). The origin of correlated variations in in-situ $^{18}O/^{16}O$ and elemental concentrations in metamorphic garnet from southeastern Vermont, USA. *Geochimica et Cosmochimica Acta* **57**, 2585–2597.

Yui, T.-F., Rumble, D. III, and Lo, C.-H. (1995). Unusually low $\delta^{18}O$ ultra-high-pressure metamorphic rocks from the Sulu Terrain, eastern China. *Geochimica et Cosmochimica Acta* **59**, 2859–2864.

CHAPTER 13

EXTRATERRESTRIAL MATERIALS

13.1 INTRODUCTION

Some of the earliest problems that physicists envisioned addressing with stable isotope chemistry involved the origin and heterogeneities of the solar system. Hyperbolic orbits (erroneously) identified for certain meteorites suggested that there might be matter from outside the solar system which ultimately reached the Earth. Early attempts to find exotic carbon and oxygen isotope ratios in meteorites failed (Manian et al., 1934; Jenkins and King, 1936), mainly because the measurements were not sufficiently precise at the time, but also partly because the "right" material was not analyzed. Hydrogen isotope analyses followed in the early 1950s, when δD values approaching 300‰ from the fresh carbonaceous chondrites Ivuna and Orgueil were measured (Boato, 1954). Additional isotopic systems were analyzed over the next several decades and included sulfur (Trofimov, 1949), nitrogen (Injerd and Kaplan, 1974), and carbon (Krouse and Modzeleski, 1970). There was a flurry of interest in lunar materials when Apollo samples were returned to Earth. Most lunar samples had $\delta^{18}O$ values very similar to those of equivalent terrestrial rock types, indicating a close genetic relationship between Earth and Moon. Then, in 1973, Clayton et al. (1973) published a paper that completely changed the meteorite community. For the first time, an "exotic" stable isotope component in meteorites had clearly been identified. Since then, stable isotope geochemistry has become one of the most important tools for constraining mechanisms of formation and history of the early solar system.

There are no distinct boundaries between traditional earth sciences, cosmochemistry, space sciences, and astrophysics, so it is difficult to limit the scope of this chapter. For example, isotope chemistry of interstellar grains (Anders and Zinner, 1993; Nittler, 2003) is important in studies of nucleosynthesis and stellar evolution. At the other extreme, average isotope chemistries of carbonaceous chondrites are used to estimate bulk Earth values, which are then applied to terrestrial studies completely unrelated to meteorite research. All of these fields spill into each other, but here we will try to limit our examinations to more traditional meteorite research. Readers interested in processes of star formation and the nucleosynthesis of elements and their isotopes are referred to Anders and Zinner (1993, in appendix) for a review and additional references. Note that the "normal" stable isotopes (H, C, N, O, S) are not used independently of other isotopic systems, such as the noble gases, nontraditional isotopes (e.g., Mg, Si), and short-lived radiogenic isotopes. But again, discussion of these systems is beyond the scope of this chapter. Further information beyond the short review here can be found in Clayton (1993) for oxygen isotopes and in McKeegan and Leshin (2001) for H, C, N, O stable isotope systems.

13.2 CLASSIFICATION OF METEORITES

The alphabet soup of meteorite names can be very confusing to the uninitiated, but a simple review reveals that it is really not complex. Table 13.1 outlines the major types of meteorites. There are basic general groups, which are then subdivided and subdivided again. ***Iron meteorites*** are the disrupted cores of asteroids or planetesimals. ***Stony iron meteorites*** are core–mantle boundary material, and ***stony meteorites*** are considered to be crustal magmatic material or relatively unprocessed material.

The stony meteorites can be further divided into chondrites and achondrites on the basis of whether they once contained chondrules or not. Chondrules are spherules of olivine or pyroxene with or without glass, together with metal inclusions formed by melting during flash heating. Chondrules are set in a fine-grained matrix of silicates, oxides, metal, sulfides, and organic matter. The chondrites are considered to be agglomerations of rocky and icy material from unprocessed remnants of planetesimals from the asteroid belt. Chondrites also contain interstellar grains, calcium–aluminum inclusions (CAIs), and other refractory objects. Achondrites are much more processed material, often representing crustal magmatic or metamorphic rocks from planets or large asteroids. The origin of many achondrites can be attributed to one of several large bodies, including Mars (the SNC meteorities), the Moon, and the asteroid 4 Vesta.

13.3 OXYGEN ISOTOPE VARIATIONS IN METEORITES

13.3.1 Introduction

The pioneers of fluorination included meteorites (or tektites) in their rock suites to fill out the general overview of natural oxygen isotope variations (Baertschi, 1950; Silverman, 1951). Vinogradov et al. (1960) measured a suite of meteorites using the method of carbon reduction, and found noticeable, but small, variations among different types. Similar conclusions were reached in a later thorough examination of different types of meteorites (Reuter et al., 1965). Silicate separates from 27 chondrites varied by only 1.2‰. Chondrules had the same values as the whole rock. Pallasites were several per mil lower than the chondrites, but overall, it appeared that meteorites were not particularly interesting from a stable isotope viewpoint. In the same year and in the same journal, Taylor et al. (1965) found much larger variations in chondrites when they concentrated on measuring only pyroxene and not olivine.[1] Now the variations observed were larger: 9‰ for pyroxenes and 13‰ between different bulk rock samples. The researchers developed a meteorite classification scheme of three groups with distinct $\delta^{18}O$ values.

---

[1]Olivine is a notoriously difficult mineral to fluorinate by conventional methods. (See Section 11.2.1 for more details.)

TABLE 13.1: Classification of different meteorite types.

Name	Mineralogy	Source	Remarks
Iron meteorites	Iron–nickel metal; iron sulfide	Cores of disrupted asteroids; impacts on asteroids	Groups distinguished by trace element abundances
Stony iron meteorites: pallasites mesosiderites	Roughly equal proportions of Fe, Ni metal, and silicate	Core–mantle boundary of disrupted asteroids; impacts on asteroids	
Stony meteorites: Achondrites			
Basaltic achondrites: howardites eucrites diogenites	Pigeonite–plagioclase-bearing basalts, orthopyroxenites or mélange of each	Thought to be a basaltic magma system from asteroid 4 Vesta	Distinguished from other achondrite groups on the basis of oxygen isotopes
Other achondrites: aubrites, ureilites, angrites, etc.	Exotic basalts and related igneous rocks	Main-belt differentiated asteroids	Each group distinguished on the basis of oxygen isotopes
Martian meteorites (SNC meteorites): shergottites nakhlites chassignites	Magmatic basalts, lherzolites, clinopyroxenites, orthopyroxenite, dunite	Mars	Young (most <1.3 Ga) (ALH84001 is 4.5 Ga). Characteristic stable and rare gas isotope ratios
Lunar	Basalts, anorthosites, and lunar breccias	Moon	Compare closely with samples returned from the Apollo and Luna missions.
Stony meteorites: Chondrites			
Agglomeratic rocks containing rocky and metallic materials. Unprocessed remnants of planetesimals from the asteroid belt. Oldest and most primitive material in the solar system. Consist of high-temperature chondrules hosted in a fine-grained matrix that represents low-T fraction of nebular material. Also contain interstellar grains, calcium–aluminum inclusions (CAIs), and other refractory objects. Groups are distinguished by chemistry and oxygen isotope ratios. Divided into types 1–6, where 3 is least altered, 2 and 1 are higher levels of low-T aqueous alteration, and 4–6 are increasing levels of thermal metamorphism.			
Carbonaceous or C chondrites	8 groups: CI, CM, CR, CV, CO, CK, CH, and CB; named after type, fall, and locality.		
Ordinary or O chondrites	3 groups: H, L, and LL: H = high total Fe content; L = low total Fe content; LL = low metallic Fe		
Enstatite or E chondrites	2 groups: EH and EL: EH = high total iron; EL = low total iron		
Other classes	Rumuruti-like (R) and Kakangari-like (K)		
Other extraterrestrial material			
Interplanetary dust particles (IDPs)	Divided into chondritic and nonchondritic types	Active periodic comets from Oort cloud and Kuiper belt.	Primordial "icy dust balls" (Rietmeijer, 1998). Average size is 10 μm.

(Brearley and Jones, 1998; Shearer et al., 1998.)

The return of the Apollo lunar samples led to a flurry of activity by a number of stable isotope laboratories in the United States (e.g., Epstein and Taylor, 1970; Friedman et al., 1970; Onuma et al., 1970; Kaplan and Petrowski, 1971). It was quickly recognized that the $\delta^{18}O$ values of lunar basalts and anorthosites were the same as the $\delta^{18}O$ values of their terrestrial equivalents.[2] These data demonstrated that the Earth and Moon were closely related, but said little about the genesis of the solar system.

Onuma and colleagues then returned to meteorites, determining temperatures of early ordinary chondrites by using oxygen isotope thermometry between the coexisting minerals plagioclase, olivine, and pyroxene (Onuma et al., 1972b). At the same time, they proposed a

---

[2] Later analyses confirmed that $\delta^{17}O$ values were also identical to terrestrial ones (Clayton and Mayeda, 1975).

model for condensation, chondrule formation, and planetary accretion based on the equilibration of solid material with a large, isotopically homogeneous nebular gas whose $\delta^{18}O$ value was $-1 \pm 2‰$ (Onuma et al., 1972a). Interestingly, some of the most refractory, presumably early formed mineral separates had $\delta^{18}O$ values that were too low to be explained by this model (e.g., melilite and spinel aggregates with $\delta^{18}O$ values as low as $-11.5‰$).

13.3.2 Discovery of an ^{17}O Anomaly

The cosmothermometer model had to be reevaluated (Onuma et al., 1974) when Clayton et al. (1973) made one of the most striking discoveries in the field of planetary science. These scientists analyzed anhydrous high-temperature phases from carbonaceous chondrites, mostly Allende (which had fallen to Earth only on Feb. 8, 1969, in Chihuahua, Mexico).[3] The $\delta^{18}O$ values obtained were extraordinary, covering a range of 30‰. This in itself was far more than anyone had seen previously, but through an extremely careful analytical procedure, they found something else that was even more surprising: The "apparent" $\delta^{13}C$ values of the CO_2 gas that they were analyzing were not constant. The reason and significance are explained in the next subsection.

The Chicago group analyzed silicates by reacting them with BrF_5 at high temperatures.[4] The oxygen that was evolved was converted to CO_2 by reacting with spectroscopic graphite at high temperatures in the presence of a platinum catalyst. This is standard procedure in most laboratories.[5] What was unusual is that they noticed that there was a systematic variation in the apparent $\delta^{13}C$ value of the CO_2 that could be correlated with the measured $\delta^{18}O$ value. In fact, the two should be completely independent of each other. The carbon comes from the graphite rod, the oxygen from the fluorinated silicate. The three isotopologues of CO_2 have masses of 44 ($^{12}C^{16}O_2$), 45 ($^{13}C^{16}O_2$ and $^{12}C^{16}O^{17}O$), and 46 ($^{12}O^{16}O^{18}O$).[6] In almost all known chemical fractionation mechanisms, the following relationship is valid:

$$\left(\frac{^{17}O}{^{16}O}\right) = \left(\frac{^{18}O}{^{16}O}\right)^{0.52}. \qquad (13.1)$$

This equation translates into $\delta^{17}O = 0.52(\delta^{18}O)$ for all but the most extreme compositions (e.g., Matsuhisa et al., 1978). Normally, there is no reason to measure the $\delta^{17}O$ value of a substance, because it varies with $\delta^{18}O$ according to equation (13.1). By subtracting out the ^{17}O contribution (from equation (13.1)) to the mass 45 isotopologue of CO_2—the so-called ^{17}O correction—we can calculate the $\delta^{13}C$ value of CO_2 gas from the 45/44 ratio. What Clayton et al. had found when they analyzed the high-temperature inclusions from chondrites was that the calculated $\delta^{13}C$ value was changing in relation to the $\delta^{18}O$ values. What they were actually seeing, and quickly realized, were variations in the $^{17}O/^{16}O$ ratio that did not conform to equation (13.1). When they plotted $\delta^{17}O$ vs. $\delta^{18}O$ (assuming that the $^{13}C/^{12}C$ ratio was constant), they found that the chondrite data plotted, not on the normal *terrestrial fractionation line* (TFL) with slope of 0.52, as is seen for almost all materials in the Earth and Moon, but rather on a line of slope 1, below the TFL line (Fig. 13.1). (See Box 13.1 for a discussion of $\delta^{18}O$ vs. $\delta^{17}O$ plots.)

---

[3]The other important carbonaceous chondrite meteorite that was analyzed was Murchison, which fell in Australia only a few months later.
[4]A typical reaction would be $2\ MgSiO_3 + 12/5\ BrF_5 \longrightarrow 2\ MgF_2 + 2\ SiF_4 + 6/5\ Br_2 + 3\ O_2$.
[5]However more and more laboratories have started analyzing O_2 directly.
[6]Other isotopologues, such as $^{12}C^{17}O^{17}O$, are so rare that they can be ignored. (See Craig, 1957.)

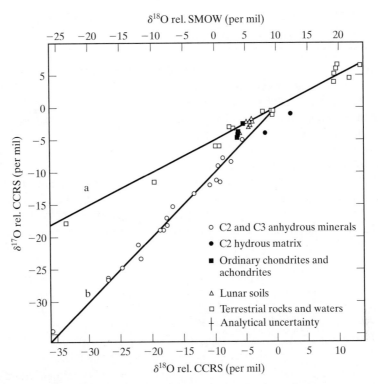

FIGURE 13.1: Plot of $\delta^{17}O$ vs. $\delta^{18}O$ values for anhydrous (high-T) chondritic samples (open circles), aqueously altered chondritic samples (filled circles), and terrestrial material, which, unlike chondrites, plots on the the terrestrial fractionation line (TFL) (slope 1/2). All data are plotted relative to a carbonaceous chondrite reference standard (CCRS) defined as the intersection of the slope 1 line from chondrite data with the TFL. Plotted in this way, the chondrite data are explained in terms of mixing between two components: a nebular gas near the CCRS and a very ^{16}O enriched source, possibly from outside the solar system. (Reprinted from Clayton et al., 1973, with permission from AAAS. Note that, because their data are defined in terms of the CCRS, the mixing line b in the figure meets the criterion that $\delta^{17}O = \delta^{18}O$.)

13.3.3 Possible Explanations: Mixing of Two Distinct Reservoirs

The results of the analysis of the anhydrous high-temperature material were interpreted as representing mixing between two different reservoirs: a nebular reservoir with $\delta^{18}O = \sim 30$ ‰, $\delta^{17}O = \sim 24$‰ (Clayton, 1993), and a primitive, presolar ^{16}O-rich dust source with $\delta^{18}O = \sim -40$‰ and $\delta^{17}O = \sim -42$‰ that apparently originated outside of the solar system. Only once before had any geochemical evidence been found suggesting a component derived from outside of our own solar system (Black, 1972). Clayton et al. postulated that the ^{16}O-rich component was probably extremely impoverished in ^{17}O and ^{18}O, as might be expected from oxygen formed from a young helium-burning star, in which only ^{16}O is produced.

Additional analyses in coming years refined the meteorite data and allowed for classifications of different types of components on the basis of oxygen isotope values (Clayton et al., 1976; Clayton et al., 1977). A mixing line between an ^{16}O-rich and a nebular component was defined on the basis of carbonaceous chondrite anhydrous mineral

BOX 13.1 The three-isotope plot for oxygen

All chemical fractionation processes (except for certain photochemical effects) fractionate the three isotopes of oxygen according to equation (13.1). On the three-isotope plot of $\delta^{17}O$ vs. $\delta^{18}O$, this leads to all data defining a $\delta^{17}O/\delta^{18}O$ slope of 0.52 (the *mass fractionation line*, or mfl). All terrestrial samples plot on an mfl that, by definition, intersects $\delta^{17}O = \delta^{18}O = 0$‰ (vs. SMOW) and is called the *terrestrial fractionation line (TFL)*. Any object that undergoes mass-dependent fractionation will spread out on an mfl line, illustrated in Fig. 13.B1 by the dashed line with slope 0.52 emanating from point *a*. Mixing between two reservoirs with distinct $\delta^{17}O - \delta^{18}O$ values will result in a straight line whose slope is defined by the compositions of the two end members. The mixing line of slope 1 could be explained by mixing between an average nebular composition given by CCRS in Fig. 13.1 and a second component rich in ^{16}O. How far a sample plots off the TFL is defined in terms of $\Delta^{17}O$ by the equation $\Delta^{17}O = \delta^{17}O - 0.52\delta^{18}O$. The inset shows the $\delta^{17}O - \delta^{18}O$ values of samples from Mars. They plot on an mfl with a $\Delta^{17}O$ value of +0.3‰

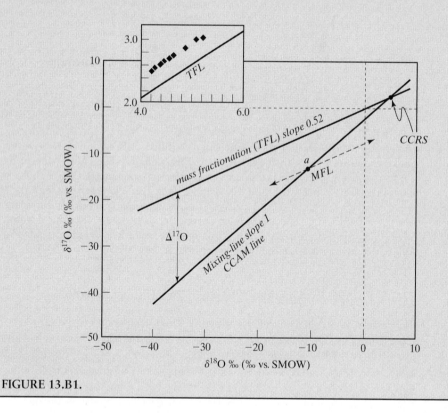

FIGURE 13.B1.

data (CCAM line) with a slope of 0.96. Later, Young and Russell (1998) analyzed some inclusions from Allende and proposed that the slope of the mixing line was exactly 1.0. Data lying to the high $\delta^{18}O$ side of this line (such as hydrous matrix phases in chondrites) were explained by mass-dependent reactions. Implied, but not explicitly stated in their study, is that it is unlikely that mixing between two distinct reservoirs would fortuitously

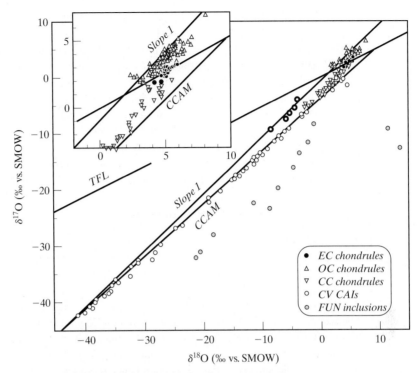

FIGURE 13.2: $\delta^{17}O-\delta^{18}O$ **plot for chondrite data.** The high-temperature anhydrous inclusion data plot on the carbonaceous chondrite anhydrous mineral (CCAM) line with slope 0.96. A slope 1 line is suggested on the basis of laser fluorination data (thick-outlined open circles at $\delta^{18}O \sim -5‰$). The ordinary (OC) and enstatite (EC) chondrites plot to the right of the slope 1 line, indicating alteration with a high $\delta^{18}O-\delta^{17}O$ reservoir or mass-dependent fractionations. (Modified from McKeegan and Leshin, 2001, with additional laser data from Young and Russell, 1998.) Shaded circles are fractionated unidentified nuclear (FUN) inclusions, which have a complex, as yet poorly understood, history.

have a slope of *exactly* unity. An alternative explanation based on mass-independent fractionation (see next subsection) was supported by the Young and Russell data.

Ordinary chondrites plot at or above the TFL, with a slope close to 1. The ordinary chondrite data plot either to the left or slightly to the right of the CCAM line, depending upon whether a slope of 0.96 or 1.0 is chosen (Fig. 13.2). If the Young and Russell mixing line is chosen, no additional low-$\delta^{18}O$ mixing reservoirs are required to explain the data, as had been proposed earlier (Clayton et al., 1991). Enstatite chondrites plot on the TFL from $\delta^{18}O = 4$ to 6‰. Other types of meteorites (not shown) have distinct fields in the $\delta^{17}O-\delta^{18}O$ diagram that can be thought of in terms of mixing and the original composition of the source planetesimal.

The original mixing-of-two-reservoirs explanation for the oxygen isotope trends in early formed material has always had some inconsistencies, which alternative models have sought to explain. Following are some of the concerns:

- If the ^{16}O-rich component formed by stellar nucleosynthesis of helium burning, then the $\delta^{18}O-\delta^{17}O$ values should correlate with those of other isotope systems, such as Mg and Si. However, no such correlations are found.

- The presolar component should be nearly pure ^{16}O. However, all measurements of CAI inclusions have $\delta^{18}O-\delta^{17}O$ values that "bottom out" around −50 to −40‰ (McKeegan and Leshin, 2001). Only one ^{16}O-rich oxide presolar grain has been found (Nittler et al., 1998), but the estimated abundance of such grains is less than 0.25 part per billion. If the exotic ^{16}O rich source was not pure ^{16}O, then why is the $^{18}O/^{17}O$ ratio the same as the solar nebular value (thereby giving the slope-1 line)?
- If an ^{16}O-rich melt exchanged with the solar nebular gas during crystallization, the degree of ^{17}O and ^{18}O enrichment should be correlated with the order of crystallization, yet this is not the case. Melilite has the highest $\delta^{18}O-\delta^{17}O$ values, but should crystallize before pyroxene (Stolper, 1982). Alternatively, exchange with the heavy nebular gas might have occurred by diffusion. Appropriate diffusion data are not sufficient to evaluate this possibility.

13.3.4 Mass-Independent Fractionation

An alternative to the two-reservoir-mixing hypothesis was born when Thiemens and Heidenreich (1983) discovered that the production of ozone (O_3) from molecular oxygen (O_2) in a high-frequency discharge occurred with a mass-independent fractionation (also called non-mass-dependent, or NoMaD, fractionation). The $\delta^{18}O-\delta^{17}O$ values of the residual oxygen and newly formed ozone plot on a slope 1 line (Fig. 13.3), nearly identical to the slope of the line for carbonaceous chondrite mineral inclusions found by Clayton et al. (1973).

Thiemens and Heidenreich (1983) suggested a mechanism of optical self-shielding to explain the mass-independent fractionation. O_2 gas undergoes strong absorption of ultraviolet light by the Schumann–Runge absorption bands between 1.76×10^{-7} m and 1.926×10^{-7} m, causing photodissociation of O_2. The absorption bands are slightly different for $^{16}O-^{16}O$ and the other isotopologues $^{17}O-^{16}O$ and $^{18}O-^{16}O$. Because more than 99.5% of O_2 consists of $^{16}O_2$, radiation corresponding to $^{16}O_2$ will strongly be absorbed (or self-shielded), so that only the wavelengths absorbed by $^{17}O-^{16}O$ and $^{18}O-^{16}O$ will filter through to the center of the reaction chamber. These UV rays will then cause mass-independent dissociation of the rare isotopologues $^{17}O-^{16}O$ and $^{18}O-^{16}O$. The dissociated ^{17}O and ^{18}O ultimately react with $^{16}O_2$ to form ozone. Figure 13.3 shows the ozone enriched in the heavy isotopes and the residual O_2 with preferential removal of ^{18}O and ^{17}O.

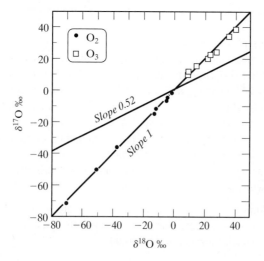

FIGURE 13.3: $\delta^{18}O-\delta^{17}O$ values of O_2 and O_3 during ozone formation by high-frequency discharge. The product ozone is enriched in both ^{17}O and ^{18}O by a mass-independent mechanism that is believed to be due to self-shielding. (Reprinted from Thiemens and Heidenreich, 1983 with permission from AAAS.)

Isotopic fractionation in interstellar clouds by a similar process of self-shielding had been predicted from astronomical observations. Langer (1977) suggested that self-shielding could produce CO gas in interstellar clouds that appeared to be enriched in ^{12}C by a factor of 2.3 compared with solar abundances. Later, Bally and Langer (1982) showed that the CO gas of interstellar clouds is enriched in ^{16}O as well (contrary to previous ideas). In other words, the $C^{18}O$ (and, presumably, $C^{17}O$) is preferentially disassociated relative to $C^{16}O$. Stated yet another way, the abundant isotopologue $^{12}C^{16}O$ strongly absorbs the appropriate radiation bands in the outer portions of the molecular clouds, so that only $^{13}C^{16}O$ and $^{12}C^{18}O$ are strongly dissociated in the interior of the cloud, leaving the residual CO enriched in ^{12}C and ^{16}O. The oxygen produced by the dissociation would combine with hydrogen to produce ^{17}O- and ^{18}O-enriched H_2O. Later estimates suggested that the photodissociation rates of $^{13}C^{16}O$ and $^{12}C^{18}O$ would be one to two orders of magnitude larger than that of $^{12}C^{16}O$ inside molecular clouds (van Dishoeck and Black, 1988).

Mass-independent reactions have been identified for the breakdown of CO (Heidenreich and Thiemens, 1985) that are compatible with the astrophysical models predicted earlier. One criticism of calling on mass-independent reactions to produce the ^{16}O-enriched CAIs is that photodissociation should produce reactive material that is isotopically heavy (e.g., H_2O). Thus, it was difficult to envision a process whereby the light CAIs could be formed if the nebular cloud was isotopically heavy (Clayton, 1993).

Recently, a molecular cloud origin for oxygen heterogeneities in the solar system has been proposed (Yurimoto and Kuramoto, 2004). Fundamental to this idea is that the solar nebula has a low $\delta^{18}O - \delta^{17}O$ value, not a high one as had been proposed earlier. According to the model, preferential dissociation of $C^{18}O$ and $C^{17}O$ occurs in the solar system molecular cloud, because the UV intensity at wavelengths of dissociation lines for the abundant $C^{16}O$ rapidly attenuate near the surface layer of the cloud. ^{17}O- and ^{18}O-rich oxygen produced from CO photodissociation combines with H_2 to form H_2O. Temperatures at this stage are as low as 10 K. Calculated fractionations are such that the CO has $\delta^{18}O$ and $\delta^{17}O$ values from -60 to $-400‰$ relative to the overall molecular cloud, while H_2O has $\delta^{18}O$ and $\delta^{17}O$ values $+100$ to $+250‰$ relative to the cloud. In the cold, dense core of the molecular cloud, CO would be frozen onto dust grains, along with H_2O. As long as the core is not heated, the large fractionations between CO and H_2O will be preserved.

A protoplanetary disk then forms as the core of the molecular cloud collapses. In the outer regions, CO sublimes, preserving its unique low $\delta^{18}O - \delta^{17}O$ values. In the inner, higher temperature core region, the heavy water evaporates and reacts with other solid components (e.g., iron metal).[7] Early formed high T grains that form in equilibrium with the nebular cloud would have low $\delta^{18}O - \delta^{17}O$ values, similar to those of the overall nebular cloud. Dust and "heavy" H_2O ice migrate towards the midplane of the nebular disk. As they approach the protosun, the H_2O melts, mixing with the dust and ultimately forming an $^{17}O - ^{18}O$ enriched region relative to the overall solar system. The inner planets reflect this anomalously heavy oxygen isotope composition.

The new model eliminates several of the problems inherent in the older mixing model. Oxygen is not expected to correlate with other isotopic systems, and no extremely ^{16}O-rich component (consistent with a supernova origin) is expected. In fact, no external material is required at all. The self-shielding model implies that the inner solar system is enriched in ^{17}O and ^{18}O relative to the solar nebula, *exactly the opposite* of what had been proposed in earlier models, according to which the solar nebula was the heavy component.

---

[7] Oxidation of Fe–Ni metal by H_2O has produced magnetite from ordinary chondrites (Choi et al., 1998), the most $^{18}O - ^{17}O$-enriched component of the solar system, consistent with the model.

With the molecular cloud model, we would expect the solar wind to have $\delta^{18}O$–$\delta^{17}O$ values of ~ -50‰, in contrast to the Clayton mixing model, in which the solar wind should have the heaviest values ($> +20$‰). Researchers expect to resolve this issue when samples from the Genesis space mission are analyzed. During this mission, solar wind was implanted on oxygen-free plates exposed to the sun for several years. The data should be analyzed soon, but in the meantime, evidence for the low $\delta^{18}O$–$\delta^{17}O$ solar wind has been identified. Hashizume and Chaussidon (2005) measured the $\delta^{18}O$ and $\delta^{17}O$ values of solar energetic particles that had been implanted in metal grains on the lunar regolith. The measured $\Delta^{17}O$ value of the solar particles was -33 ± 16‰, corresponding to a protosolar nebula of ~ -67‰ for $\delta^{18}O$ and $\delta^{17}O$ (when appropriate fractionations are taken into account).

13.4 HYDROGEN

13.4.1 Introduction

The D/H ratios of materials in our galaxy span such an enormous range as to strain the utility of the delta notation. Solar wind is virtually deuterium free, with δD values approaching $-1{,}000$‰. Cold molecular clouds, at the other extreme, have spectroscopically measured D/H ratios as high as 1×10^{-2}, corresponding to a whopping 63,000‰. (See Robert et al., 2000, for a review of galactic values.) Hydrogen isotope systematics differ significantly from those of oxygen, and the information available from D/H ratios is quite different as well. Hydrogen has only two stable isotopes in place of oxygen's three, so mixing of sources can be tracked only in "one dimension" instead of two. While far and away the most abundant phase in molecular clouds and our solar system, hydrogen is rare in solids preserved in meteorites. Finally, the isotopes of hydrogen have a very different formation history than does oxygen, which forms during different star-forming events, with each isotope identified with specific nucleosynthetic processes. Deuterium, in contrast, was all formed in the Big Bang (Burbidge et al., 1957). There is no additional stellar input to the primordial mix. For oxygen, a knowledge of the solar abundances would place firm constraints on our theories of the composition and evolution of the solar system. For the hydrogen–deuterium system, the Sun is of no help in giving us average solar system values, as all deuterium was "burned" during the contraction of our proto-Sun.[8] Therefore, D/H ratios of protosolar matter are estimated from $^3He/^4He$ abundances of solar wind.

The final difference between hydrogen and oxygen is that the relative mass difference between ^2H and ^1H is far larger than for any other isotopic system. Fractionations can be huge, and when these are considered in light of the extremely low temperatures of the solar nebula, the effect is greatly enhanced. Even slight degrees of dehydrogenation or dehydration can dramatically affect the D/H ratios, so obtaining meaningful hydrogen isotope data is challenging.

The average protostellar D/H ratio is estimated as 20 ppm ($\delta D \sim -870$‰), compared with Earth's 140 ppm ($\delta D = -100$‰). The D/H ratio of Jupiter is thought to reflect that of the proto–solar nebula, because it has incorporated hydrogen with minimal fractionation due to its immense size. The spectroscopically measured D/H ratio of 26 ± 7 ppm agrees with other independent estimates for the solar nebula. The D/H ratios measured in comets is higher, approximately 1,000‰.

---

[8] Of all the isotopes of all the elements, deuterium is the most easily destroyed in thermonuclear reactions.

Compared with the Earth, Venus and Mars have significantly elevated D/H ratios of 20,000 ppm ($\delta D = 125,000‰$) and 800 ppm ($\delta D = 4,000‰$), respectively, due presumably to the preferential loss of H_2 by escape from the atmosphere. In contrast, the loss of H_2 from the atmosphere of Earth is not thought to have been important (Lécuyer et al., 1997; Robert et al., 2000). Bulk Martian meteorites have been analyzed by step-heating. Low-temperature heating steps clearly reflect terrestrial contamination. Water extracted at higher temperatures gives δD values of $+50‰$ (Chassigny) to $+2,100‰$ (Shergotty) (both cited in Leshin et al., 1996; the low δD Chassigny data probably reflect exchange with terrestrial water).

Ion microprobe studies of hydrous phases in Martian meteorites indicate two distinct sources of hydrogen (Fig. 13.4). The first is interpreted as magmatic water, with a δD value of $\sim 900‰$, a D/H ratio 10 times that of Earth. The second is interpreted as being the Martian atmosphere, with a δD value of $4,000‰$, indicative of an evolved atmosphere with loss of H_2 (e.g., Leshin, 2000; Sugiura and Hoshino, 2000).

The δD value of hydrogen extracted from lunar soils is between $-890‰$ and $-1,000‰$, explained as being entirely sourced by H_2 from solar wind (Epstein and Taylor, 1970). The measurable deuterium in solar wind implanted samples may be due to terrestrial contamination.

13.4.2 Meteorites

The first hydrogen isotope measurements of meteorites gave "unearthly" high δD values up to $+300‰$ (Boato, 1954). The range was greatly expanded in the 1970s and 1980s. For example, Kolodny et al. (1980) measured δD values between -70 and $+771‰$ from seven carbonaceous chondrites. A fundamental question was addressed in this study: Were the deuterium-rich phases hydrous minerals or organic matter? After low-temperature plasma oxidation (which removed organic matter), the δD values dropped to near the range of typical terrestrial values. The authors concluded that the organic matter (removed during the oxidation) had D/H ratios in excess of $1,600‰$. To explain such high D/H ratios, they proposed that the organic matter formed at temperatures of only 180 K. Because there is no correlation of the D/H ratios with $\delta^{13}C$ values, researchers have suggested that an extremely deuterium rich phase resides *as inclusions* in the organic matter (Yang and Epstein, 1983).

During the ensuing years, δD values in excess of $+6,000‰$ were measured in organic macromolecules from carbonaceous chondrites (e.g., McNaughton et al., 1981; Robert and Epstein, 1982). The nature of such high-δD material could not be explained by simple

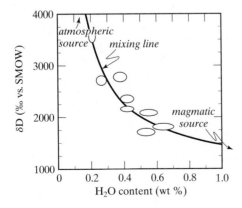

FIGURE 13.4: Ion probe δD values of apatite grains from Martian meteorite QUE94201. The data are interpreted as mixing between a magmatic source of $\sim 900‰$ with an atmospheric source in excess of $4,000‰$. (After Leshin, 2000.)

low-temperature chemical exchange reactions. In order for fractionations to be large enough to get such huge enrichments, hydrogen isotope exchange reactions between gas phases such as CH_4 and H_2 or H_2O and H_2 would have to occur at extremely low temperatures. However, at such low temperatures, reaction rates are simply too slow for any appreciable exchange to occur. An alternative mechanism for producing the extreme deuterium enrichments observed in macromolecules is ion exchange reactions (Watson, 1973).

Ion exchange reactions could occur in molecular clouds or solar nebula, where UV radiation breaks apart molecules, creating ionic radicals that can then recombine with pre-existing molecules. Pressures need to be low to prevent the recombination of the dissociated radicals. The rates of ion exchange reactions are mostly independent of temperature, and such reactions can occur with huge fractionations at extremely low temperatures. In doing so, they can effectively fractionate deuterium into water preferentially, explaining the high δD values found in organic macromolecules. This mechanism is compatible with astrophysical observations. D/H ratios are very high in carbon-bearing molecules in dense molecular clouds, where temperatures are on the order of <50 K. Unfortunately, the D/H ratio for H_2O in molecular clouds cannot be measured because the partial pressure of the H_2O is too low.

Results from ion microprobe analyses changed some of the basic assumptions (but not mechanisms) regarding deuterium-rich sources in meteorites. All earlier results pointed to high D/H ratios in organic matter, but normal terrestrial δD values in phyllosilicates (e.g., clay minerals). Deloule and Robert (1995) measured D/H ratios with an ion microprobe from an achondrite (Semarkona) known to be relatively free of terrestrial alteration. The pure phyllosilicates were calculated to have δD values of $+3,800 \pm 800‰$, far higher than previously found in such material. The authors recognized that water (now incorporated in phyllosilicates), in addition to organic material, could have high δD values. The high concentration of deuterium was explained in terms of the ion exchange reaction

$$D + H_3O^+ \rightleftharpoons H + H_2DO^+ \quad (13.2)$$

13.5 CARBON

Carbon isotope analyses of meteorites began extremely early on in the game, due to the identification of a distinct spectroscopic band found in N-type stars attributable to high concentrations of $^{12}C^{13}C$. In 1936, several attempts were made to measure the $^{13}C/^{12}C$ ratio of graphite from the Cañon Diablo iron meteorite by means of spectrographic measurements of graphite discharge. Unfortunately, the spectrographic image was identical to that obtained from terrestrial graphite (e.g., Jenkins and King, 1936), suggesting that no ^{13}C anomaly existed in meteorites. Additional analyses in the coming decades also found that meteoritic carbon had $\delta^{13}C$ values that overlapped with terrestrial values. The first anomalous values were found in the carbonate fraction of two carbonaceous chondrites (Orgueil and Ivuna) in which $\delta^{13}C$ values of $\sim+60‰$ (PDB) were measured on the dolomite fraction (Clayton, 1963). Along with Boato's early high δD values in meteorites, this was one of the first findings of anomalous stable isotope ratios in extraterrestrial material.

In the same year, Briggs (1963) reported $\delta^{13}C$ values of solvent-extractable organic material from carbonaceous chondrites that were only slightly less than 0‰, higher than terrestrial equivalents. The anomalous $\delta^{13}C$ values provided some of the strongest evidence to date that organic material in meteorites was indeed of extraterrestrial origin.

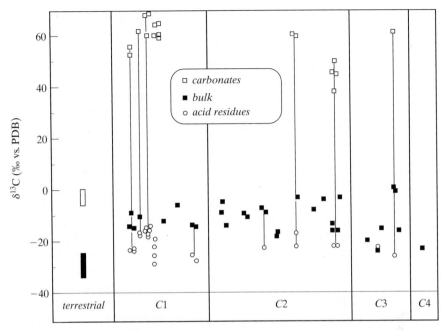

FIGURE 13.5: Carbon isotope composition of organic and carbonate material from carbonaceous chondrites contrasted with typical values for terrestrial material chemically treated in the same manner. The most striking difference are the high $\delta^{13}C$ values of carbonate in the meteorites. However, the organic matter is also elevated compared with its terrestrial equivalents. (Reprinted from Krouse and Modzeleski, 1970, with permission from Elsevier.)

Two papers appearing in 1970 significantly expanded the database for carbonaceous chondrites (Krouse and Modzeleski, 1970; Smith and Kaplan, 1970). Both showed very high $\delta^{13}C$ values of carbonates and less striking but still high, $\delta^{13}C$ values of organic extracts (Fig. 13.5). The data were clearly incompatible with a terrestrial carbon origin. Instead, the "endogenous" origin (Smith and Kaplan, 1970) of carbon in carbonaceous meteorites was firmly established.

The emerging picture was that the $\delta^{13}C$ values of organic carbon averaged around -16‰, while the much less abundant carbonate had $\delta^{13}C$ values as high as $+60$ or $+70$‰. The fractionation of greater than 80‰ between the oxidized and reduced form was far larger than what was observed on Earth, where the differences were only on the order of 25 to 30‰. Urey (1967) considered two distinct carbon synthesis events in the solar system as a possible explanation. Several year later, Lancet and Anders (1970) proposed an alternative explanation that did not require an exotic source of carbon. They showed experimentally that oxidation–reduction reactions of CO (Fischer–Tropsch synthesis) given by

$$2n\,CO + (n+1)\,H_2 \rightleftharpoons C_nH_{2n+2} + n\,CO_2 \tag{13.3}$$

could produce large carbon isotope fractionations. In this reaction, both oxidized and reduced carbon are formed abiogenically. Fractionations between organic species and CO_2 and between CH_4 and CO_2 were up to 60‰ and 100‰, respectively, more than enough to explain the variations that were encountered in the meteorities.

The isotopic composition of individual amino acids from CM chondrites have been measured (Engel et al., 1990). The $\delta^{13}C$ values are as high as $+30$‰, clearly of extraterrestrial origin. Interestingly, the measured samples are not racemic,[9] a property that had been used as previous evidence for terrestrial contamination. The high $\delta^{13}C$ values eliminate contamination as a possibility of their origin. Hydrogen isotope analyses of the hydroxy acids from Murchison are also extraterrestrial, topping out at higher than 500‰ (Cronin et al., 1993).

In a number of stepwise combustion analyses of carbonaceous chondrites, an extremely high $\delta^{13}C$ value of 1,100‰ was measured (Swart et al., 1983). Later work using high-spatial-resolution ion microprobe analyses revealed $\delta^{13}C$ values in interstellar residual grains in excess of 7,000‰ (Zinner and Epstein, 1987). These are primitive and rare grains of "stardust."

13.6 NITROGEN

Nitrogen isotope geochemistry of meteorites lagged behind the isotope geochemistry of the other "traditional" elements: H, C, O, and S. Some of the first analyses of carbonaceous chondrites found a range considerably broader than existed on Earth. Injerd and Kaplan (1974) measured $\delta^{15}N$ ratios between -20 and $+46$‰ for four chondrites. Kung and Clayton (1978) expanded the number of analyses and extended the total range observed for carbonaceous chondrites from lows of -40 to -30‰ for enstatite chondrites to highs of $+30$ and $+50$‰ (Renazzo with an anomalous 170‰). These researchers felt that the variations were too large to be explained by mass-dependent fractionation and instead suggested that the meteorites were recording nitrogen isotope heterogeneities in the solar nebula.

In a step-heating experiment of Allende chondrite, larger variations were seen than in all the other meteorites analyzed previously (except Renazzo). Nitrogen released below 900°C had low $\delta^{15}N$ values of -90 to -60‰, whereas the gas released above 1,000°C had $\delta^{15}N$ values averaging -20‰ with some values as high as 0‰ (Thiemens and Clayton, 1981). The spread of the data were explained in terms of spatial and temporal variations in $\delta^{15}N$ values of the solar system.

The conclusions of Thiemens and Clayton were in line with earlier ideas that the $\delta^{15}N$ values of our solar system have changed dramatically over time. The inorganic phases in the enstatite chondrites were thought to have formed early and at high temperatures. Little nitrogen isotope fractionation would be expected under these conditions, so the early solar nebula likely had a $\delta^{15}N$ value of -30‰, consistent with the measured chondrite values of -40 to -30‰. The low $\delta^{15}N$ values of gas released in the lower temperature step-heating experiments is postulated to be of presolar origin. This argument is based partly on rare-gas data obtained from the same step-heating interval (Thiemens and Clayton, 1981). It had been suggested that the solar wind has had variable $^{15}N/^{14}N$ ratios through time (Kerridge, 1975). The early solar wind had a very low $\delta^{15}N$ value of -210‰, which increased to its present-day value of $+120$‰. Removal of ^{14}N to the early formed organic material, together with "diffusive loss" of N_2, can explain the high $\delta^{15}N$ values in the high-temperature extracted phase.

The secular variation in solar wind $\delta^{15}N$ values was based on measurements of lunar soils (Kerridge, 1975). The nitrogen isotope values of soils were found to differ by over 200‰, related to the age of exposure. It was thought that the $\delta^{15}N$ value of the solar wind had changed over the its lifetime due to spallation of ^{16}O. The results obtained from these data were completely reasonable, but are not correct. The authors did not have the benefit of high-spatial-resolution analyses to explain the range of values that they were measuring. Recently,

---

[9]A racemic amino acid is an amino acid that has equal abundances of left- and right-handed optical isomers (enantiomers). The Murchison sample has a D/L value of 0.85, significantly different from the racemic value of 0.5. It is normally assumed that nonracemic amino acids are due to biological synthesis.

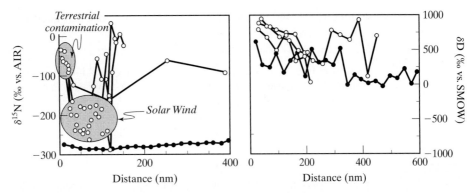

FIGURE 13.6: **Ion microprobe depth profiling for nitrogen (open circles and left scale) and hydrogen (filled circles and right scale) of two samples of lunar soil.** The sample at the left has dD values as low as −900‰ clearly of solar wind origin. The implanted solar wind nitrogen has $\delta^{15}N$ values of ∼−250‰. (Material close to the surface is contaminated, and samples deep within the grain have nitrogen concentrations that are too low to be measured reliably.) The sample at the right has both high dD and $\delta^{15}N$ values, consistent with a meteorite grain origin. (After Hashizume et al., 2000.)

Hashizume et al. (2000) made depth profile measurements of nitrogen and hydrogen isotope ratios in *individual* lunar soil grains (Fig. 13.6). They found that, for some grains, the δD values were extremely low, indicative of solar wind. The $\delta^{15}N$ values in these samples were correspondingly low, demonstrating that the $\delta^{15}N$ solar wind value was not heavy, as had been proposed earlier. Samples with high $\delta^{15}N$ values also had high δD values. These were clearly meteoritic fragments without input from solar wind implantation. What the earlier researchers had been measuring was a mixture of two components: the original planetary component (heavy for both hydrogen and nitrogen isotope ratios), and the light component from the solar wind. The different proportions of the two components as a function of age were simply related to differing accumulation rates of meteorite fragments on the lunar surface over time.

There have been many surprises in isotope studies of meteorites. Additional exciting work is coming to light using high-resolution techniques and from analyses of other isotope systems. For example, sulfur isotope mass-independent anomalies have been found in meteorites (Rees and Thode, 1977; Farquhar et al., 2000), and nontraditional isotopes are being applied to all types of extraterrestrial materials. Ion microprobe measurements are finding much larger fractionations over much smaller scales than had been observed earlier. Finally, solar wind material has been brought back to Earth from the Genesis mission. Perhaps a Mars mission in the not-too-distant future will also be undertaken, returning fresh material for analysis.

PROBLEM SET

1. Large isotopic variations have been found in hydrogen, carbon, nitrogen, and oxygen in meteoritic material. The explanation for the large range differs for each element. Review the mechanisms that have been proposed to explain the large isotopic range for each element.
2. What is self-shielding, and how can it explain the CCAM mixing line?
3. It has been proposed that the $\delta^{18}O$ value of the Sun is either very heavy ($\delta^{18}O > 20‰$) or very light ($\delta^{18}O < -60‰$). On what basis have these values been obtained, and what are the implications in terms of explaining the CCAM line?

Chapter 13 Extraterrestrial Materials

4. If the CCAM line intersected the TFL at $\delta^{18}O = +50‰$ *and* passed through pure ^{16}O at its lower end, what is its slope?

5. We generally think of the relationship $\left(\dfrac{^{17}O}{^{16}O}\right) = \left(\dfrac{^{18}O}{^{16}O}\right)^{0.52}$ from equation (13.1) and the equation $\delta^{17}O = 0.52 \times (\delta^{18}O)$ as equivalent, but strictly speaking, they are not. Graph $\delta^{17}O$ vs. $\delta^{18}O$ over a $\delta^{18}O$ range of 0 to 1,000‰, assuming first that equation (13.1) is correct and then that $\delta^{17}O = 0.52 \times (\delta^{18}O)$ is correct.

ANSWERS

4. It is generally considered that the CCAM line has a slope of unity and passes through $\delta^{17}O = \delta^{18}O = -1{,}000‰$ (pure ^{16}O) and $\delta^{17}O = \delta^{18}O = 0‰$ (SMOW). In fact, the line passes through the carbonaceous chondrite reference standard (CCRS) with a $\delta^{18}O$ value of ~9‰ (vs. SMOW) and a $\delta^{17}O$ value of ~5‰ (vs. SMOW). The slope of the CCAM line on the SMOW scale is therefore $1005/1009 = 0.996$. If the CCAM line intersected the TFL at +50, it would have a $\delta^{18}O$ value of +50 and a $\delta^{17}O$ value of $0.52 \times 50 = 26‰$. The CCAM slope (on the SMOW) scale would then be $1026/1050 = 0.977$.

5. A plot of $\delta^{17}O = 0.52 \times (\delta^{18}O)$ is trivial: It is a straight line with a slope of 0.52. For the second part, we need to rearrange two equations. First, the definition of $\delta^{18}O = (R_{sa}/R_{std} - 1) \times 1000$ is rearranged to $R^{18}_{sa} = R^{18}_{std} \dfrac{1000 + \delta^{18}O}{1000}$. Next, we know that $R^{17}_{sa} = (R^{18}_{sa})^{0.52}$. Then

$$R^{17}_{sa} = \left(R^{18}_{std}\dfrac{1000 + \delta^{18}O}{1000}\right)^{0.52}$$

and $R^{17}_{std} = (R^{18}_{std})^{0.52}$. From the definition of δ, we therefore have

$$\delta^{17}O = \left(\dfrac{R_{sa}}{R_{std}} - 1\right) \times 1000 = \left[\dfrac{\left(R^{18}_{std}\dfrac{1000 + \delta^{18}O}{1000}\right)^{0.52}}{R^{18}_{std}} - 1\right] \times 1000$$

$$= \left[\left(1 + \dfrac{\delta^{18}O}{1000}\right)^{0.52} - 1\right] \times 1000,$$

which plots as follows:

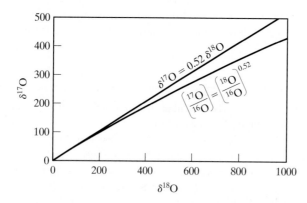

FIGURE 13.q.5

REFERENCES

Anders, E., and Zinner, E. (1993). Interstellar grains in primitive meteorites: Diamond, silicon carbide, and graphite. *Meteoritics* **28**, 490–514.

Baertschi, P. (1950). Isotopic composition of the oxygen in silicate rocks. *Nature* **166**, 112–113.

Bally, J., and Langer, W. D. (1982). Isotope-selective photodestruction of carbon monoxide. *The Astrophysical Journal* **255**, 143–148.

Black, D. C. (1972). On the origins of trapped helium, neon and argon isotopic variations in meteorites—II. Carbonaceous meteorites. *Geochimica et Cosmochimica Acta* **36**, 377–394.

Boato, G. (1954). The isotopic composition of hydrogen and carbon in the carbonaceous chondrites. *Geochimica et Cosmochimica Acta* **6**, 209–220.

Brearley, A., and Jones, R. H. (1998). Chondritic meteorites. In *Planetary Materials,* Vol. 36 (ed. J. J. Papike), pp. 3.01–3.370. Mineralogical Society of America, Washington, DC.

Briggs, M. H. (1963). Evidence of an extraterrestrial origin for some organic constituents of meteorites. *Nature* **197**, 1290.

Burbidge, E. M., Burbidge, G. R., Fowler, W. A., and Hoyle, F. (1957). Synthesis of the elements in stars. *Reviews of Modern Physics* **29**, 547–650.

Choi, B. G., McKeegan, K. D., Krot, A. N., and Wasson, J. T. (1998). Extreme oxygen-isotope compositions in magnetite from unequilibrated ordinary chondrites. *Nature* **392**, 248–250.

Clayton, R. N. (1963). Carbon isotope abundance in meteoritic carbonates. *Science* **140**, 192–193.

Clayton, R. N. (1993). Oxygen isotopes in meteorites. *Annual Review of Earth and Planetary Sciences* **21**, 115–149.

Clayton, R. N., Grossman, L., and Mayeda, T. K. (1973). A component of primitive nuclear composition in carbonaceous meteorites. *Science* **182**, 485–488.

Clayton, R. N., and Mayeda, T. K. (1975). Genetic relations between the moon and meteorites. *Proceedings of the Lunar Science Conference* **6**, 1761–1769.

Clayton, R. N., Mayeda, T. K., Goswami, J. N., and Olsen, E. J. (1991). Oxygen isotope studies of ordinary chondrites. *Geochimica et Cosmochimica Acta* **55**, 2317–2337.

Clayton, R. N., Onuma, N., Grossman, L., and Mayeda, T. K. (1977). Distribution of the pre-solar component in Allende and other carbonaceous chondrites. *Earth and Planetary Science Letters* **34**, 209–224.

Clayton, R. N., Onuma, N., and Mayeda, T. K. (1976). A classification of meteorites based on oxygen isotopes. *Earth and Planetary Science Letters* **30**, 10–18.

Craig, H. (1957). Isotopic standards for carbon and oxygen and correction factors for mass-spectrometric analysis of carbon dioxide. *Geochimica et Cosmochimica Acta* **12**, 133–149.

Cronin, J. R., Pizzarello, S., Epstein, S., and Krishnamurthy, R. V. (1993). Molecular and isotopic analyses of the hydroxy acids, dicarboxylic acids, and hydroxydicarboxylic acids of the Murchison Meteorite. *Geochimica et Cosmochimica Acta* **57**, 4745–4752.

Deloule, E., and Robert, F. (1995). Interstellar water in meteorites? *Geochimica et Cosmochimica Acta* **59**, 4695–4706.

Engel, M. H., Macko, S. A., and Silfer, J. A. (1990). Carbon isotope composition of individual amino acids in the Murchison meteorite. *Nature* **348**, 47–49.

Epstein, S., and Taylor, H. P., Jr. (1970). $^{18}O/^{16}O$, $^{30}Si/^{28}Si$, D/H, and $^{13}C/^{12}C$ studies of lunar rocks and minerals. *Science* **167**, 533–535.

Farquhar, J., Jackson, T. L., and Thiemens, M. H. (2000). A ^{33}S enrichment in ureilite meteorites; evidence for a nebular sulfur component. *Geochimica et Cosmochimica Acta* **64**, 1819–1825.

Friedman, I., O'Neil, J. R., Adami, L. H., Gleason, J. D., and Hardcastle, K. (1970). Water, hydrogen, deuterium, carbon, carbon-13, and oxygen-18 content of selected lunar material. *Science* **167**, 538–540.

Hashizume, K., and Chaussidon, M. (2005). A non-terrestrial ^{16}O-rich isotopic composition for the protosolar nebula. *Nature* **434**, 619–622.

Hashizume, K., Chaussidon, M., Marty, B., and Robert, F. (2000). Solar wind record on the Moon: Deciphering presolar from planetary nitrogen. *Nature* **290**, 1142–1145.

Heidenreich, J. E. I., and Thiemens, M. H. (1985). The non-mass-dependent oxygen isotope effect in the electrodissociation of carbon dioxide: A step toward understanding NoMaD chemistry. *Geochimica et Cosmochimica Acta* **49**, 1303–1306.

Injerd, W. G., and Kaplan, I. R. (1974). Nitrogen isotope distribution in meteorites. *Meteoritics* **9**, 308–309.

Jenkins, F. A., and King, A. S. (1936). A test of the abundance of the heavy isotope of carbon in a graphite meteorite. *Publications of the Astronomical Society of the Pacific*, 323–325.

Kaplan, I. R., and Petrowski, C. (1971). Carbon and sulfur isotope studies on *Apollo 12* lunar samples. *Proceedings of the Lunar Science Conference* **2**, 1397–1406.

Kerridge, J. F. (1975). Solar nitrogen: Evidence for a secular increase in the ratio of nitrogen-15 to nitrogen-14. *Science* **188**, 162–164.

Kolodny, Y., Kerridge, J. F., and Kaplan, I. R. (1980). Deuterium in carbonaceous chondrites. *Earth and Planetary Science Letters* **46**, 149–158.

Krouse, H. R., and Modzeleski, V. E. (1970). C^{13}/C^{12} abundances in components of carbonaceous chondrites and terrestrial samples. *Geochimica et Cosmochimica Acta* **34**, 459–474.

Kung, C. C., and Clayton, R. N. (1978). Nitrogen abundances and isotopic compositions in stony meteorites. *Earth and Planetary Science Letters* **38**, 421–435.

Lancet, M. S., and Anders, E. (1970). Carbon isotope fractionation in the Fischer–Tropsch synthesis and in meteorites. *Science* **170**, 980–982.

Langer, W. D. (1977). Isotopic abundance of CO in interstellar clouds. *The Astrophysical Journal* **212**, L39–L42.

Lécuyer, C., Gillet, P., and Robert, F. (1997). The hydrogen isotope composition of seawater and the global water cycle. *Chemical Geology* **145**, 249–261.

Leshin, L. A. (2000). Insights into Martian water reservoirs from analyses of Martian meteorite QUE 94201. *Geophysical Research Letters* **27**, 2017–2020.

Leshin, L. A., Epstein, S., and Stolper, E. M. (1996). Hydrogen isotope geochemistry of SNC meteorites. *Geochimica et Cosmochimica Acta* **60**, 2635–2650.

Manian, S. H., Urey, H. C., and Bleakney, W. (1934). An investigation of the relative abundance of the oxygen isotopes O^{16}/O^{18} in stone meteorites. *Journal of the American Chemical Society* **56**, 2601–2609.

Matsuhisa, Y., Goldsmith, J. R., and Clayton, R. N. (1978). Mechanisms of hydrothermal crystallization of quartz at 250°C and 15 kbar. *Geochimica et Cosmochimica Acta* **42**, 173–182.

McKeegan, K. D., and Leshin, L. A. (2001). Stable isotope variations in extraterrestrial materials. In *Stable Isotope Geochemistry,* Vol. 43 (ed. J. W. Valley and D. R. Cole), pp. 279–318. Mineralogical Society of America, Washington, DC.

McNaughton, N. J., Borthwick, J., Fallick, A. E., and Pillinger, C. T. (1981). Deuterium/hydrogen ratios in unequilibrated ordinary chondrites. *Nature* **294**, 639–641.

Nittler, L. R. (2003). Presolar stardust in meteorites: Recent advances and scientific frontiers. *Earth and Planetary Science Letters* **209**, 259–273.

Nittler, L. R., Alexander, C. M. O. D., and Wang, J. (1998). Meteoritic oxide grain from supernova found. *Nature* **393**, 222.

Onuma, N., Clayton, R. N., and Mayeda, T. K. (1970). Oxygen isotope fractionation between minerals and an estimate of the temperature of formation. *Science* **167**, 536–538.

Onuma, N., Clayton, R. N., and Mayeda, T. K. (1972a). Oxygen isotope cosmothermometer. *Geochimica et Cosmochimica Acta* **36**, 169–188.

Onuma, N., Clayton, R. N., and Mayeda, T. K. (1972b). Oxygen isotope temperatures of "equilibrated" ordinary chondrites. *Geochimica et Cosmochimica Acta* **36**, 157–168.

Onuma, N., Clayton, R. N., and Mayeda, T. K. (1974). Oxygen isotope cosmothermometer revisited. *Geochimica et Cosmochimica Acta* **38**, 189–191.

Rees, C. E., and Thode, H. G. (1977). A ^{33}S anomaly in the Allende meteorite. *Geochimica et Cosmochimica Acta* **41**, 1679–1682.

Reuter, J. H., Epstein, S., and Taylor, H. P., Jr. (1965). O^{18}/O^{16} ratios of some chondritic meteorites and terrestrial ultramafic rocks. *Geochimica et Cosmochimica Acta* **29**, 481–488.

Rietmeijer, F. J. M. (1998). Interplanetary dust particles. In *Planetary Materials*, Vol. 36 (ed. J. J. Papike), pp. 2-1–2-95. Mineralogical Society of America, Washington, DC.

Robert, F., and Epstein, S. (1982). The concentration and isotopic composition of hydrogen, carbon and nitrogen in carbonaceous meteorites. *Geochimica et Cosmochimica Acta* **46**, 81–95.

Robert, F., Gautier, D., and Dubrulle, B. (2000). The solar system D/H ratio: Observations and theories. *Space Science Reviews* **92**, 201–224.

Shearer, C. K., Papike, J. J., and Rietmeijer, F. J. M. (1998). The planetary sample suite and environments of origin. In *Planetary Materials,* Vol. 36 (ed. J. J. Papike), pp. 1-1–1-28. Mineralogical Society of America, Washington, DC.

Silverman, S. R. (1951). The isotope geology of oxygen. *Geochimica et Cosmochimica Acta* **2**, 26–42.

Smith, J. W., and Kaplan, I. R. (1970). Endogenous carbon in carbonaceous meteorites. *Science* **167**, 1367–1370.

Stolper, E. (1982). Crystallization sequences of Ca-, Al-rich inclusions from Allende: An experimental study. *Geochimica et Cosmochimica Acta* **46**, 2159–2180.

Sugiura, N., and Hoshino, H. (2000). Hydrogen-isotopic compositions in Allan Hills 84001 and the evolution of the Martian atmosphere. *Meteoritics Planetary Sciences* **35**, 373–380.

Swart, P. K., Grady, M. M., Pillinger, C. T., Lewis, R. S., and Anders, E. (1983). Interstellar carbon in meteorites. *Science* **220**, 406–410.

Taylor, H. P., Jr., Duke, M. B., Silver, L. T., and Epstein, S. (1965). Oxygen isotope studies of minerals in stony meteorites. *Geochimica et Cosmochimica Acta* **29**, 489–512.

Thiemens, M., and Clayton, R. N. (1981). Nitrogen isotopes in the Allende meteorite. *Earth and Planetary Science Letters* **55**, 363–369.

Thiemens, M. H., and Heidenreich, J. E. I. (1983). The mass-independent fractionation of oxygen: A novel isotope effect and its possible cosmochemical implications. *Science* **219**, 1073–1075.

Trofimov, A. (1949). Isotopic constitution of sulfur in meteorites and in terrestrial objects. *Doklady Akademii Nauk SSSR* **66**, 181–184.

Urey, H. C. (1967). The abundance of the elements with special reference to the problem of the iron abundance. *Quarterly Journal of the Royal Astronomical Society* **8**, 23–47.

van Dishoeck, E. F., and Black, J. H. (1988). The photodissociation and chemistry of interstellar CO. *The Astrophysical Journal* **334**, 771–802.

Vinogradov, A. P., Dontsova, E. I., and Chupakhin, M. S. (1960). Isotopic ratios of oxygen in meteorites and igneous rocks. *Geochimica et Cosmochimica Acta* **18**, 278–293.

Watson, W. D. (1973). Formation of the HD molecule in the interstellar medium. *The Astrophysical Journal* **182**, L73–L76.

Yang, J., and Epstein, S. (1983). Interstellar organic matter in meteorites. *Geochimica et Cosmochimica Acta* **47**, 2199–2216.

Young, E. D., and Russell, S. S. (1998). Oxygen reservoirs in the early solar nebula inferred from an Allende CAI. *Science* **282**, 452–455.

Yurimoto, H., and Kuramoto, K. (2004). Molecular cloud origin for the oxygen isotope heterogeneity in the Solar System. *Science* **305**, 1763–1766.

Zinner, E., and Epstein, S. (1987). Heavy carbon in individual oxide grains from the Murchison Meteorite. *Earth and Planetary Science Letters* **84**, 359–368.

APPENDIX A

STANDARD REFERENCE MATERIALS FOR STABLE ISOTOPES

Isotopic composition of selected reference materials after the compilation of Coplen et al. (2002). Standards not presented here are not official IAEA-approved. Additional IAEA standards are developed continually. Refer to the IAEA website for up-to-date delta values.

Coplen, T. B., Hopple, J. A., Böhlke, J. K., Peiser, H. S., Rieder, S. E., Krouse, H. R., Rosman, K. J. R., Ding, T., Vocke, R. D. J., Révész, K. M., Lamberty, A., Taylor, P., and DeBièvre, P. (2002) *Compilation of Minimum and Maximum Isotope Ratios of Selected Elements in Naturally Occurring Terrestrial Materials and Reagents,* United States Geological Survey, Reston, VA.

Hut G. (1987) Consultants' group meeting on stable isotope reference samples for geochemical and hydrological investigations. International Atomic Energy Agency. Vienna

Reference Standard	Substance	$\delta^{18}O$ (SMOW)	$\delta^{18}O$ (PDB)	$\delta^{13}C$ (PDB)	δD (SMOW)
VSMOW (Standard Mean Ocean Water)	water	$\equiv 0.00$			$\equiv 0.00$
SLAP (Standard Light Antarctic Precipitation)	water	-55.50			$\equiv -428.00$
GISP (Greenland Ice Sheet Precipitation)	water	-24.78			-189.73
NGS1	CH_4 in natural gas			-29.0	-138
NGS2	CH_4 in natural gas			-44.7	-173
NGS2	C_2H_6 in natural gas			-31.7	-121
NGS3	CH_4 in natural gas			-72.7	-176
NBS-19	calcite	28.64	$\equiv -2.20$	$\equiv +1.95$	
NBS-18	calcite	7.20	-23.00	-5.01	
NBS-17	CO_2 tank gas	21.99	-8.65	-4.41	
IAEA-CO-1	calcite	28.39	-2.44	2.48	
IAEA-CO-8	calcite	7.54	-22.67	-5.75	
IAEA-CO-9	$BaCO_3$	15.16	-15.28	-47.12	
LSVEC	Li_2CO_3	3.63	-26.46	-46.48	
USGS-24	graphite			-15.99	
NBS-22	oil			-29.74	
IAEA-CH-7	polyethylene			-31.83	-100.3
IAEA-C-6	sucrose			-10.43	
NBS-28	quartz	9.58			
NBS-30	biotite	5.24			-65.7
NIST RM 8562	CO_2	22.20	-8.45	-3.76	
NIST RM 8563	CO_2	6.46	-23.72	-41.56	
NIST RM 8564	CO_2	31.11	0.19	-10.45	

Reference Standard	Substance	$\delta^{18}O$ (SMOW)	$\delta^{15}N$ (air)	$\delta^{34}S$ (CDT)	$\delta^{37}Cl$ (SMOC)
NSVEC	N_2 gas		−2.77		
NBS-14	N_2 gas		−1.18		
IAEA-N-1	$(NH_4)_2SO_4$		0.43		
IAEA-N-2	$(NH_4)_2SO_4$		20.32		
IAEA-NO-3	KNO_3	25.3	4.69		
IAEA-305A	$(NH_4)_2SO_4$		39.8		
IAEA-305B	$(NH_4)_2SO_4$		375.3		
IAEA-310A	$CO(NH_2)_2$		47.2		
IAEA-310B	$CO(NH_2)_2$		244.6		
IAEA-311	$(NH_4)_2SO_4$		4693		
USGS25	$(NH_4)_2SO_4$		−30.25		
USGS26	$(NH_4)_2SO_4$		53.62		
USGS32	KNO_3		179.2		
IAEA-S-1	Ag_2S			≡ −0.30	
IAEA-S-2	Ag_2S			22.67	
AEA-S-3	Ag_2S			−32.55	
NBS-123	sphalerite			17.44	
NBS-127	$BaSO_4$	8.7		21.1*	
IAEA-SO-5	$BaSO_4$	12.0		0.49	
IAEA-SO-6	$BaSO_4$	−11.0		−34.05	
Soufre de lacq	Sulfur			16.90	
SRM 975	NaCl				0.43
SRM 975a	NaCl				0.2
ISL 354	NaCl				0.05

*The original published $\delta^{34}S$ value of NBS-127 was 20.32‰, with SO_2 gas (Hut, 1987). A revised value of 21.1‰ was determined when SF_6 gas was used.

APPENDIX B

SAMPLE CALCULATION OF THE CORRECTION PROCEDURE FOR ADJUSTING MEASURED ISOTOPE DATA TO ACCEPTED IAEA REFERENCE SCALES

There are a number of analytical factors that will tend to reduce the measured difference between two samples with very different isotopic compositions. Leaking of changeover valves, memory, blanks, etc., could all cause the apparent isotopic difference between two samples to be less than what it actually is. A correction for compression effects can be made by "stretching" all data so that they conform to a universally accepted scale. The problem is most severe for hydrogen isotope determinations. Two standards with very different compositions, VSMOW $\equiv$ 0‰ and SLAP $\equiv$ -428‰ were developed to allow each laboratory to correct for the compression unique to its own system. For the hydrogen example, the difference between VSMOW and SLAP *must be* 428‰. If the difference measured in the laboratory is less than 428‰, a stretching factor needs to be applied to the data. Similar stretching factors can be applied to all isotope systems. The example that follows is for sulfur.
Consider the measured $\delta^{34}S$ values of SO_2 gas samples given in Table A2-1.

Step 1. The accepted difference ($\Delta_{accepted}$) between IAEA-S-2 and IAEA-S-1 is $(22.67 - -0.3) = 22.97$‰ The measured difference ($\Delta_{measured}$) between IAEA-S-2 and IAEA-S-1 in this example is $(21.5 - -0.5) = 22.0$‰ The measured isotope scale is compressed relative to the accepted one by a ratio of $\Delta_{accepted}/\Delta_{measured}$, or $22.97/22.0 = 1.0441$. This is the "stretching" factor required to bring the scale of the measured values into agreement with the accepted one. All measurements are multiplied by the stretching factor (column 3).

Step 2. The data now need to be shifted by a constant amount to bring them into agreement with the VCDT scale. The difference between the accepted and measured values, "stretched"

Sample	Measured $\delta^{34}S$ value	Measured × stretching factor*	Shifted data corrected to the VCDT scale	Accepted $\delta^{34}S$ value VCDT
1	−14.2	−14.83	−14.61	
2	+2.3	2.40	2.62	
IAEA-S-1	−0.5	−0.52	−0.3	−0.3
IAEA-S-2	21.5	22.45	22.67	22.67

*Two decimal places are recorded in column 3. This does not imply that the data have a higher level of precision than the measured data in column 2. The extra decimal place is added to avoid rounding errors. Reported data should have the level of decimal places appropriate to their uncertainty.

to the IAEA scale, is $(-0.3 - -0.52) = 0.22$ or $(22.67 - 22.45) = 0.22$. Thus, 0.22 is added to all data in column 3 to bring them into agreement with the VCDT scale (column 4). Note that the two measured IAEA samples have been corrected so that they now yield values identical to the accepted IAEA values. The two samples have changed from their measured values of −14.2 and +2.3 to −14.61 and +2.62, respectively. Note that the extra decimal place (higher apparent precision) in the corrected data is only included to avoid rounding errors. Such higher precision should not be reported, as it gives the impression that the isotopic values are known to two decimal places.

The two steps just outlined are illustrated graphically in Fig. A2. In the top row, the measured delta values of two references and two samples are given. In step 1, all delta values are multiplied by a constant to increase the difference between them, so that the measured

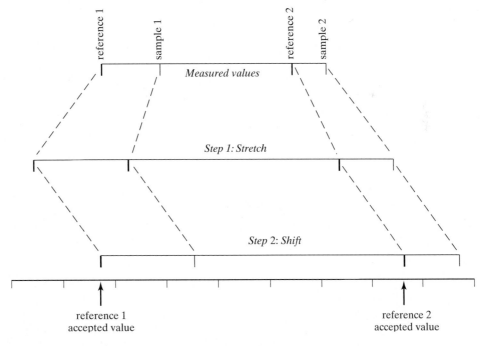

FIGURE A2: Graphical representation of the steps necessary to bring measured isotope ratios into agreement with accepted values defined by two different standards.

difference between the two standards is the same as the accepted difference. (The difference between the measured values of the references could be larger than the accepted values. In such a case, step 1 involves multiplication by a constant less than 1, and the delta values are compressed). In the step 2, a constant is added to each value to bring the measured values of the two references into agreement with the accepted values. At this point, the corrected delta values of the samples will be on the IAEA scale.

INDEX

Page references followed by italicized *f*, *t*, *b*, and *n* refer to figures, tables, boxes, and notes, respectively.

A

AABW (Antarctic Bottom Waters)
   delta value, 106, 106*f*
   delta versus salinity, 107*f*
Absolute zero, 43
Absolute isotope ratios, 18
Acid
   decomposition, 121
   fractionation factor, 23, 123, 124*t*
   formula for making phosphoric acid, 122, 122*b*
AFC (Assimilation–Fractional Crystallization) process, 263, 264*f*
Altitude effect, 83, 84*f*
Amount effect, 83, 83*f*
Ampoule, NBS-1, 27*f*
Anatexis, 258
Animal, nitrogen isotope ratio, 216
Antarctic Bottom Waters (AABW)
   delta value, 106, 106*f*
   delta versus salinity, 107*f*
Apatite, mammalian, 187
Aquatic plant, 155, 158, 158*f*
Archean sulfate, 234
Aragonite, paleotemperature equation, 129
Assimilation
   nitrogen, 207, 210, 210*f*
   wall-rock material, 263
Assimilation–Fractional Crystallization (AFC) process, 263
Assimilatory reduction, 229
Atom, definition, 5
Atomic mass, 11, 20*b*
Atomic number, 5
Atomic weight, boron in minerals, 1

B

β-(beta) factor, 23
Bacterial sulfate reduction, 229

Basalt
   continental and oceanic, 245, 245*f*
   lunar, 244
Basaltic glass, carbon presence, 251
Basinal brine
   delta value, 90, 90*f*
   formation water, 89
Batch
   end-member, 277
   fractionation, 278*f*
   volatilization, 275
Belemnite, PDB, 128*f*
Bellows system, mass spectrometer, 31
Benthic foraminifera
   environment, 132
   oxygen isotope analysis, 140
   vital effect, 163
Biogenic calcite and water, 126
Biogenic carbonate
   carbon isotope equilibrium, 162, 163
   marine calcareous algae, 164
   oxygen, 120
   vital effect, 162
Biological pump, 166*b*
Bitumen, total amount in Earth, 1
Boron, atomic weight, 1
Brachiopods, 137
Brines, 89, 90*f*
Brucite–water fractionation, 54*f*

C

C_3-C_4 plants, 153–155
Calcite–graphite thermometry, 295, 295*f*
Calcite–water fractionation curve, 129*f*
Calibration, fractionation
   empirical, 57
   experimental, 47

335

Calibration, fractionation (*Continued*)
  semi-theoretical, 47
  theoretical, 41
Camp Century ice core, 94, 95f, 96f
Campanian Mishash formation, 182, 182f
Carbon, 28
  ^{12}C as official reference mass, 20b
  composition, 248
  cycle, 149–153, 228
  dioxide, 149
  organic, 157
  oxidation state, 149
  reservoirs, isotopic composition, 152f
  versus oxygen isotope, 143
Carbon–hydrogen isotope, 160f
Carbon isotope
  air values, 160
  analysis, 1, 187, 320
  carbonates, general rules, 162
  composition, 188f
  fractionation, 155, 166, 166b, 166f, 248f, 282f, 321
  isotopologues, 33
  mantle values, 247
  methane values, 160, 160f
  Proterozoic values, 168
  secular variation, 167f
  shift, 170, 171f
  systematic, 247, 282
  typical values, 152f
  value of diamond, 250f
  value of mantle, 249
Carbon reservoir
  mantle, 153
  methane, 159–160, 160f
  plants, 153–157
Carbonate
  change of value, 283
  characterizing, 162
  marine, 167, 167f
  oxygen isotope composition of, 182f
  Proterozoic, 168f
  stable isotope analysis, 122b
  terrestrial, 171
Carbonate analysis, 121
Carbonate-water paleothermometer, 104, 125, 127f
Carbonatite, mantle, 250
Caribbean carbonate cores, isotopic composition, 131
Catastrophic event, 169
Cementation, diagenetic process, 133
Cenozoic isotope record, 140
Cetacean, osmoregulatory system, 186
Chemical fractionation, 314b
Chemical process, physical, 1
Chemistry
  Nobel Prize, 1
  stable isotope publications, 2t
  See also geochemistry
Chert
  ancient, 189f
  application to recent sediments, 193
  diagenesis, 191, 192f
  isotope composition, 285
  nodule of Jurassic age, 192f
  oxygen isotope composition of, 182f
  Phanerozoic, 190, 191f
  Precambrian, 189
  variations within single nodule, 192f
Clay mineral
  attaining equilibrium, 197
  bulk study, 195–199
  equilibrium, 198f
  grain-size considerations, 196
Closed system, 273
Closure temperature, 292, 296, 297
Coal, total amount in Earth, 1
Collector
  Faraday cup, 32
  ion, 33
Common mistakes in terminology, 16t
Condensation
  closed-system (batch) isotopic fractionation, 75
  isotopic equilibrium, 77
  open-system (Rayleigh) isotopic fractionation, 78
Continental carbonate, 141
Continentality effect, 82
Continuous-flow method, 33
Converting between standards, 18
Cooling rates,
  estimating, 297
  retrograde reequilibration, 296f
Correction equations, 332
Crystallization
  fractionation, 243, 244f, 263
  isotope equilibrium, 40
Cycle
  carbon, 150
  nitrogen, 207
  sulfur, 228, 229f

D

Decarbonation, 277, 283
Decomposition, acid and thermal, 121, 123
Delta
  big, 23
  symbol, 18
Delta value
  definition, 17
  North Atlantic Deep Waters (NADW), 107f
  variations in space and time of ocean water, 130
Denitrification, 208, 210
  isotope fractionation, 211
  poorly oxidized soil, 212
Deuterium
  absolute ratios of standards, 28t
  excess parameter, 71, 72, 72f
  excess value distribution, 73f
  hydrogen isotope, 5
Devils Hole, 143
Devolatilization, hydrogen isotope, 262, 280
Diagenesis
  burial, 195
  carbonate, 133
  diagenetic effect, 191
  increasing, 157
  meteoric, 135

procedures for identifying, 138f
strategies circumventing and exploiting effects of, 136, 137
tooth enamel, 186
Diagenetic potential, 134
Diamond
  carbon isotope value, 249f, 250f
  mantle, 248
  nitrogen isotope value, 252f
Diatomic hydrogen, potential-energy curve, 44f
Diatomic molecule, 45
Diet, carbon and nitrogen value, 217, 217f
Dietary reconstruction, tooth enamel, 187
Diffusion rate, oxygen, 299
Dissimilatory reduction, 229, 230
Distance effect, temperature, 82
Dolomite-water system, 50f

E

Early publications, 2t
Earth science, 9
Eclogites
  oxygen isotope anomalies, 247
  oxygen isotope value, 247f
Ecology, 138
Effect (isotope)
  altitude, 83
  amount, 83, 83f
  continuality, 82
  diagenetic, 191
  distance, 82
  distillation, 80f
  equilibrium isotope, 10, 11
  kinetic isotope, 10, 11, 40
  latitude, 83
  metabolic, 162, 163
  salinity, 104, 105
  seasonal, 86
  speciation, 165
  vital, 132, 162–163, 165
Electron
  definition, 5
  mass, 5t
Element
  changing order of mass number and symbol, 20
  isotopic abundances and atomic masses, 8
  isotopic fractionation, 7, 8
  rare isotopes, 7
Empirical calibration, 57
Empirical determination, 52
Empirical semi-theoretical method, 46
Energy, vibrational, 44
Enrichment factor, 209
Equation, Rayleigh, 277
Equilibration, scale of, 284
Equilibrium
  calcite–water fractionation, 48
  clay mineral, 197, 198f
  condensation, 77
  constant, 21, 22
  exchange reaction, 20, 40, 42
  fractionation, 40, 52, 75, 225, 287

kinetic fractionation, 74
mineral–water exchange reaction, 49
reaction, 20, 40
resetting during cooling, 292
testing, 293
two-directional approach, 49f
value with change in temperature, 76
Equilibrium isotope effect
  atomic mass on bond energy, 11
  compared to kinetic isotope effect, 10
Evaporation
  below-cloud air temperature, 85
  kinetic process, 70, 74
  seawater, 91
Event, catastrophic, 169
Extraterrestrial material, 309

F

Fischer–Tropsch synthesis, 321
Fixation, nitrogen, 207
Fluid
  calculating fluid/rock ratio, 259
  flow, controls on, 286
  loss during volatilization, 275
  sources in metamorphic rocks, 280
Fluid–rock
  interaction, 280, 287
  mass-balance equation, 286
  ratio, 286
Foraminifera
  carbon isotope values across
    K/T boundary, 169f
    Paleocene–Eocene boundary, 170f
  coeval and planktic, 131
  variation in delta value versus age, 131f
  speciation effect, 165
  vital effect, 132
Formation water, basinal brine, 89
Fractionation
  anhydrous and hydrous minerals, 48
  batch, 77, 278f
  biogenic calcite and water, 126
  brucite-water, 54f
  carbon isotope, 155, 248f, 321
  chemical, 314b
  crystallization, 243, 244f, 263
  empirical determination, 52
  equilibrium, 40, 52, 75, 225, 287
  evaporation, during, 70
  experimental determinations, 47
  factors, 21, 47, 58, 75, 81, 123, 124t
  general rules governing, 40
  goethite, 200
  high pressure experimental exchange, 51
  isotopic, 1, 293, 317
  mass-independent, 235, 316
  mineral, 292f
  nitrogen, 208, 208f, 209f, 210f
  oxygen isotope, 55

Fractionation (*Continued*)
  oxygen in sulfates, 237
  pressure effect, 53, 54f
  phosphate–water, 181
  polymorphism, effect, 55
  quartz–calcite, 56, 57f
  Rayleigh, 278f
  stable isotope, 120, 292
  statistical mechanics, 41
  sulfur isotope, 225, 229
  temperature variation, 41
  theoretical determination, 41
Freshwater, salinity, 288, 288f
Friedman, Irving, 35

G

Gas, polyatomic, 46
Gas chromatography–isotope ratio mass spectrometry (GC-IRMS), 33
  schematic, 34f
GC-IRMS (gas chromatography–isotope ratio mass spectrometry)
  advantages, 33
  analyzing isotope ratio, 35
  schematic, 34f
Geochemical tracer, 273
Geochemistry
  goethite, 200
  isotopic abundances and atomic masses of elements, 8t
  nitrogen isotope, 207, 322
  publications, 2t
  stable isotope, 1, 120, 284, 300
  sulfur isotope, 273
Geological thermometer, 3
Geospeedometry, 296
Geothermal system, 88–89
Glacial ice
  age, 93
  paleoclimatology, 92
  reservoir of freshwater, 91
  thinning, 94
Glacial–interglacial cycle, 130
Glacial paleoclimatology, 92
Glacial period, 72
Glaciation, 116
Glass changeover valve, 32f
Global Meteoric Water Line (GMWL)
  deuterium excess parameter, 72, 72f
  example of, 68f
  general features, 69
  slope value, 70
GMWL (*See* Global Meteoric Water Line)
Geothermal systems, 88
Goethite
  fractionation, 200
  iron oxide, 199
Goethite geochemistry, 200
Granitoid, characterizing, 256f
Granodiorite, 261
Greenland Ice Core Project (GRIP), 97, 97f
Greenland Ice Sheet Project (GISP), 93
Ground vibrational state, 43

Groundwater
  isotopic composition, 86
  mixing with other water, 88f

H

Harmonic oscillator, 44
Hermatite, iron oxide, 199
Heterogeneities, 317
High-temperature exchange
  country rock, 258
  post-emplacement, 259
Homogenization, 285
Hydrogen
  diatomic, 44f
  equilibrium, 197
  ion microprobe study, 319
  isotopes of, 5, 6f
  SMOW, 25, 26
Hydrogen isotope
  analysis, 35
  composition, 252, 253
  devolatizing, 262
  fractionation, 1, 51, 55, 76
  ratio, 110, 323
  systematic, 318
Hydrosphere, 64
Hydrothermal alteration
  low-temperature, 259
  shallow-level, 256
Hydrothermal system, fossil, 257

I

Ice core, 94
  variations with depth, 95f, 96f
  volcanic eruption, 93, 93f
Igneous rock, 242
  alteration, 256
  fluid/rock ratios, 259
  degassing, 261
  high oxygen isotope value, 258
  hydrothermally altered, 257
  oxygen isotope composition, 255
  typical isotope values, 255f
  unaltered, 256
Illite–smectite, 197f, 198f
Immobilization, 207
Indian Deep Waters (IDW), delta value, 106, 106f
Interglacial period, 72
Internal energy, 42
International Atomic Energy Agent (IAEA), 19, 24
Interstellar cloud, isotopic fractionation, 317
Ion, charging, 32
Ion microprobe
  general, 36
  profiling for nitrogen and hydrogen, 323f
  study of hydrogen, 319, 319f, 320
Iron oxide, hematite and goethite, 199
Isochemical process, 134

Isotope
  chemistry, 2t
  composition of, 64, 65t
  data across lithological contact, 291b
  effects
    equilibrium, 11
    kinetic, 10, 85
  exchange reaction, 20, 40, 74
  homogenization, 285
  mass-independent sulfur anomaly, 235
  partitioning, 53–55
  profile, 290, 290f
  rare, 7
  ration, 223
  research, 4
  stable measurement, 1
  See also Stable isotope
Isotope ratio mass spectrometry, 30
Isotopic abundances of light elements, 8t
Isotopic composition, ocean water, 105f, 109, 110
Isotopic equilibration, 286
Isotopic equilibrium
  condensation, 77
  disturbing, 292
Isotopic fractionation
  closed-system, 75
  coexisting minerals, 293
  denitrification, 211
  factor between two substances, 21
  hydrogen, 51
  interstellar clouds, 317
  light element, 1
  oxygen, 123
  physical chemical processes, 1
  semiempirical calculation, 3
  speciation effect, 165
  sulfur, 21, 225, 225f, 229
  velocity differences, 10
Isotopic geochemistry, stable, 1
Isotopically substituted molecules, 15
Isotopologue
  chemical and physical properties, 6t
  defining, 6, 16
  diffusion rate, 75
  masses measured, 34t
  measuring, 33, 35
  natural abundances in water, 65, 66t
  vapor pressure, 71
  water abundance of, 65. 66t
  zero-point energies, 44f
Isotopomer, 17

J

Jupiter, 318
Juvenile water, 88

K

Kaolinitization, 195
Kimberlite, mantle, 250

Kinetic effect, 162, 163
Kinetic fractionation, equilibrium, 74, 80f
Kinetic isotope effect
  defining, 10
  high-temperature processes, 11
  reaction, 40
Kinetic process, evaporation, 74
K/T boundary, 169, 169f

L

Latitude effect, 83
Light element, isotopic fractionation, 1
Limestone, marine, 3
Liquid, isotopic exchange, 74
Lithological contact, isotopic data, 291b
Local Meteoric Water Line (LMWL),
  slope value, 70, 70f
  variations in slopes and intercepts, 69
Low-temperature mineral, phosphate, 179
Lunar basalt, 244
Lunar soil
  ion microprobe profiling, 323f
  solar wind nitrogen isotope value, 322

M

Mafic lava, 244, 245
Magmatic volatile, 261
Mammal
  apatite, 187
  carbon isotope composition, 188f
  marine paleothermometry, 183
  bone phosphate, 184
Mantle
  carbon isotope values, 249
  carbonatite and kimberlite, 250
  composition, 242, 243
  diamond, 248
  eclogite, 247, 247f
  hydrogen isotope values, 252
  Mid Ocean Ridge Basalt (MORB), 252
  sulfur isotope values, 254, 254f
  xenolith, 246
Marble
  carbon isotope value, 274f
  isotope composition, 285
  oxygen isotope value, 274f, 286
Marine carbonate
  alteration by meteoric water, 136f
  carbon value vs. atmospheric oxygen concentration, 167, 167f, 168f
  catastrophic event causing variation in value, 169
Marine paleothermometry, 181, 183
Marine phosphate, analysis, 183f
Marine sediment, secular change, 111
Mars, 314b, 319
Mass-balance equation, fluid–rock, 276b, 286
Mass-independent fractionation, 235, 316
Mass ratio, 40

Mass spectrometer, 30, 31t
　gases commonly analyzed, 34t
Material balance, 276b
Measurement
　precise density, 1
　spectrometric, 1
　stable isotope, 1, 4, 9
Melting, isotope variation, 1
Mercury piston, 30
Metabolic effect, 162, 163
Metamorphic fluid isotope values, 281f
Metamorphic rock
　carbon isotope variations, 282
　fluid/rock ratios, 286
　geological feature, 272
　open system, 274
　oxygen isotope thermometry, 291–295
　sulfur isotope variations, 283
　volatilization, 277
Metamorphism, 284
Metastable
　aragonite and vaterite, 135
　calcite, 134
Meteoric diagenesis, 135
Meteoric water
　distillation effect, 80f
　fluxes, 67f
　glacial versus interglacial, 72
　hydrothermal alteration, 256
　precipitation on Earth, 67
Meteoric Water Line (MWL), 68, 68f
Meteorite, 244
　carbon isotope values, 320, 321f
　classifying, 310, 311t
　hydrogen isotope values, 318, 319f
　nitrogen isotope values, 322, 323f
　oxygen isotope values, 310
Methane, 159–160, 160f
Mid Ocean Ridge Basalt (MORB)
　high-temperature alteration, 113
　low-temperature alteration, 112
　mantle, 243, 252
　oxygen isotope composition, 245
　sulfur, 254
Mineral
　clay, 195–199
　diagenetic potential, 134
　fractionation, 292f
　low-temperature, 179
　refractory, 294
Mineral–calcite exchange reaction, 50
Mineral–CO_2 exchange reaction, 51
Mineral–water exchange reaction, 49
Mineral–water experimental method, 50
Mineralization, 209
Mistakes in terminology, 16t
Molecular flow, 30
Molecule
　diatomic, 45
　internal energy, 42
　isotopically substituted, 15
　vapor pressure, 74
　vaporizing, 85

Mollusc, 131
Monterey formation, 192f
MORB (Mid Ocean Ridge Basalt)
　high-temperature alteration, 113
　low-temperature alteration, 112
　mantle, 243, 252
　oxygen isotope composition, 245
　sulfur, 254
MWL (Meteoric Water Line), 68
　See also Global meteoric water line, Local Meteoric water line

N

NADW (North Atlantic Deep Waters), 107f
National Institute of Standards and Technology (NIST), 19, 24
Natural substance, isotopic fractionation, 1
Neogene isotope record, 140
Neomorphism, 134
Neutron, 5
Nier, A., 1
Niggli, Paul, 3
Nitrate, 211f
Nitrification, 207, 208, 208f, 210
Nitrobacter, 208
Nitrogen, 28, 29
　composition, 206
　cycle, 207, 208f
　fixation, 207, 209
　ocean, 213–214
　plants and soil, 212
　terrestrial reservoir, 212–213
　value in solar system, 322
Nitrogen isotope
　analysis, 35
　composition of marine plants and animals, 216f
　cycle, 207, 208f
　fixation, 209
　fractionation, 208, 210f
　geochemistry, 322
　ocean values, 211f, 213
　ratio, 33, 323
　ratio in animals, 216
　systematic, 251
　typical values, 211, 213f
Nitrosomona, 208
Nobel Prize in Chemistry
　Soddy, Frederick, 5n
　Urey, Harold C., 1
North Atlantic Deep Waters (NADW), 106, 106f, 107f
Nuclide, 6

O

^{17}O anomalies in meteorites, 309–318
Ocean water
　ancient, 109, 111
　buffering of isotopic composition, 112
　carbon isotope depth profiles, 109
　depth profile, 108
　isotopic composition, 105f
　nitrogen in, 213–214

oxygen dissolved in, 108
oxygen isotope composition, 103
oxygen isotope variation, 104
salinity effect, 105
variations in delta value across Atlantic, 76, 107
variations in delta value in space and time, 130
Open system
    Rayleigh model, 276
    volatilization and fluid infiltration process, 274
Organic carbon, 157
Organic matter
    contamination, 122, 124
    formation, 109
    oxidation, 109
Organism
    ecology, 138
    oxygen source, 184
    source for water, 184, 184f
Orthogneisses, 273
Osmoregulatory system
    cetacean, 186
Oxidizing, 234
Oxygen, 29
    biogenic carbonate, 120
    equilibrium, 197
    factors affecting paleotemperature, 129
    heterogeneities, 317
    organism source for, 184
    three-isotope plot, 314b
    versus carbon isotope, 143
Oxygen isotope
    analysis, 3, 132, 139, 140, 180
    composition, 103, 182, 243, 255
    data, 292
    fractionation, 55, 77, 123, 165
    paleotemperature, 125
    paleothermometry, 137, 139
    ratio, 110, 111, 135
    ratio variation of shell carbonate, 142f
    thermometry, 293, 311
    value, 247f, 286
    variation, 76, 237, 310

P

Pacific Deep Waters (PDW), 106, 106f, 108f
Paleocene–Eocene boundary, 170, 170f
Paleoclimate reconstruction, 195
Paleoclimatology, 3
    cave deposit, 143
    defining, 9
    glacial, 92
Paleogene–Neogene time, 140
Paleotemperature
    estimating, 4
    factors affecting oxygen isotope, 129
    salinity effect, 104
    scale, 125, 131
Paleothermometer, 104
Paleothermometry, 181
Pacific Deep Waters, 106, 108f
Paragneisses, 273

Partition function
    complete, 46
    rotational, 45
    translational, 45
    vibrational, 43
PDW (Pacific Deep Waters), 106, 106f, 108f
PeeDee Belemnite (PDB), 24
    vs. SMOW, 126f
Per mil fractionation, 22
PET (Potential Evapotranspiration Loss), 73f
Petrology, 242, 255
Phanerozoic chert
    application, 190
    delta plot, 191f
Phosphate
    analytical techniques, 180
    carbon isotope values, 188f
    low-temperature mineral, 179
    mammalian bone, 183
    marine analysis, 183f
    oxygen isotope analysis, 180
    primates, 188f
    thermometer, 181
Phosphoric acid method
    carbonate analysis, 121
    recipe, 122
Phosphorite, 182f
Photorespiration, 154
Photosynthesis, 153–157, 163
Physical chemical process, 1
Planetary science, ^{17}O anomaly, 312
Planktic foraminifera
    measuring, 131
    oxygen isotope analysis, 140
    oxygen isotope equilibrium, 132
Plant
    aquatic, 155, 158, 158f
    C_3-C_4, 153, 155f
    nitrogen-fixing, 212
    terrestrial, 153, 155f
Plutonic rock
    emplacement, 255, 256
    oxygen isotope value, 258f
Polyatomic gas, 46
Potential-energy curve
    diatomic hydrogen, 44f
    harmonic oscillator, 44
Potential evapotranspiration loss (PET), 73f
Precambrian chert, 189
Precipitation, 80
Precipitation-PET, 73f
Precise density measurement, 1
Proterozoic carbonate, 168f
Protium, 5
Proton, 5, 5t
Publications, 2t

Q

Quantum mechanical technique, 43
Quartz
    burial diagenesis, 195
    crystallizing from silica, 193, 194

Quartz–calcite fractionation, 56
Quartz–water fractionation, 56
Quaternary isotope record, 139

R

Ratio
   analyzing stable isotopes, 35
   calculating, 259
   element in natural materials, 1
   hydrogen and nitrogen, 323
   hydrogen and oxygen, 1, 110
   measured and desired, 35
   oxygen isotope, 135
   stable isotope, 65
Rayleigh equation, 277
Rayleigh fractionation, 78–80, 211, 278f
Rayleigh volatilization, 275, 278
Reaction mechanisms, 9
Recrystallization
   metamorphic rock, 272
   Precambrian chert, 189
   problem of, 57
   solution and redeposition, 134
Reduced mass, 44
Reference standards, 24, 25t, 329–331
   carbon, 28
   hydrogen, 25
   nitrogen, 28
   oxygen, 29
   sulfur, 30
Refractory mineral, 294
Regional-scale exchange, 284
Relative difference, 18
Reservoir, isotopic composition, 276b
Retrograde exchange, 296
Robinson, P., 1
Rock
   calculating fluid/rock ratio, 259
   chert, 285
   country, 258
   igneous, 242, 256, 257, 258
   marble, 285
   metamorphic, 272, 277, 293
   metamorphism, 272
   origin and history, 4
   plutonic, 255
   volcanic, 256
Royal Society of London, 3

S

Sahara desert, 71
Salinity, 288, 288f
Salinity effect
   -isotope relationship, 104, 105f
   deep ocean water, 105
   paleotemperature, 104
Scale
   paleotemperature, 131
   PDB versus SNOW, 127

Science
   planetary ^{17}O anomaly, 312
Scientist
   Briscoe, H., 1
   Friedman, Irving, 35
   Nier, A., 1
   publications by, 2t
   Robinson, P., 1
   Soddy, Frederick, 5n
   Urey, Harold C., 1, 3, 15, 120, 121
   Vernadsky, V., 1
   Wickman, F., 1
Seasonal effect, 86
Seawater
   evaporation, 91
   salinity, 288, 288f
Seawater–basalt interaction
   buffering competing types, 116
   low temperature alteration, 112, 113
Sedimentation, 89
Sediments, organic carbon, 157
Semiempirical calculation, 3
Sensitive high-resolution ion microprobe (SHRIMP), 35
Silica
   application, 193–195
   phytolith, 195
SLAP (Standard Light Antarctic Precipitation), 27
SMOW (Standard Mean Ocean Water)
   fictitious reference standard, 25–29
   hydrogen, 25, 26
   ice core, 95f, 96f
   vs. PeeDee Belemnite (PDB), 126f
   stable isotope analysis, 64
   Vienna, 26
Soddy, Frederick, Nobel Prize in Chemistry, 5n
Solar system
   discovery of an anomaly, 312, 313
   nitrogen value, 322
   origin, 309
   oxygen heterogeneity, 317
   solar wind, 318
Solar wind
   nitrogen value, 322
   solar system, 318
Solution and redeposition
   recrystallization and neomorphism, 134
   speciation effect, 165
Spectral Mapping Project (SPECMAP), 139, 140
Spectrometer
   60-degree sector mass, 2f
   gases commonly measured, 34t
   isotope ratio mass, 30, 33
   masses measured, 223
   modern mass, 31
   schematic, 31f
Spectrometric measurement, 1
Spectrometry
   analyzing isotope ratios, 35
   gas chromatography–isotope ratio mass spectrometry (GC-IRMS), 33
   isotope ratio mass, 30
Speleothem, 143
Stable isotope
   chemistry, 2t

data for photosynthetic and nonphotosynthetic coral, 165*f*
defining, 5
fractionation, 41, 292
geochemistry, 7, 120, 273, 284, 300
nitrogen, 206
oxygen and hydrogen, 255
ratio, 65, 320
sulfur, 222
thermometry, 47, 291, 295
zero-dimensional system, 289
Stable isotope analysis
application to mammal, 186
carbonate, 122*b*
Standard Light Antarctic Precipitation (SLAP), 27
Standard Mean Ocean Water (SMOW), 73*f*
fictitious reference standard, 25–29
hydrogen, 25, 26
ice core, 95*f*, 96*f*
PeeDee Belemnite (PDB), 126*f*
stable isotope analysis, 64
Vienna, 26
Standards, *See* Reference Standards
Strangelove Ocean, 169
Sulfate
archean, 234
oxidizing, 234
oxygen isotope variation, 237
Sulfur
absolute isotope abundance, 223
analytical techniques, 223
Archean sulfates, 234
barite analyses, 229
composition, 222, 230*f*, 235*f*
exchangeability, 284
forming at low temperature, 226
fractionation, equilibrium, 225*f*
isotope ratio, 34*t*, 223, 236
isotope thermometry, 226
Mid Ocean Ridge Basalt (MORB), 254
pH, effect of, 226, 227*f*
reduction, 230
secular variation, 231
Sulfur cycle, 228
Sulfur isotope
anomaly, 235
Archean values, 234
composition, 254*f*
curve, 232
fractionation, 225, 225*f*, 229
geochemistry, 273
mass independent fractionation, 235
ratio, 34*t*, 223, 236
secular variations, 231
systematic, 283
thermometry, 226
Symmetry, 45, 66

T

Temperature
changes across Paleocene–Eocene boundary, 170, 170*f*
closure, 292, 296, 297
cooling rate, 297

equilibrium value, 76
evaporation, 85
forming sulfur, 226
fractionation, 41
isotopic composition of precipitation, 80
oxygen isotope composition, 82
seawater–basalt interaction, 112, 113
value of basinal brine, 90
Terrestrial
carbonate, 171
climate, 183
environment, 236
plant, 153, 155*f*, 158*f*
reservoir, 212–213
Thermal decomposition, 121
Thermodynamic equilibrium constant, 22
Thermometer
geological, 3
phosphate, 181
Thermometry
calcite–graphite, 295, 295*f*
defining, 9
oxygen isotope, 293, 311
stable isotope, 47, 291, 295
sulfur isotope, 226
Tooth enamel
diet, 187
stable isotope analysis, 186
Tracer, 9
Tritium, 5

U

University of Minnesota, 1
Urey, Harold C.
isotope exchange reaction, 1
isotopically substituted molecules, 15
oxygen isotope analysis, 121
Royal Society of London lecture, 3
stable isotope geochemistry, 120

V

Valve, 32*f*
Vaporizing, 74, 85
Venus, 319
Vibrational partition function ratio, 45
Vienna Standard Mean Ocean Water (VSMOW), 26
Viscous flow, 30
Vital effect
carbon, 164*f*
organisms secreting shells, 131, 162–163
speciation effect, 165
Volatile
degassing, 262
magmatic, 261
Volatilization
batch, 275
isotopic composition of minor elements, 278
metamorphic rock, 277
open system, 274
Rayleigh, 275, 278

Volatilizing, isotope variation, 1
Volcanic
    eruption, 93, 93*f*
    rock, 256
VSMOW (Vienna Standard Mean Ocean Water), 26

W

Wall-rock material, assimilation, 263
Water, 184
    factors controlling precipitation, 80–86
    input and output by mammal, 184, 184*f*
    meteoric, 67
    reservoir size and isotopic composition, 65*t*
Water-rock interaction model
    one-dimensional (directional), 287
    zero-dimensional, 286, 287

Wave number, 44
Weight, atomic, 1
Wickman, F., 1
Working standard (WS), 18, 19

X

Xenolith, mantle, 246

Z

Zero-point energy, 11*f*, 43